Conference Proceedings of the Society for Experimental Mechanics Series

For other titles published in this series, go to
www.springer.com/series/8922

Tom Proulx
Editor

Mechanics of Biological Systems and Materials, Volume 2

Proceedings of the 2011 Annual Conference on Experimental and Applied Mechanics

Springer

Editor
Tom Proulx
Society for Experimental Mechanics, Inc.
7 School Street
Bethel, CT 06801-1405
USA
tom@sem1.com

ISSN 2191-5644 e-ISSN 2191-5652
ISBN 978-1-4614-0218-3 e-ISBN 978-1-4614-0219-0
DOI 10.1007/978-1-4614-0219-0
Springer New York Dordrecht Heidelberg London

Library of Congress Control Number: 2011929511

Printed on acid-free paper

Springer is part of Springer Science+Business Media (www.springer.com)

Preface

Mechanics of Biological Systems and Materials—The 1st International Symposium on the Mechanics of Biological Systems and Materials—represents one of eight volumes of technical papers presented at the Society for Experimental Mechanics Annual Conference & Exposition on Experimental and Applied Mechanics, held at Uncasville, Connecticut, June 13-16, 2011. The full set of proceedings also includes volumes on Dynamic Behavior of Materials, Mechanics of Time-Dependent Materials and Processes in Conventional and Multifunctional Materials, MEMS and Nanotechnology; Optical Measurements, Modeling and, Metrology; Experimental and Applied Mechanics, Thermomechanics and Infra-Red Imaging, and Engineering Applications of Residual Stress.

Each collection presents early findings from experimental and computational investigations on an important area within Experimental Mechanics. The 1st International Symposium on the Mechanics of Biological Systems and Materials was organized by: Bart Prorok, Auburn University; Francois Barthelat, McGill University; Chad Korach, State University of New York (SUNY) at Stony Brook; K. Jane Grande-Allen, Rice University; Elizabeth Lipke, Auburn University.

This symposium was organized to providing a forum to foster the exchange of ideas and information among scientists and engineers involved in the research and analysis of how mechanical loads interact with the structure, properties and function of living organisms and their tissues. The scope includes experimental, imaging, numerical and mathematical techniques and tools spanning various length and time scales. Establishing this symposium at the Annual Meeting of the Society for Experimental Mechanics provides a venue where state-of-the-art experimental methods can be leveraged in the study of biomechanics. A major goal of the symposium is for participants to collaborate in the asking of fundamental questions and the development of new techniques to address bio-inspired problems in society, human health, and the natural world.

The current volume on The 1st International Symposium on the Mechanics of Biological Systems and Materials includes studies on:

Simulation and Modeling in Biomechanics
Mechanics of Tissue Damage
Cell Mechanics
Mechanics of Cardiovascular Tissues

Advanced Imaging Methods Applied to Biomechanics
Mechanics of Hydro Gels and Soft Materials
Mechanics of Hard Tissues
Mechanics of Biocomposites
Nanomechanics in Nature
Indentation Methods for Biological and Soft Materials

The Biological Systems and Materials TD would like to thank the presenters, authors and session chairs for their participation.

The opinions expressed herein are those of the individual authors and not necessarily those of the Society for Experimental Mechanics, Inc.

Bethel, Connecticut

Dr. Thomas Proulx
Society for Experimental Mechanics, Inc

Contents

Analysis and Simulation of PTMA Device Deployment into the Coronary Sinus: Impact of Stent Strut Thickness

Thuy M Pham[1], Milton DeHerrera[2], Wei Sun[1]

[1]Tissue Mechanics Lab
Biomedical Engineering & Department of Mechanical Engineering, University of Connecticut, Storrs, CT 06269
[2]Advanced Technology Group, Edwards Lifesciences, Irvine, CA, 92614

ABSTRACT

Surgical repair and replacement of mitral valve for functional mitral regurgitation (MR) are often limited due to high operative mortality. Recently, a new non-surgical intervention, percutaneous transvenous mitral annuloplasty (PTMA), is being investigated as an endovascular alternative to invasive open-heart surgery. Excellent short-term results have been reported in animal and several human clinical trials, proving the device is feasible. However, device fracture was observed. It is postulated that PTMA device failure is associated with its design (e.g., material, structural geometry) and interactive coupling effects between the device and hosting tissues. In this study, we developed a computational model to investigate the impact of PTMA design on its performance and fatigue life by simulating the deployment of a variety of anchor stents into human coronary sinus (CS) vessel. Peak stresses, strains, interaction forces (shear, normal) of CS wall and stent, as well as device fatigue life and safety factor, were examined, offering insights for a better PTMA design. Results showed that a stiffer Nitinol stent induced high stresses on the vessel wall. Consequently, using a stiffer stent should be coupled with an alternation of stent geometry (e.g. strut thickness) in order to reduce vessel stress as well as radial structural stiffness.

INTRODUCTION

Surgical repair and replacement of mitral valve for functional mitral regurgitation (MR) are often limited due to high operative mortality [1, 2]. Recently, a new non-surgical intervention, percutaneous transvenous mitral annuloplasty (PTMA), utilizing the trans-coronary sinus (CS) vein approach, is being investigated as an endovascular alternative to invasive open-heart surgery. A PTMA device, as illustrated in **Fig. 1**, is consisted of three parts - the proximal and distal anchors and the "spring-like" bridge elements. The anchors are made of self-expanding Nitinol material that is now frequently used for stent grafts [3]. The proximal and distal anchors are deployed into the CS ostium and the great cardiac vein (GCV), respectively. The bridge element is also a Nitinol material and intertwined with a biodegradable suture, which dissolves over a period of 1 month. Once the suture completely absorbs, the bridge element will shorten and induce tension, drawing the proximal and distal anchors together, hence the septo-lateral mitral annulus distance can be reduced.

The initial human clinical study of the first generation of PTMA device was reported by Webb et al [4]. Four of the five patients received the implants successfully, and an acute reduction in MR (35 ± 1 mm vs. 36 ± 3 mm at baseline) was observed in three out of these four patients. However, a reduction in contraction force of the device following bridge separation occurred in all three patients several weeks implantation, leading to the recurrence of MR. Currently, another study of 72 patients reported results of a second generation PTMA device with a reduction in MR by$\geq$ 1 grade in 50% of 22 implanted patients and 85.7% of 7 severe MR patients [5]. The device was modified with a non-biodegradable suture to reinforce the bridge element. No bridge separation but fractures in the proximal anchor in 4 cases were observed. We postulate that the device fracture is associated with the design (e.g., material and structural geometry) and the interactive coupling effects between the device and the hosting tissues. Numerous computational studies of stent (both self-expandable and balloon expansion) deployment into the arterial vessels, including stenotic coronary arteries [6, 7], carotid arteries [8, 9], iliac arteries [10, 11], and other arteries [12-14], have shown that stent induces mechanical stress on the vessel wall [7, 14] and reversely undergoes stress itself during the pulsatile constriction of the vessel resulted from the cardiac cycle [15].

T. Proulx (ed.), *Mechanics of Biological Systems and Materials, Volume 2*, Conference Proceedings of the Society for Experimental Mechanics Series 9999, DOI 10.1007/978-1-4614-0219-0_1, © The Society for Experimental Mechanics, Inc. 2011

Therefore, it is necessary to be able to predict the behavior (e.g. reaction forces and changes in the mechanical factors such as stress and strain fields) of both the device and the host tissue prior to the procedure, to minimize the negative outcomes and reduce the procedural cost.

Currently, biomechanics involved in the PMTA procedure has not been well studied. In our previous work [16, 17], we have studied the deployment of proximal anchor of the PTMA device using computational models to predict the interaction between the human/porcine derived material CS vessel models and the stent. We observed that stiffer stent material tends to have a better safety factor when deployed in a relatively softer tissue (porcine). Although human tissue offered an adequate radial forces, another concern arises that is high stresses on the vessel wall may induce a vascular injury.

In this study, we attempted to alter the stent design to reduce radial interactive forces. The approach includes maintain the same material properties of the stent and varying the design parameters, e.g. the strut thickness. Two stents models with different strut thickness values were developed. The results from these models include peak stresses and strains, interaction forces (shear and normal) of the CS wall and the stent, as well as the device fatigue life and safety factor were compared to the original design, and differences between these design can provide insights into the stent designs that may help to reduce future device failure in the PTMA intervention.

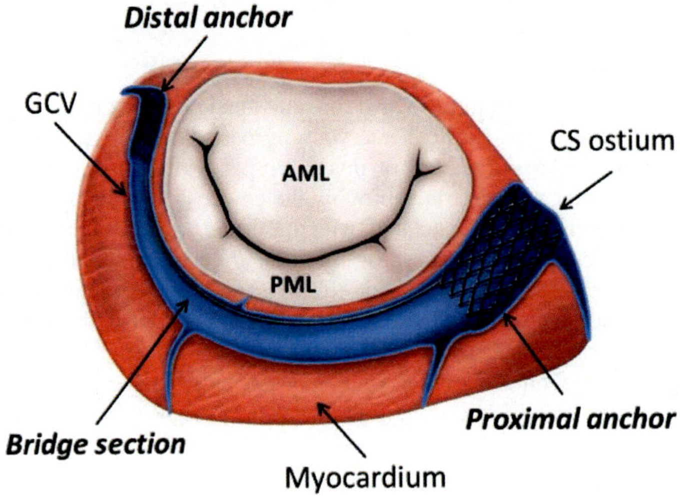

Figure 1 Illustration of a Monarch™ PTMA being deployed into the CS vessel, which is adjacent to the posterior mitral annulus. **ANT** - anterior mitral leaflet. **PML** - posterior mitral leaflet. **GCV** - great cardiac vein.

METHODS/MATERIALS

Stent models. Typical mechanical properties of superelastic Nitinol material are illustrated in **Fig. 2a**, and the material parameters used in this study are listed in **Table 1**. Three stent designs were developed, 1) the original design with a stent strut thickness of 0.30 mm, 2) modified stents with a thickness of 0.23 mm and 3) thickness of 0.18 mm, denoted as, Original, Mod1 and Mod2, respectively. **Fig. 3** illustrates the stent in the undeformed and deformed configurations. Details of design parameters of these stent models are listed in **Table 2**. The original PTMA device is composed of two and twelve strut cells in the axial and circumferential directions, respectively.

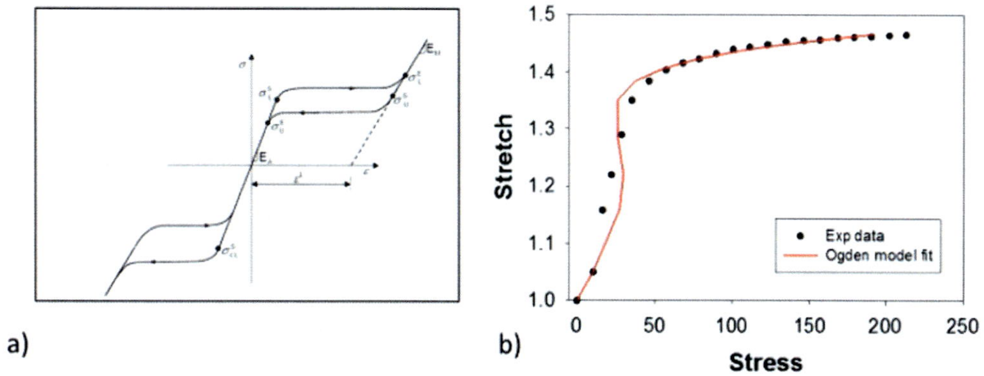

Figure 2 a) A stress-strain curve for Nitinol material, and b) a mean human CS stress-stretch response from inflation experiment fitted with the Ogden model

Table 1 The material parameters of the CS vessels and stents

	μ_1(GPa)	a_1	μ_2(GPa)	a_2	μ_2(GPa)	a_2				
Human	127.653	5.928	-63.082	11.851	4.674	19.317				
	E_A(MPa)	v_A/v_M	E_M (MPa)	ε_L	σ_L^S	σ_L^E	σ_U^S	σ_U^E	σ_{CL}^S	T
Nitin	70,000	0.3	47,800	0.063	600	670	288	254	900	37

Details of the proximal anchor design in 2D drawings and 3D crimped geometry were reported previously [18]. All stents are consisted of brick elements (ABAQUS element type C3D8R, the 8-node linear brick, reduced integration with hourglass control element). One end of the anchor is attached to the bridge-to-anchor connector, see Fig. 3. Two unit cells are joined together by a strut bridge, and two adjacent cells share a strut bar. Other FE models for the simulation system include the expanding and crimping sleeves. The expanding sleeve was used in the expansion-annealing process and the crimping sleeve acted as a restraining sheath which is removed to release the stent into contact with the CS wall. Both the expanding and crimping sleeves are consisted of 2,231 4-node quadrilateral membrane elements.

Characterization of tissue model. The CS material parameters were obtained from fitting the experimental pressure-inflation data collected from four cadaver CS vessels to the nonlinear hyperelastic Ogden model [19]. Briefly, the CS vessels were subjected to the pressure inflation test while they were intact (i.e. CS was not dissected out from its surrounding myocardium) and submerged in the small tank of phosphate buffered saline (PBS) solution at room temperature. After 10 preconditioning cycles, vessels were incrementally dilated up to 80 mmHg of pressure, the dilated CS diameters were measured to obtain the CS pressure-radius curve. The hoop/axial stress-stretch relation were calculated [20]. We chose the Ogden constitutive model [21] to characterize the mean experimental data,

$$W = \sum_{i=1}^{N} \frac{2\mu_i}{a_i^2}(\lambda_1^{a_i} + \lambda_2^{a_i} + \lambda_3^{a_i} - 3) \tag{1}$$

where μ_i and a_i are material constants and λi are the principal stretches. The goodness of the fit was determined using the R-square value based on the Levenberg-Marquardt nonlinear regression algorithm using SYSTAT 10 (Systat Software Inc., Chicago, IL). The Ogden model curve fitting results are illustrated in Fig. 3, with the parameters listed in Table 2 below.

For simplicity the geometry of the CS, at the section that is in contact with the proximal anchor stent, is assumed to be a cylindrical tube with an ID of 12 mm and a thickness of 0.74 mm for the CS vessel. A total of 43,200 brick elements were used to model the CS.

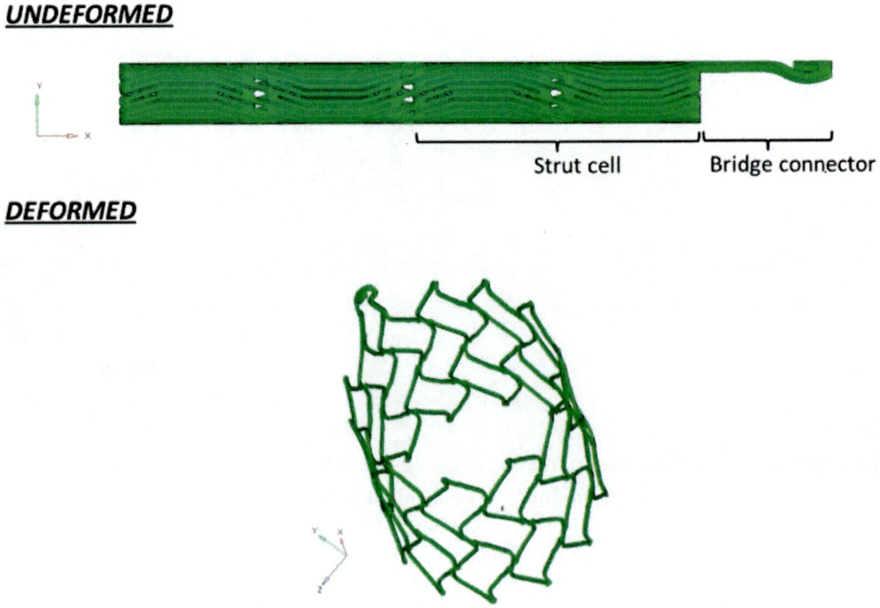

Figure 3 Stent designs in the undeformed (top) and deformed configurations: top – undeformed original, bottom – deformed original and Mod1 after released to come in contact with the vessel wall

Table 2 Values of undeformed design parameters of different stents

			No. of strut		
	Length	Thickness	Circ.	Axial	No. of elements
Original	26	0.30	12	2	114,000
Mod1	26	0.23	12	2	91,370
Mod2	26	0.18	12	2	68,532

Finite Element (FE) modeling of stent expansion and crimping. Prior to the deployment, each stent model underwent the expansion and annealing process to reach a targeted 15.50 mm ID. This is achieved by applying a series of displacement-controlled expansion of the sleeve, allowing it to contact and expand radially the inner surface of the stent. Afterwards, stents were crimped to a final ID of 11.98 mm, which is slightly less than the CS's ID of 12 mm. Crimping was performed by radially displaced the crimping sleeve inward.

PTMA implantation FE modeling. The PTMA procedure consists of a series of steps including stent crimping, stent release, interaction with the vessel wall, contraction of the bridge element, and displacement of the mitral annulus. This study focuses on the tissue-stent interaction (TSI). The simulation is divided into a three-step loading procedure:

Step 1 – Release of the stent anchor. The initial contact between the stent and the CS wall is accomplished by removing the crimping sleeve to release the proximal anchor into contact with the CS inner wall.

Step 2 – Contraction force at the connector end of the anchor. An axial load with a magnitude of 2.45 N in the axial direction is applied to the connector elements to simulate the pulling force generated by the contraction of the bridge section.

Step 3 – Simulation of the pressure in the vessel wall. The incremental pressure, from 0 to 10 mmHg, is imposed directly on the luminal surface of the CS wall to mimic internal blood pressure.

The coefficient of friction between the CS inner wall and the stent for the Step-1 and -2 is 0.1. For Step-3, the contact definition is set as a rough surface (no slipping) condition (an ABAQUS option: ROUGH) to imitate the condition that is equivalent to tissue in-growth over time resulting in the attachment between the stent and the CS wall. The analysis was run with ABAQUS/Explicit release 6.9 on a Linux cluster using typically 12 2.8GHz CPUs.

FE analysis of the tissue-stent interaction (TSI). Each step in the TSI simulation after crimping was analyzed, and the following output variables were obtained:

1) Stress and strain. Distribution of von Mises stress and maximum values were investigated, which can provide insight into the potential injury induced by stents on the vessel wall. Stent strain measurements were analyzed with the maximum tensile strains (SDV24). Stent strain is zero at the initial un-crimped configuration and is at the maximum after crimping. When release to the vessel, stent will undergo unloading deformation or expanding to its pre-set shape or approaching its zero strain. The maximum principal strains (LE) were used for the CS wall strains.

2) Contact forces. Contact forces between the outer stent and inner CS wall were measured. This is an important measure because elevated contact forces in the vicinity of stent struts may lead to injury of the CS wall, which may increase neo-intimal hyperplasia formation, or ultimately causes tears of the posterior CS wall. Conversely, insufficient radial contact forces will prevent proper device anchoring; hence the stent may migrate and lost its annulus cinching function. The contact forces have normal and shear components, denoted by *CNF* and *CSF*, respectively, and are expressed as:

$$CNF = \sum_{n=1}^{n_c} NF_{n,post},$$
$$CSF = \sum_{n=1}^{n_c} SF_{n,post} \qquad (2)$$

where n_c is the total number of nodes of the stent cell that are in contact with the CS inner wall, and $NF_{i,post}$ and $SF_{i,post}$ are the normal contact force and shear contact forces, respectively, at each node after deployment.

***3) Stent fatigue analysis and safety factor* [22]**. Quantitative studies of metal fatigue often make use of the well-established Goodman-Haigh diagram [23], which is recommended by the FDA for stent fatigue analysis [24]. In Goodman diagrams, a pair of the mean stress or strain and its amplitude (or half-amplitude) at a particular point is plotted and compared with constant life curves [24] for that particular material. The mean tensile strain, ε_{mean}, and the half-amplitude oscillating strain, $\Delta\varepsilon$, at a given node are calculated by:

$$\varepsilon_{mean} = (\varepsilon_{\max} + \varepsilon_{\min})/2$$
$$\Delta\varepsilon = (\varepsilon_{\max} - \varepsilon_{\min})/2 \quad , \qquad (3)$$

where ε_{max} and ε_{min} are the maximal and minimal strains, upon the application of 10 mmHg at a node on the PTMA stent after deployment. For a Nitinol material, the constant life curves from Pelton *et al.* 2008 [15, 25, 26] were used. Using the 0.4% strain amplitude delineated by the constant life line [15], the stent fatigue safety factor can be predicted using the equation: safety factor = 0.4%/half-amplitude strain.

RESULTS

Stress and strain analysis. It can be seen in **Fig. 4** that the stent strut was twisted after expansion and annealing. The stent outer surface is not smooth. Thus, it is necessary to perform the expansion and annealing procedure to obtain accurate per-deployment stent geometries. **Table 3** lists all the maximum Von mises stresses and strains of the vessel wall, as well as three stent models. Both Mod1 and Mod2 stents are softer than the original stent model, resulted in higher maximum strains on the stent after releasing the stents. Stresses on the stent increased as strains increased.

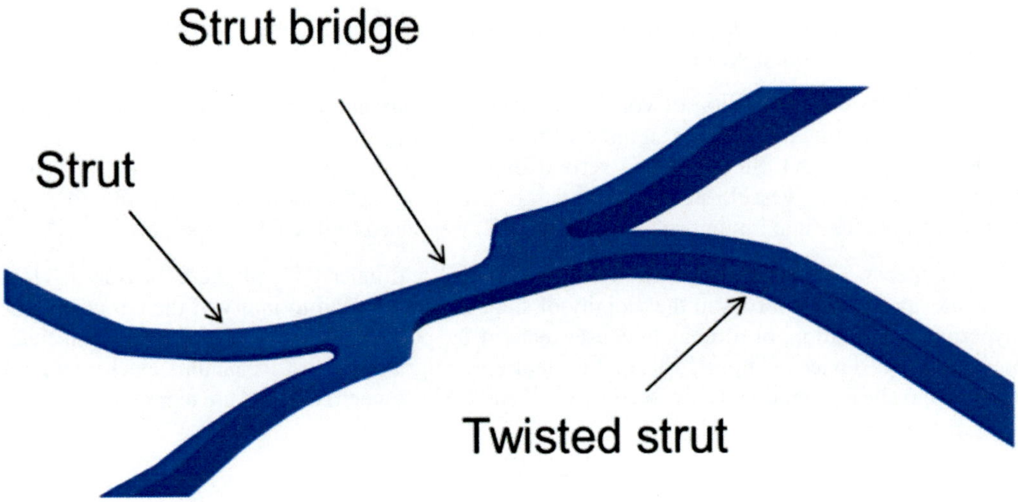

Fig. 4 The illustration of a twisted strut after expansion

Table 3 Maximum von Mises stresses and strains of CS wall and stent models

		STRESS (MPa)			STRAIN		
	Steps	Original	Mod1	Mod2	Original	Mod1	Mod2
CS WALL	1	3.47E-02	3.44E-02	3.30E-02	0.23	0.23	0.23
	2	3.66E-02	3.55E-02	3.52E-02	0.26	0.23	0.28
	3	3.58E-02	3.55E-02	3.46E-02	0.27	0.25	0.25
STENT	1	1001.00	955.00	947.40	2.49E-02	2.40E-02	2.38E-02
	2	358.60	380.00	371.60	4.96E-03	5.40E-03	5.32E-03
	3	444.70	480.30	525.90	6.37E-03	6.88E-03	7.54E-03

After releasing the stent, stresses induced by the stents on the CS wall were similar. However, stresses on the CS wall were lowest when interacted with the Mod2 stent compared to the other two stents. **Fig. 5** illustrates the stress distribution on the CS wall induced by each stent. By reducing the strut thickness, stresses were lower at the region of contact in both Mod1 and Mod2 stents.

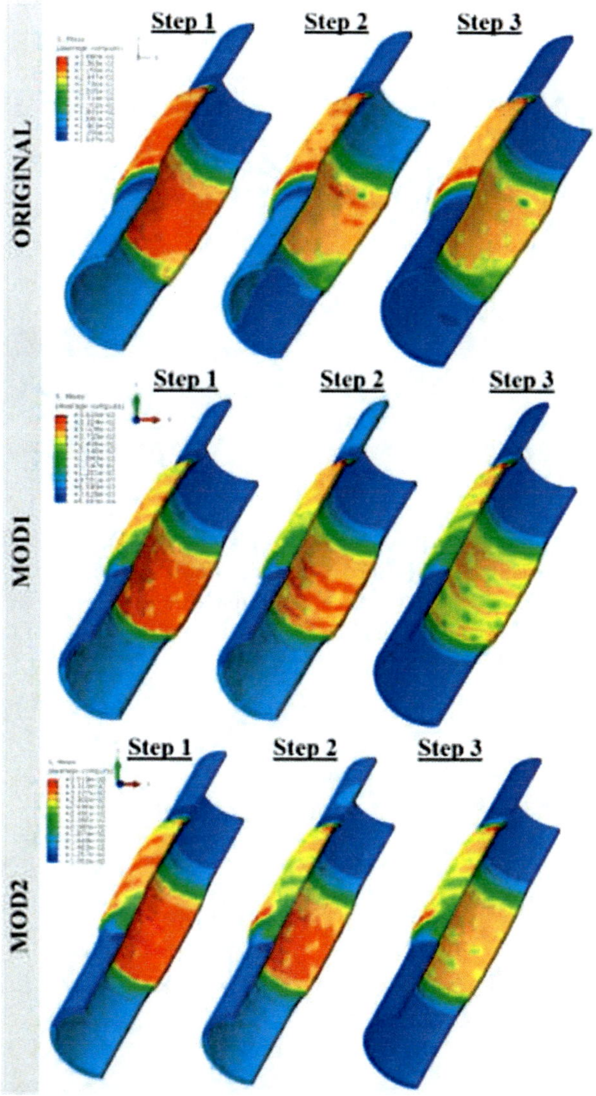

Figure 5 – Stress distributions on CS walls induced by the three stent models in 3 steps.

Normal and shear forces. Because stiffer stent generated a higher normal force, Mod1 and Mod2 stents exerted less forces compared to the original stent, as shown in **Fig. 6a**. There is a slight difference in forces between the stent models in steps 1 and 2 but no difference was observed in step 3. No device slippage was observed in step 2. Shear forces, illustrated in **Fig. 6b**, increased in step 3demonstrating the tissue in-growth over time.

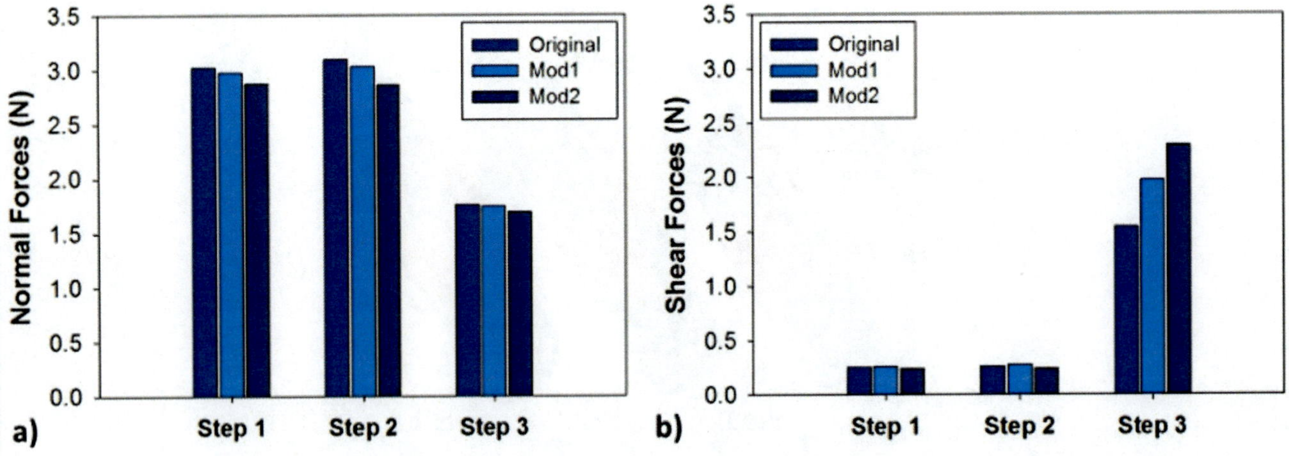

Fig. 6 a) Normal and b) shear forces of the original, Mod1 and Mod2 stents in 3-step TSI simulation

Stent fatigue analysis. The strain amplitude of the original model is 8.43E-2% and the safety factor is 4.75. Decreasing in strain amplitudes of Mod1 and Mod2 (6.61E-2% and 6.68E-2%) resulted in higher safety factors of 6.05 and 5.98, respectively. Changing stent thickness does not reduce the fatigue life of the stent in the PTMA intervention. As shown in **Fig. 7**, strain points on the Goodman diagram of Mod1 and Mod2 are similar to the strain points of the original design, and all are condensed and below 0.1% strain amplitude and 1% mean strain.

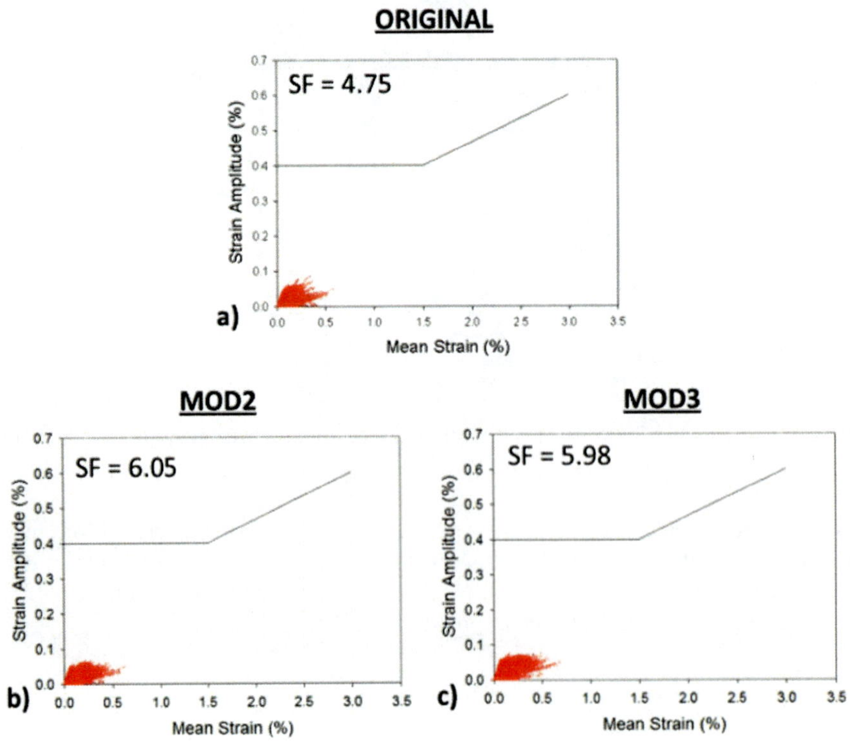

Figure 7 – Goodman diagrams of the stent models.

DISCUSSION/CONCLUSION

The evaluation of PTMA device performance was performed based on two criteria: 1) inducing a minimal vessel injury and 2) enabling stent mechanical function (i.e., anchoring and fatigue life). The mechanical stress on the vessel wall induced by the stent strut cells often provokes vascular wall injury that stimulates the intimal hyperplasia [27]. Therefore, to minimize the vessel wall stress, the structural compliance of the stent needs to be low to meet the compliance of the vessel wall. In the meantime, the stent itself also needs to have enough structural strength to withstand the cyclic mechanical changes induced by cardiac pulsatile pressures, and have enough radial force for its anchoring mechanism. This computational study indicated that PTMA device performance can be improved by altering the stent thickness. By reducing the stent thickness at a particular value, stress on the vessel wall can be reduced without compromising the stent anchoring forces and fatigue life. These numerical results need to be further validated by a future experimental study.

ACKNOWLEDGEMENTS

Research for this project was funded in part by the AHA SDG grant #0930319N and a NIH Pre-doctoral Fellowship to TP.

REFERENCES

[1] Wu, A. H. e. a., 2005, "Impact of mitral valve annuloplasty on mortality risk in patients with mitral regurgitation and left ventricular systolic dysfunction.," J Am Coll Cardiol, 45, pp. 351-354.
[2] Grigioni, F., Enriquez-Sarano, M., Zehr, K. J., Bailey, K. R., and Tajik, A. J., 2001, "Ischemic mitral regurgitation: long-term outcome and prognostic implications with quantitative Doppler assessment," Circulation., 103(13), pp. 1759-1764.
[3] Duda, S. H., Bosiers, M., Lammer, J., Scheinert, D., Zeller, T., Tielbeek, A., Anderson, J., Wiesinger, B., Tepe, G., Lansky, A., Mudde, C., Tielemans, H., and BÃ©rÃ©gi, J. P., 2005, "Sirolimus-eluting versus bare nitinol stent for obstructive superficial femoral artery disease: The SIROCCO II trial," Journal of Vascular and Interventional Radiology, 16(3), pp. 331-338.
[4] Webb, J. G., Harnek, J., Munt, B. I., Kimblad, P. O., Chandavimol, M., Thompson, C. R., Mayo, J. R., and Solem, J. O., 2006, "Percutaneous transvenous mitral annuloplasty: initial human experience with device implantation in the coronary sinus," Circulation., 113(6), pp. 851-855. Epub 2006 Feb 2006.
[5] Harnek, J., Webb, J. G., Kuck, K. H., Tschope, C., Vahanian, A., Buller, C. E., James, S. K., Tiefenbacher, C. P., and Stone, G. W., 2010, "Transcatheter implantation of the MONARC coronary sinus device for mitral regurgitation," JACC: Cardiovascular Interventions, 4(1), pp. 115-122.
[6] Lally, C., Dolan, F., and Prendergast, P. J., 2005, "Cardiovascular stent design and vessel stresses:a finite element analysis," Journal of Biomechanics, 38, pp. 1574-1581.
[7] Zahedmanesh, H., John Kelly, D., and Lally, C., 2010, "Simulation of a balloon expandable stent in a realistic coronary artery-Determination of the optimum modelling strategy," Journal of Biomechanics, 43(11), pp. 2126-2132.
[8] Auricchio, F., Conti, M., De Beule, M., De Santis, G., and Verhegghe, B., 2010, "Carotid artery stenting simulation: From patient-specific images to finite element analysis," Medical Engineering and Physics.
[9] Wu, W., Qi, M., Liu, X. P., Yang, D. Z., and Wang, W. Q., 2007, "Delivery and release of nitinol stent in carotid artery and their interactions: A finite element analysis," Journal of Biomechanics, 40(13), pp. 3034-3040.
[10] Holzapfel, G. A., Stadler, M., and Gasser, T. C., 2005, "Changes in the mechanical environment of stenotic arteries during interaction with stents: Computational assessment of parametric stent designs," Journal of Biomechanical Engineering, 127(1), pp. 166-180.
[11] Kiousis, D. E., Gasser, T. C., and Holzapfel, G. A., 2007, "A numerical model to study the interaction of vascular stents with human atherosclerotic lesions," Annals of Biomedical Engineering, 35(11), pp. 1857-1869.
[12] Timmins, L. H., Meyer, C. A., Moreno, M. R., and Moore Jr, J. E., 2008, "Mechanical modeling of stents deployed in tapered arteries," Annals of Biomedical Engineering, 36(12), pp. 2042-2050.
[13] Early, M., Lally, C., Prendergast, P. J., and Kelly, D. J., 2009, "Stresses in peripheral arteries following stent placement: a finite element analysis," Computer Methods in Biomechanics and Biomedical Engineering, 12(1), pp. 25-33.
[14] Bedoya, J., Meyer, C. A., Timmins, L. H., Moreno, M. R., and Moore Jr, J. E., 2006, "Effects of stent design parameters on normal artery wall mechanics," Journal of Biomechanical Engineering, 128(5), pp. 757-765.
[15] Pelton, A. R., Schroeder, V., Mitchell, M. R., Gong, X. Y., Barney, M., and Robertson, S. W., 2008, "Fatigue and durability of Nitinol stents," Journal of the Mechanical Behavior of Biomedical Materials, 1(2), pp. 153-164.
[16] Sun, W., and Deharrera, M., 2010, "Modeling the interaction between the coronary sinus and the proximal anchor stent in percutaneous transvenous mitral annuloplasty," ASME Summer Bioengineering ConferenceProceedings of the ASME 2010 Summer Bioengineering Conference (SBC2010), June 16-19, Grande Beach Resort, Naples Florida, USA, pp. SBC2010-19715.
[17] Pham, T., Deharrera, M., and Sun, W., 2011, "Simulated deployment of percutaneous transvenous mitral annuloplasty device into human coronary sinus vessel," ASME Summer Bioengineering ConferenceProceedings of the ASME 2011 Summer Bioengineering Conference, SBC2011, June 22-25, Nemacolin Woodlands Resort, Famington, Pennsylvania, USA, pp. SBC2010-19715.
[18] Deharrera, M., and Sun, W., 2007, "Simulation of the Forming of a Superelastic Anchoring Stent and its Deployment in the Coronary Vein," ABAQUS annual conferencePairs, France.
[19] Ogden, R., 2003, "Nonlinear elasticity, anisotropy, material stability, and residual stresses in soft tissue," Biomechanics of soft tissue in cardiovascular system, R. Ogden, ed., Springer, New York.

[20] Pham, T., and Sun, W., 2010, "Characterization of the mechanical properties of the coronary sinus for percutaneous transvenous mitral annuloplasty," Acta Biomaterialia, 6(11), pp. 4336-4344.
[21] Ogden, R. W., 1997, Non-linear elastic deformations, Dover, Toronto.
[22] Kleinstreuer, C., Li, Z., Basciano, C., Seelecke, S., and Farber, M., 2008, "Computational mechanics of Nitinol stent grafts," J Biomech, 41(11), pp. 2370-2378.
[23] FDA, 2005, "Guidance for Industry and FDA Staff: Non-Clinical Tests and Recommended Labeling for Intravascular Stents and Associated Delivery Systems," U.S. Department of Health and Human Services.
[24] Cavanaugh Jr., KJ, Holt, V., Goode, J., and Anderson, E., 2006, "FDA Recommendations for Nitinol Stent and Endovascular Graft Fatigue Characterization and Fracture Reporting," Journal of ASTM International, 3(6), pp. 8-18.
[25] Pelton, A. R., Gong, X. Y., and Duerig, T., 2003, "Fatigue testing of diamond-shaped specimens," Medical Device Materials - Proceedings of the Materials and Processes for Medical Devices Conference 2003, pp. 199-204.
[26] Wick, A., Gong, X. Y., Fino, J., Sheriff, J., and Pelton, A. R., 2005, "Bending fatigue characteristics of Nitinol," Medical Device Materials II - Proceedings of the Materials and Processes for Medical Devices Conference 2004, pp. 15-20.
[27] Rogers, C., and Edelman, E. R., 1995, "Endovascular stent design dictates experimental restenosis and thrombosis," Circulation, 91(12), pp. 2995-3001.

Mechanical identification of hyperelastic anisotropic properties of mouse carotid arteries

Badel Pierre[(a)], PhD, Avril Stéphane[(a)], Pr, Lessner Susan[(b)], PhD, Sutton Michael[(c)], Pr

[(a)] Center for Health Engineering – Ecole des Mines de Saint Etienne – 158, cours Fauriel – 42023 Saint Etienne – France

[(b)] Department of Cell Biology and Anatomy - University of South Carolina School of Medicine - Columbia, South Carolina 29208, USA

[(c)] Department of Mechanical Engineering - University of South Carolina - Columbia, South Carolina 29208, USA

ABSTRACT

The role of mechanics is known to be of primary order in many arterial diseases; however determining mechanical properties of arteries remains a challenge. This paper discusses the identifiability of a Holzapfel-type material model for a mouse carotid artery, using an inverse method based on a finite element model and 3D digital image correlation measurements of the surface strain during an inflation/extension test. Layer-specific mean fiber angles are successfully determined using a five parameter constitutive model, demonstrating good robustness of the identification procedure. Importantly, we show that a model based on a single thick layer is unable to render the biaxial mechanical response of the artery tested here. On the contrary, difficulties related to the identification of a seven parameter constitutive model are evidenced; such a model leads to multiple solutions. Nevertheless, it is shown that an additional mechanical test, different in nature with the previous one, solves this problem.

1. Introduction

Identification of mechanical and structural properties of the arteries is a major topic in cardiovascular research. Many arterial disorders involve significant changes in vascular mechanical properties. Not only do the structure and mechanical response of arteries vary according to many factors such as the location in the vasculature and age, but also their properties may alter under various physiological conditions and during the development of diseases [1,2]. Accurate mechanical identification of the arteries can therefore provide helpful information for clinical diagnoses and treatments. To improve the contribution of solid mechanics, a lot of effort has been undertaken to develop constitutive models of the arterial wall as well as experimental and numerical methods to identify these models.

T. Proulx (ed.), *Mechanics of Biological Systems and Materials, Volume 2*, Conference Proceedings of the Society for Experimental Mechanics Series 9999, DOI 10.1007/978-1-4614-0219-0_2, © The Society for Experimental Mechanics, Inc. 2011

Several constitutive models intended to describe the mechanical response of arterial tissues at finite strains have been developed, see [3] for an extensive review of these models. Most of anisotropic non-linear models considering the passive response of arteries are hyperelastic. Fung [1] first introduced a phenomenological exponential strain energy function. More recently, structurally-motivated models including fiber reinforcements have been developed. Bischoff [4] suggested representing the nonlinear orthotropic material response of the arterial wall, with a homogeneous orthotropic model. This model has proven adequate to capture the nonlinear orthotropic response of vascular tissues although the physical meaning of its parameters is not clear [5]. Holzapfel [6] introduced a two-fiber family model to account for the helically-oriented distribution of collagen fibers within the arterial wall. To ideally describe the arterial wall from the mechanical point of view, two separate layers of this material are required for medial and adventitial layers.

Originally, Holzapfel [6] grounded the choice of two symmetrically- and helically-oriented fiber families per layer on histological observations. Since the compositions of medial and adventitial layers (elastin, collagen and cell contents) are different [2], two separate layers must be distinguished. Note also that some authors introduced the active response of the arterial wall due to smooth muscle activity [7], which is not the concern of this paper.

Correct identification of the parameters of the constitutive model is a key issue in considering the reliability of interpretation for medical purposes or subsequent utilization in numerical models, for instance. The process of identification requires experimental data obtained from mechanical testing. Most of the previous biomechanical studies on mouse carotid arteries have been conducted using global or average data such as pressure-diameter and/or force-length measurements [8,9]. In addition, the identification of layer-specific parameters of the Holzapfel model has been performed, at this time, only from dissected layers on large arteries [10]. The data which are used in this study are collected with a 3D-DIC stereo-microscopy system on a mouse carotid artery, which is, to our knowledge, unique at this time. See [11] for a description of this previous experimental work.

From experimental data, the identification of constitutive models relies, most of the time, on inverse approaches because establishing response curves from the model may involve complex non-linear relations between the parameters. Classical inverse approaches are based on updating methods [5, 7, 10, 12] using optimization algorithms such as the Levenberg-Marquardt algorithm to find the best-fit parameters in a least-square sense with respect to a given cost function. In these approaches, previous authors have most often used analytical developments to derive their modeled data from the given constitutive equations, which presupposed multiple assumptions. Among these, note the widely-accepted assumption of axisymmetry and that of a single homogeneous layer. The latter may be relevant when experimental data is global or averaged over the arterial wall, though two separate layers would be closer to reality and would emphasize the distinction between the layer properties. In addition, for the study of mouse carotid arteries, Gleason [12] performed analytical developments within the frame of thin-tube elasticity theory. This may be a strong assumption in cases of thick arteries like mouse carotid arteries where the ratio of thickness to inner radius was reported to be about 0.6 [8, 12].

Finite element (FE) simulations have rarely been used to recover those modeled data and perform inverse identification. Yet, this kind of approach allows using experimental tests capable of providing richer or otherwise unavailable data and modeling complex problems. Regarding non-linear anisotropic vascular properties, the study of Ning [5] was focused on stress and strain distributions within the arterial wall and how they are influenced by axial pre-stretch. Using the same data as the present study, they identified the parameters of the constitutive model of Bischoff [4], thereby not considering heterogeneity between media and adventitia, which would likely affect these distributions. From our point of view, the advantage of using finite element based identification approaches is to model complex mechanical tests and/or complex structures (see, for instance, [13]), like a thick multi-layer artery presented in this paper.

The question of whether an identification method is relevant with respect to the problem to be treated is seldom addressed. Introducing multiple assumptions and parameters may lead to improper identification or multiple solutions. The objective of the present paper is to address the feasibility of the simultaneous inverse identification of mean fiber angles in both medial and adventitial layers using DIC surface strain measurements.

2. Methods

2.1. Experimental considerations

The experimental data referenced to in this study were described in deep details in [12] where three-dimensional digital image correlation (3D-DIC) is used to obtain full-field surface strain measurements on mouse carotid arteries at the micro-scale during an inflation/extension test (Fig. 1). The mechanical test performed here allows both pressurization loading and extension loading at the same time (see the schematic principle of the setup in Fig. 1). This test is relevant as it provides biaxial loading conditions close to physiological conditions.

To briefly describe the experimental setup, both ends of a freshly-dissected carotid artery are cannulated with Luer stubs. For image processing and local deformation measurements, a high contrast speckle pattern is incorporated into the vessel structure thanks to ethidium bromide nuclear staining. The experiments are performed with one end of the artery attached to the pressure controller and pressure source, while the other capped end is free in the axial direction, thereby allowing axial translation. The artery is pressurized from 5 to 150 mmHg in steps of 9 mmHg with a flow rate of 0.2 ml/min and an average pressurization rate of 1.8 mmHg/s. After each pressurization step, synchronized images are acquired from two cameras and analyzed using existing commercial software, VIC-3D.

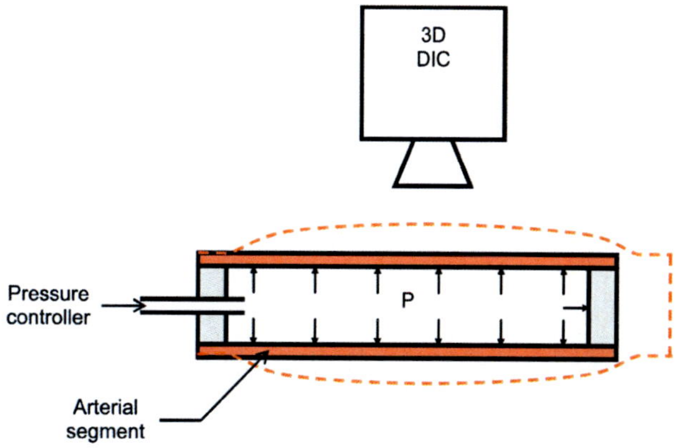

Figure 1. Schematic of the inflation/extension test of the mouse carotid artery showing one fixed end (on the left) linked to the pressure controller and one end (on the right) free to translate axially. Dashed lines represent the schematic shape of the deformed segment.

The region of interest being small (about 200 by 140 μm²), due to the depth of field of the system, and displaying very little heterogeneity, only the strains averaged over this region are considered in the analysis. Note that usual 2D-DIC is not suitable for this problem due to the non planar nature of the specimen and possible out-of-plane deformation. More details about the experimental setup and procedure can be found in (Sutton et. al. 2008).

2.2. Numerical and constitutive model

The development of the FE model of the inflation/extension test in Abaqus® is based on the experimental considerations and measurements described in Sutton et al. (2008). The geometry of the artery is assumed to be perfectly cylindrical with one end of the artery being capped. Due to axial symmetry, only one quarter of the geometry is meshed with 4280 8-node brick elements resulting in 22070 degrees of freedom. The element type chosen here, called C3D8RH in Abaqus® (hybrid formulation with constant pressure), is recommended for nearly incompressible constitutive models.

The open end of the cylinder is blocked in the axial direction whereas symmetry boundary conditions are applied on the surfaces of the quarter cylinder. Pressure is applied onto the inner surface of the artery, with values ranging up to 140 mmHg.

The constitutive model used in this study is implemented in Abaqus® and based on the developments of Holzapfel [6]. This hyperelastic incompressible model was developed to describe the passive mechanical response of arterial tissues at finite strains. The material considered is a collagen-fiber-reinforced material with two fiber directions being symmetrically arranged with respect to the axis of the artery. This theoretical basis provides a strong physical meaning to the constitutive parameters involved in the model. The simplest form of its isochoric strain energy function consists of two terms (note that incompressibility of the tissue is a well-known characteristic). The first term represents the isotropic response of the medium, related to the ground substance and elastin content, and the other two terms represent the response of the collagenous fiber network, each fiber direction having its own contribution:

$$\psi = \frac{C}{2}(I_1 - 3) + \frac{k_1}{2k_2}\left[\exp\left(k_2 (I_4 - 1)^2\right) - 1\right] + \frac{k_1}{2k_2}\left[\exp\left(k_2 (I_6 - 1)^2\right) - 1\right] \quad (1)$$

where C is the parameter of the isotropic neo-Hookean term, k_1 and k_2 are the parameters for the exponential response of the collagen fiber networks. The structural anisotropy induced by the fiber network arises from both I_4 and I_6. These terms are pseudo-invariants of the right Cauchy Green tensor $\underline{\underline{C}}$ and the fiber directions $\underline{\mathbf{f}}_1$ and $\underline{\mathbf{f}}_2$. Therefore they are driven by β, the mean fiber angle, in the medium, with respect to the circumferential direction. I_4 and I_6 give the squares of stretch for the two fiber families.

Due to histological differences between media and adventitia, two separate strain energy functions are assigned to each of these mechanically-relevant layers of an artery, the contribution of the intima being commonly considered negligible. Thus, two sets of parameters are to be identified for each layer, yielding eight material parameters. In this study, two variants of this model are considered: a simplified five parameter model and a seven parameter model. The simplified five parameter model includes the assumption that media and adventitia have identical exponential parameters. In addition, it is assumed that the value of C is the same in both layers. Hence, it features the following five parameters: C for the Neo-Hookean isotropic term (elastin and ground substance), two parameters k_1 and k_2 for the exponential response of collagen fibers, and two parameters β^{media} and $\beta^{adventitia}$ for the fiber angles in media and adventitia.

However, a full two-layer Holzapfel-type model considers that the exponential terms used in medial and adventitial layers are different [6], necessitating two additional parameters. Therefore, the seven parameter model releases the constraint on exponential parameters, making them different in each arterial layer.

2.3. Identification procedure

Given a set of experimental pressure and surface strain measurements, the principle of the present identification method is to minimize the following cost function:

$$J(\vec{\chi}) = \frac{1}{2}\left[\sum_i \left(E_{11}^{sim}(p_i) - E_{11}^{exp}(p_i)\right)^2 + \left(E_{22}^{sim}(p_i) - E_{22}^{exp}(p_i)\right)^2\right] \quad (2)$$

where $\vec{\chi}$ is the vector of parameters to be identified (ie. the constitutive parameters), p_i is the pressure applied during the inflation test, with index i ranging over the available experimental data points., E_{11} and E_{22} are Green Lagrange circumferential and axial strain components on the surface of the artery, superscripts 'sim' and 'exp' standing respectively for the simulated and experimental data. In this study, synthetic data generated by a FE calculation are also used in order to avoid noise issues when identifying the seven parameter model.

This cost function is minimized using an in-house Levenberg-Marquardt algorithm with bounds handling.

To asses the robustness of the identification method, multiple identification runs with random starting points are performed in order to compare the obtained results. However in the case of the seven parameter model, since noise in data is a major source of identification errors, especially when identifying a lot of parameters, we choose to use noise-free data which are obtained by finite element simulation with an arbitrary set of parameters. Here the set of parameters obtained with the first identification run is used.

3. Results

Using the five parameter model, convergence of the optimization algorithm is obtained after 46 iterations. We report in Table 1 the results of this first identification run. The pressure/strain curves are shown in Fig. 2. To further test the identification method, a second set of experimental data obtained by performing a second identical test with the same arterial segment is also used, the aim being to compare the results. We also report in Table 1 the results of this identification run (curves are not shown here). Note that they are very close to those obtained with the first set of data. In the following developments, only the first set of data is used.

In addition to these results, the robustness of the identification method is assessed with the method mentioned in section 2.3. The range of spanned starting points is chosen according to values found in the literature for this type of artery [12]. The range of the obtained results is reported in Table 1. Note that C shows quite a large standard deviation because these multiple runs showed that there exist two close minima in the space of parameters, the influence on the response being negligible.

Regarding the seven parameter model, synthetic noise-free data are generated using the first parameter set identified with the algorithm, referred to as the "true" set. Again, to evaluate the reliability of identifying this constitutive model, we try to identify its seven constitutive parameters using multiple identification random starting points. The results obtained through these tests are presented in Table 1. They raise several noteworthy comments.

The first comment to be made is that all of the runs lead to a very low value of the cost function J, which means that the algorithm is practically always able to find a solution. However, only 20% runs lead to the true set of parameters and the dispersion among the other results is very large.

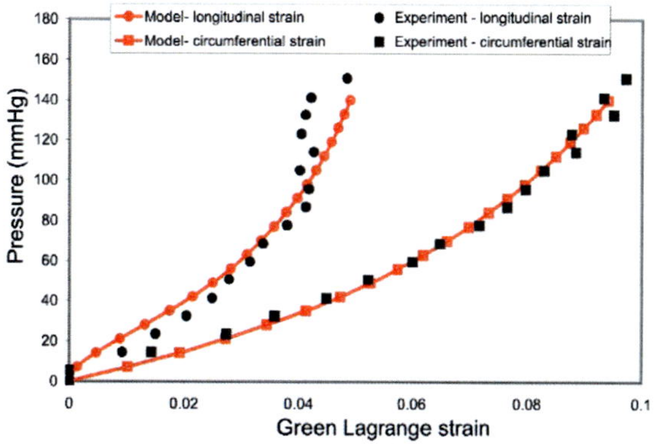

Figure 2. Pressure/strain curves obtained with the five parameter constitutive model.

Table 1. Results of parameter identification for each model and each type of procedure (single run or multiple random starting points run).

-	C (kPa)	k_1^{media} (kPa)	$k_1^{adventitia}$ (kPa)	k_2^{media} (kPa)	$k_2^{adventitia}$ (kPa)	β^{media} (°)	$\beta^{adventitia}$ (°)
Parameter bounds	[0.3;50]	[0.5;100]	[0.5;100]	[0.1;100]	[0.1;100]	[5;60]	[5;60]
-	Five parameter identification						
Results with the first data set	0.5	33		12.8		46.4	27.2
Results with the second data set	0.86	31.3		14		43.3	27.3
Results with random starting points (mean ± SD). $J \approx 1.10^{-4}$	0.45±0.37	30.5±1.07		15.4±0.51		46.7±0.2	26.8±0.17
-	Seven parameter identification						
True solution	8.8	13.9	11.8	21.6	0.242	41.9	5.14
Results with random starting points (mean ± SD). $J < 1.10^{-5}$	8.8±3	12.8±7.9	19.6±10	29±21.7	18.7±16	41.3±18	13.3±13
-	Seven parameter identification with enriched cost function						
Solution #1 Occurrence 85.7% and $J < 1.10^{-8}$	8.8	13.7	12.3	21.8	5.4	5.4	8
Solution #2 Occurrence 14.3% and $J < 1.10^{-5}$	8.8	8.11	14.2	0.1	64.7	5	38.9

4. Discussion / conclusion

4.1. Five-parameter two-layer model

In this study, firstly, two separate thick layers are considered, with the further assumption that the material parameters in the exponential terms related to the response of fiber bundles are identical in each layer. This assumption seems to be reasonable as long as only the passive mechanical behavior is considered. The reason for this is that the passive mechanical response is mainly driven by elastin and collagen fibers. The response of elastin, as well as that of the ground substance, is included in the neo-Hookean term of the strain energy function. On the other hand, the response of collagen fibers is included in the exponential terms. Whether they are in media or adventitia, it is assumed in this model that collagen fibers of the arterial wall have the same behavior.

This parameterization of the constitutive model has been shown to be correctly identified. With any initial values, the method provides a single set of parameters leading to a good agreement with the axial and circumferential experimental data, which proves that the model is well parameterized with respect to the available data and that the method is robust. These results confirm that identifying layer-specific fiber angles, based on the constitutive model of Holzapfel, is possible for such arteries using data from only one experimental test of inflation with free axial movement.

In our results, the mean fiber angle in the media is close to 45° while fibers are, in average, more circumferentially oriented in the adventitia. This means that the adventitia is found to be circumferentially stiffer than the media. Though it cannot be generalized based on a single example, this result brings some knowledge about the mechanical behavior of these arteries at the layer-scale. The relationship with the actual microstructure remains, however, a pending question [14]. From a global point of view the overall anisotropy of the artery, with the circumferential direction being stiffer, is typical for arteries in general, and is in qualitative agreement with previous studies [11, 12, 14].

4.2. Seven-parameter model

In contrast with the five parameter model, the full two-layer Holzapfel-type model has failed in providing a unique solution to the problem of identification. Our results (see Table 1) show the typical trend of an over-parameterized problem. They confirm that identifying such a model on these experimental data is very easy because the algorithm is always able to capture the experimental data with a close fit. Unfortunately, robustness is very poor because seven families of different possible solutions are obtained. Yet, this issue is not unexpected because the problem of inflating a tube made up of two fiber-reinforced layers potentially presents two solutions: one solution with a large fiber angle for the inner layer and small angle for the outer layer, and vice versa. Both solutions are possible as long as their constitutive properties can be different to compensate 3D through-thickness effects. Interestingly, our results clearly show these two alternatives, with a noticeable preference for large media angle and low adventitia angle (80% solutions). Note, also, that the mechanism of compensation is clearly illustrated by the ratio of k_1 parameters between media and adventitia, which is inversed in these two situations.

To correctly identify the parameters for this model, more abundant and/or relevant experimental data are required. In order to confirm this hypothesis, additional synthetic data are generated by simulating a simple axial tension test on the arterial segment, without any pressurization, recording axial force versus axial stretch. Accordingly, the cost function (see Eq. 2) is enriched with these data:

$$J(\vec{\chi}) = \frac{1}{2}\left[\sum_i \left(E_{11}^{sim}(p_i) - E_{11}^{exp}(p_i)\right)^2 + \left(E_{22}^{sim}(p_i) - E_{22}^{exp}(p_i)\right)^2 + \sum_i \left(F^{sim}(\lambda_i) - F^{exp}(\lambda_i)\right)^2\right] \tag{3}$$

where λ_i is the axial stretch applied during the tension test, with index i ranging over the available experimental (synthetic) data points, F^{sim} and F^{exp} are simulated and experimental (synthetic) axial force. Values are scaled so that both sum terms in J are of the same order of magnitude.

Using this updated cost function, the same multiple identification runs are performed again. Instead of obtaining seven different families of solutions, the algorithm provides only two (see Table 1). Among these two solutions, the first one, obtained only 14.3% times, reaches two boundaries of the allowed range for the parameters. This solution must obviously be rejected, as the procedure would have diverged or found inconsistent values. The other solution is considered as the true solution.

These results show that additional relevant data may easily make the identification procedure robust, discriminating thus the true solution. In this case, a simple tension test added to the inflation/extension test is sufficient.

Circumventing the problem of non-uniqueness emphasized above would also be made possible by using different types of additional data, or other strategies. For instance, separate mechanical tests of medial and adventitial layers can be used as was done in [10], or inversion tests in which the arterial segment is turned inside out to reverse the spatial locations of the media and adventitia and then redistribute through-thickness stresses in a different manner (theoretically studied in [2]). Another possible way to access useful experimental information would be to acquire through-thickness data, thanks to the use of optical coherence tomography or confocal microscopy, for instance. Such layer-specific data would help discriminate the true solution. Otherwise, in order to reduce the number of dependent parameters to be identified and help make the solution unique, additional information regarding the fiber angle distributions within each layer (from histology for instance) would be helpful. A part of the work in progress at this time is related to these last two aspects.

5. References

[1] Fung YC, Biorheology of soft tissues, Biorheology, 10, 139-55, 1973.

[2] Humphrey JD, Cardiovascular solid mechanics: cells, tissues, organs, New York: Springer-Verlag, 2002.

[3] Vito RP, Dixon SA, Blood vessel constitutive models, Annu Rev Biomed Eng, 5, 413–39, 2003.

[4] Bischoff JE, Arruda EA, Grosh K, A Microstructurally Based Orthotropic Hyperelastic Constitutive Law, J Appl Mech, 69(5), 570-579, 2002.

[5] Ning J, Xu S, Wang Y, Lessner SM, Sutton, MA, Anderson K, Bischoff JE, Deformation measurements and material property estimation of mouse carotid artery using a microstructure-based constitutive model, J Biomech Eng, 132, DOI:10,1115/1,4002700, 2010.

[6] Holzapfel GA, Gasser TC, Ogden RW, A New Constitutive Framework for Arterial Wall Mechanics and a Comparative Study of Material Models, J Elasticity, 61, 1-48, 2000.

[7] Masson I, Boutouyrie P, Laurent S, Humphrey JD, Zidi M, Characterization of arterial wall mechanical behavior and stresses from human clinical data, J Biomech, 41, 2618– 2627, 2008.

[8] Dye WW, Gleason RL, Wilson E, Humphrey JD, Altered biomechanical properties of carotid arteries in two mouse models of muscular dystrophy, J Appl Physiol, 103, 664-672, 2007.

[9] Guo X, Oldham MJ, Kleinman MT, Phalen RF, Kassab GS, Effect of cigarette smoking on nitric oxide, structural, and mechanical properties of mouse arteries, Am J Physiol Heart Circ Physiol 291, H2354–H2361, 2006.

[10] Holzapfel GA, Sommer G, Gasser TC, Regitnig P, Determination of the layer-specific mechanical properties of human coronary arteries with non-atherosclerotic intimal thickening, and related constitutive modelling, Am J Physiol Heart Circ Physiol, 289, 2048-2058, 2005.

[11] Sutton MA, Ke X, Lessner SM, Goldbach M, Yost M, Zhao F, Schreier HW, Strain field measurements on mouse carotid arteries using microscopic three-dimensional digital image correlation, J Biomed Mater Res Part A, 84, 178-190, 2008.

[12] Gleason RL , Dye WW, Wilson E, Humphrey JD, Quantification of the mechanical behavior of carotid arteries from wild-type, dystrophin-deficient, and sarcoglycan-δ knockout mice, J Biomech, 41, 3213–3218, 2008.

[13] Avril S, Badel P, Duprey A, Anisotropic and hyperelastic identification of in vitro human arteries from full-field measurements, J Biomech, 43 (15), 2978-2985, 2010.

[14] Haskett D, Johnson G, Zhou A, Utzinger U, Vande Geest J, Microstructural and biomechanical alterations of the human aorta as a function of age and location, Biomech Model Mechanobiol, 9, 725–736, 2010.

An Inverse Method to Determine Material Properties of Soft Tissues

Leonardo Ruggiero [a,b], Hugo Sol [a], Hichem Sahli [b], Sigrid Adriaenssens [a], Nele Adriaenssens [c]

[a] Vrije Univeristeit Brussel
Mechanics of Materials and Constructions (MeMC)
Pleinlaan 2, 1050 Brussel

[b] Vrije Universiteit Brussel
Electronics and Informatics (ETRO)
Pleinlaan 2, 1050 Brussel

[c] Vrije Universiteit Brussel
Lymfoedeem en Revalidatiewetenschappen en Kinesitherapie (LYMF-KINE)
Laarbeeklaan 103, 1090 Brussel

ABSTRACT

Material characterization of soft biological tissues by mixed experimental/numerical (inverse) techniques represents a powerful tool for the analysis of their complex mechanics. However, the uncertainty related to the accuracy and capability of the technique is not yet completely investigated and understood. In this work, a quasi-static indentation procedure is implemented in order to extract the Mooney-Rivlin material parameters and the equivalent Young's modulus for a hyperelastic rubber-like material. A non-standard approach based on a Finite Element (FE) inverse method is employed. An experimental set-up consisting of a dedicated micro-indentation system has been employed to extract reaction force values as a function of the penetration depth. A cost function, based on the square difference between experimental and numerical data, is optimized trough a modified Nelder-Mead direct search algorithm (MNMA). The accuracy of the identified parameters is discussed using results of a virtual benchmark case study.

INTRODUCTION

Characterization of soft biological tissues by mixed experimental/numerical (inverse) techniques is more and more often employed in bioengineering. Kevin et al. [1] used image correlation techniques Kauer M. et al. [2] used aspiration. Authors [3,4] used indentation as non-destructive experimental set-ups for ex-vivo and in-vivo material characterization.

T. Proulx (ed.), *Mechanics of Biological Systems and Materials, Volume 2*, Conference Proceedings of the Society for Experimental Mechanics Series 9999, DOI 10.1007/978-1-4614-0219-0_3, © The Society for Experimental Mechanics, Inc. 2011

The current study focuses on determination of material parameters for hyperelastic Mooney-Rivlin material model by an inverse technique employing indentation as experimental procedure. Literature shows that the methodology capabilities are not completely investigated and understood.
Different authors performed indentation studies, but few of them focused on its application in inverse techniques to execute material characterization of soft biological tissues. Lyra et al. [5] developed a small size indentation instrument to investigate the changes in the biomechanical properties of articular cartilage. Vannah et al. [6] measured the quasi-static response of bulk muscular tissue on the posterior lower legs of living human subjects evaluating the hysteresis and stress relaxation effects. Rome et al. [7] developed a hand-held device to provide an accurate, reliable, non-invasive, portable instrument for the measurement of heel pad indentation in a clinical setting. An experimental set-up with a linear actuator to perform the indentation was carried out by Miller et al. [8]. They compared the results of brain indentation experiments with numerical simulations. Han et al. [9] developed a hand-held ultrasound indentation system able to acquire force-displacement response in-vivo. Korhonen et al. [10] studied the risk of damaging for cartilage thickness lower than 2 (mm) by a commercial arthroscopic indentation instrument [5]. Hu and Desai [11] presented an experimental apparatus and the formulation to implement the experimental data trough FE simulations of the biomechanical model of a liver. Kerdok et al. [12] emphasized the manifestation of breast pathology by changes in stiffness using an inverse technique. Liu et al. [13] presented a method to determine the unique material parameters of breast tissue from indentation experiments. Gales et al. [14] validated the work conducted by [7] on hand-held heel pad indenter for determining pad properties. Recently, Kim et al. [15] collected intra-abdominal organs measurements from pigs. Jachowicz et al. [16] and Lu et al. [17] used a heuristic approach to calculate the Young's modulus of skin and thick soft tissues respectively.

It is globally accepted that soft biological tissues mechanical behaviour can be reasonably modelled by hyperelastic material models [18]. Typically, Mooney-Rivlin [2, 19, 20], Polynomial [21, 22], Ogden [18, 22], Arruda-Boyce [13], and Yeoh [23] represent the constitutive laws adopted for breast, brain, liver, cartilage or cells mechanical characterization. Recently, authors working on nano-indentation [24] employed Mooney-Rivlin models characterize human cells.

An extensive work, on breast tissues, has been conducted by Samani et al. [3]. They performed preliminary clinical studies to characterize human breast ex-vivo samples. The potential of the methodology as clinical tool to investigate pathologies, at different scales (organs, tissues, and cells), by changes in material parameters is unanimously recognized by the scientific community.

Unfortunately, authors [3] showed accuracy levels not appropriate for the application, reporting cases of poor convergence of the algorithm, and providing not congruent results [3, 25]. Recent works neglected completely the accuracy of the results and the method reliability. Particularly, Namani et al. [26] stated that indentation does not appear to have the sensitivity to extract two or more hyperelastic parameters also if the results suggest that it has the potential to evaluate the influence of processing conditions, pathology, etc., via changes in the generalized elastic modulus.

The goal of this work is to demonstrate the validity of the technique without investigate the aspect related to the correct choise of a specific material model. Assuming that hyperelastic material models can be effectively used to model the mechanical behaviour of soft biological tissues, the Mooney-Rivlin formulation is considered. A mixed experimental/numerical technique is employed and a modified Nelder-Mead optimization algorithm (MNMA) is implemented to minimize a cost function based on the square differences of reaction force values. A non-distructive indentation procedure is realized by a dedicated micro-indenter system. The validation of the device, and consequently of the methodology, is performed comparing the equivalent (globalized) Young's modulus of silicon rubber, calculated referring to Mooney-Rivlin material constants (see Appendix for the analytical description), with the tensile Young's modulus calculated from the slope stress-strain curve obtained by standard tensile tests. The Boussinesq's analytical solution of the "force on boundary of a semi-infinite body" problem [27] is also considered for validation. The results derived from an experimental and a benchmark case study are reported and compared. The benchmark allowed to considering the possible error sources coming out from the optimization procedure. The critical points of the methodology are deeply discussed and analyzed.

THEORY

In this section, the mathematical formulation of a Mooney-Rivlin (incompressible) material model and the equivalent Young's modulus calculation are discussed (see more details in the Appendix).

The Mooney-Rivlin material model (incompressible)

The classical strain energy formulation of the Mooney-Rivlin form for incompressible hyperelastic materials is often employed in the description behaviour of incompressible isotropic rubber-like materials and soft tissues [18]. Soft biological tissues and rubber may be characterized as *highly nonlinear* materials undergoing finite strains. As mentioned in Holzapefl et al. [28] the probably most widely used and established model to replicate the phenomenological behavior of rubberlike solids in the finite strain domain is the Ogden model Eq.(1). The incompressible Mooney-Rivlin material formulation is a special case of the incompressible Ogden material constitutive law,

$$\psi = \psi(\lambda_1, \lambda_2, \lambda_3) = \sum_{p=1}^{N} \frac{\mu_p}{\alpha_p}(\lambda_1^{\alpha_p} + \lambda_2^{\alpha_p} + \lambda_3^{\alpha_p} - 3) \tag{1}$$

where Ψ *is the strain energy density function,* λ_α *(α=1,2,3)* are the principal stretches, and μ denotes the classical shear modulus in the reference configuration, known from the linear theory. N is a positive integer that determinates the number of terms in the strain energy function, α_p are dimensionless constants with $p=1,\dots,N$.

The Ogden material model Eq.(1) describes the changes of the principal stretches from a reference to the current configuration. It plays a crucial role in the theory of finite elasticity [18].
The derived Mooney-Rivlin constitutive law for incompressible hyperelastic materials is written in the form Eq.(2),

$$\psi = c_{10}(\overline{I_1} - 3) + c_{01}(\overline{I_2} - 3) \tag{2}$$

where Ψ is the strain energy density function, $\bar{I}_1$ and $\bar{I}_2$ are the first and the second invariants of the deviatoric component of the left-Cauchy-Green deformation tensor, and c_{10} and c_{01} are material constants.

In this work, Eq.(2) is implemented in a numerical simulation of a quasi-static micro-indentation. The basic assumption, referring to incompressibility, is that soft tissues, are composed prevalently by water which is ideally considered incompressible. Rubber is also characterized by a Poisson's coefficients close to 0.5.

Equivalent Young's modulus for a Mooney-Rivlin solid

For the case of an incompressible Mooney-Rivlin material under uniaxial elongation (simple tension), the constitutive law Eq.(2) can be written, considering the Cauchy stress in function of the stretches as,

$$\sigma = \left(2c_{10} + \frac{2c_{01}}{\lambda}\right) \cdot \left(\lambda^2 - \frac{1}{\lambda}\right) \tag{3}$$

where σ is the Cauchy stress in the applied tensile direction, c_{10} and c_{01} are the material constants introduced in Eq.(2) and λ is the uni-axial stretch. According to Eq.(3) can be said that, considering small strain, the ratio σ/ε at infinitesimally small strain can be determined by the relation,

$$E_{eq} = 6 \cdot (c_{10} + c_{01}) \tag{4}$$

The complete description of Eq.(4) is given in Appendix.

In this work, the equivalent Young's modulus E_{eq}, given by Eq.(4) is used to compare the results with that one provided by standard tensile tests on silicon rubber and the Boussinesq's analytical solution [27].

MATERIAL & METHODS

The material characterization procedure is schematically represented in Fig.(1). The calculation of the two material constants for a Mooney-Rivlin material model, formulated under incompressibility condition, is performed employing a non-destructive indentation test. The methodology couples the results obtained by a real experiment with that one calculated numerically by a FE simulation of indentation of a deformable sample. The coupling consists in the minimization of a cost function built considering the square residuals given by,

$$S = \sum_{i=1}^{N} r_i^2 \qquad \text{with} \quad r_i = f(\delta_i)^{EXP} - f(\delta_i)^{FEM} \tag{5}$$

where S is the sum of the square residuals and r_i is given by the difference of the reaction force values $f(\delta_i)^{EXP}$ obtained by the real experiment and $f(\delta_i)^{FEM}$ numerically calculated. N is the number of sampling points for the function $f(\delta)$ expressing the reaction forces values derived from a quasi-static indentation in function of the penetration depth δ. In this study, the maximum value for δ is 1 (mm). The benchmark study is performed taking into account N=10 and N=100. The influence of the number of sampling points is analyzed.

The cost function Eq.(5) is optimized using the Modified Nelder-Mead Algorithm (MNMA) starting from an initial guess value $G_{(0)}$ (Fig.(1)). It is a direct-search optimization method for global minima determination. The description of its working principle is out of the scope of this work.

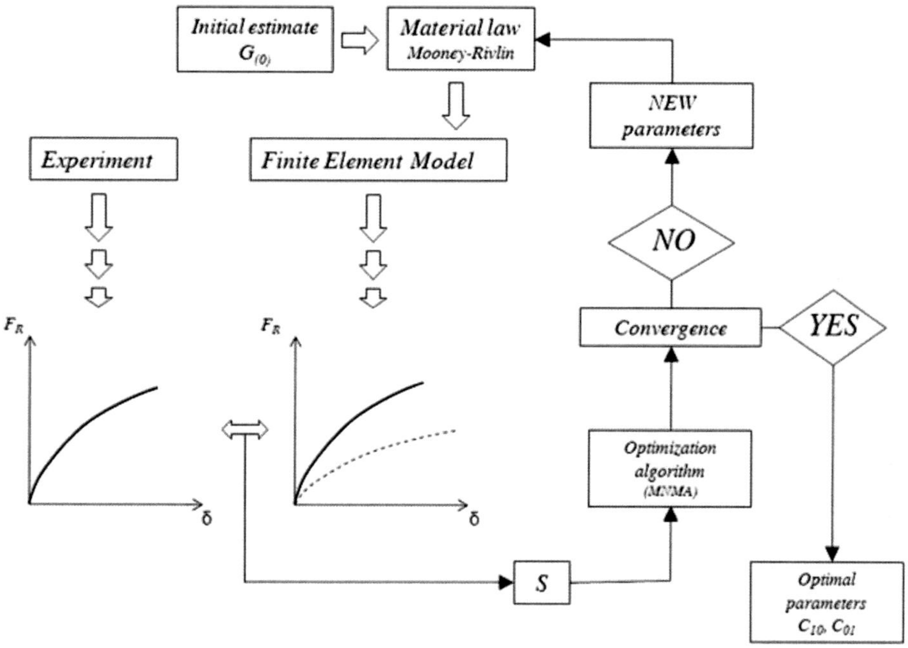

Fig. 1 Inverse method scheme

In order to evaluate the reliability of the methodology two different studies typologies have been conducted: a virtual benchmark and a real case.

The implementation of a virtual benchmark study allowed the evaluation of accuracy in the parameters (c_{10} and c_{01}) determination eliminating a possible error source due to real experimental data utilization. The benchmark "unknown" parameters have been determined by a numerical calculation of the reaction forces values at the interface of an analytical rigid surface (micro-indenter probe) and a deformable cubic specimen.

Real experiment is performed using a dedicated micro-indentation system consisting of a linear actuator, a force transducer and an acquisition data card connected to a computer. The device has a maximum penetration resolution of 220 nm and is automatically driven at speed of 0.02 (mm/s) for a measured distance of 1 (mm) in a cubic silicon rubber specimen of 1000 ml volume. A flat cylindrical punch of d=2 (mm) diameter is mounted. Ten different measurements of the force-displacement curves of the same sample are taken at distance of 5 min. The analogue signal coming out from the transducer has been sampled at frequency of 200 (MHz). A quadratic curve fitting and an average of the ten force-displacement curves is performed to obtain the final reaction force data.

For both, benchmark and real case study a final sampling of the reaction force curve is used for the calculation of the residuals r_i *(i=1,...,10)* Eq.(5).

The FE model of a quasi-static indentation is implemented (Fig.(2)) using the commercial software Abaqus-6.10. A cube of 8 ml volume is considered for the numerical calculations used for both benchmark and real case. The indentation penetration depth is δ=1 (mm) and the probe is flat cylindrical with a diameter d=2 (mm). The probe is modelled as an analytical rigid surface and it is tied to the cube surface. The cube is modelled as deformable solid. The mesh used for the cube is composed of 1500 hexahedral solid element coded as C3DH in Abaqus-6.10. Geometrical constraints are applied to the basis of the sample, to the probe reference point and at the interface probe-specimen.

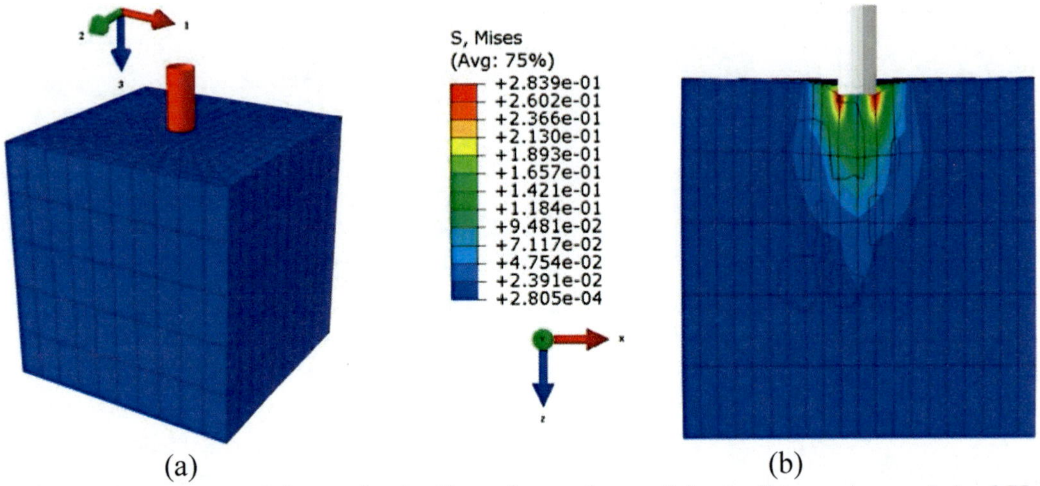

Fig. 2 (a) Finite Element model: probe (red) and specimen (blue); (b) contour plot of Von Mises (MPa) stress distribution (middle cut view): the areas far from the probe show very low stress values

The bottom of the cube is fixed in all degrees of freedom ,

$$U_1 = U_2 = U_3 = 0 \tag{6}$$

where U_1, U_2, and U_3 are the displacement in direction 1,2 and 3 respectively (Fig.(2)). The reference point at the center of the probe surface is leaved free to move only in direction 3.
At the interface probe-cube a tie constraint, and not a contact model, is applied: the analytical rigid surface is directly constrained to the surface of the elements of the deformable solid imposing a tie constraint between them (see Abaqus-6.10 Manual [29]).

The ratio δ/t (t, sample thickness) is limited to the value of 0.05 to avoid influences of boundary conditions on the reaction force values [30]. In fact, for the benchmark and the real ones (related to the real experiment) the Von Mises stress values in areas of the specimen far from the interface probe-specimen are of order 10^{-4} (MPa).

In order to test the accuracy of the method multiple starting guesses are considered for initializing the MNMA algorithm. Recent works [26] show an average error in accuracy for a Mooney-Rivlin material constants estimation of 75.08 % using starting values of 50%, 25%, and 100% smaller than the benchmark values.

The benchmark parameters considered are c_{10}=0.035 (MPa) and c_{01}=0.068 (MPa). The inverse method is tested for three different initial guesses $G_{B(0)}^1$,$G_{B(0)}^2$, and $G_{B(0)}^3$ (Tab.(1)) in the case of the benchmark study. For the real experimental case the parameters are determined using 20 different starting points $G_{R(0)}^i$ (i=1,...,20) (Tab.(2)).
Tab.(3) shows the equivalent Young's modulus $E_{eq.}^T$ calculated measuring the slope of stress-strain data obtained from standard tensile tests (ASTM D412- 06ae2 [31]), and the benchmark equivalent Young's modulus $E_{eq.}^L$ obtained by Eq.(4). This values are used as reference to determine the accuracy error. $E_{eq(B)}^{MR}$ is the equivalent Young's modulus calculated from the benchmark results and $E_{eq(R)}^{MR}$ from the results gained in the real case. Both, $E_{eq(R)}^{MR}$ and $E_{eq(B)}^{MR}$, are calculated according to Eq.(4). E_{eq}^{Num} and E_{eq}^{Exp} are the equivalent Young's moduli calculated using the Boussinesq's analytical solution of the "force on a boundary of a semi-infinite body" problem [27]. E_{eq}^{Num} is calculated from reaction force values F_R^{Max} obtained numerically performing a FE simulation using the benchmark Mooney-Rivlin constants. According to the Boussinesq's solution,

$$E_{eq} = \frac{F_R \cdot \pi \cdot r^2 \cdot \Delta l}{l} \qquad \text{with} \quad F_R^{Max} = F_R \cdot \frac{\Delta l}{l} \quad \text{for} \quad \frac{\Delta l}{l} = 1 \tag{7}$$

where E_{eq} is the equivalent Young's modulus, F_R is the reaction force value, r is the probe radius, and Δl and l are penetration depth and max. penetration depth respectively.

One can notice (Tab.(3)) that the error between E_{eq}^L and E_{eq}^{Num} (7.6%) for benchmark is in the same order of that one calculated between E_{eq}^T and E_{eq}^{Exp} (9.4%) for the real case. This means that the experimental data are reliable. The relatively high percentage in the error is due to the fact that the Boussinesq's solution is valid for linear elasticity and the data taken in consideration originate from non linear problems. In fact, the error decreases to less than 2% simulating the indentation of a linearly elastic specimen.

Tab. 1 Benchmark value for c_{10} and c_{01} and initial guesses

	c_{10} (MPa)	c_{01} (MPa)
	Benchmark Values	
	0.035	0.068
	Initial Guess Values	
$G_{B(0)}^{1}$	0.0175	0.034
$G_{B(0)}^{2}$	0.00875	0.017
$G_{B(0)}^{3}$	0.0035	0.0068

Tab. 2 Initial guess values for the real case study

	c_{10} (MPa)	c_{01} (MPa)
	Initial Guess Values	
$G_{R(0)}^{1}$	0.001	0.001
$G_{R(0)}^{2}$	0.050	0.001
$G_{R(0)}^{3}$	0.100	0.001
$G_{R(0)}^{4}$	0.100	0.050
$G_{R(0)}^{5}$	0.100	0.100
$G_{R(0)}^{6}$	0.050	0.100
$G_{R(0)}^{7}$	0.001	0.100
$G_{R(0)}^{8}$	0.001	0.050
$G_{R(0)}^{9}$	0.010	0.050
$G_{R(0)}^{10}$	0.075	0.050
$G_{R(0)}^{11}$	0.050	0.010
$G_{R(0)}^{12}$	0.050	0.075
$G_{R(0)}^{13}$	0.025	0.050
$G_{R(0)}^{14}$	0.050	0.025
$G_{R(0)}^{15}$	0.050	0.090
$G_{R(0)}^{16}$	0.090	0.050
$G_{R(0)}^{17}$	0.025	0.025
$G_{R(0)}^{18}$	0.025	0.075
$G_{R(0)}^{19}$	0.075	0.025
$G_{R(0)}^{20}$	0.075	0.075

Tab. 3 Equivalent Young's modulus calculated considering stress-strain data obtained by standard tensile tests (ASTM D412- 06ae2 [31])

	Boussinesq			$6(c_{10}+c_{01})$	Tensile test	
	F_R^{Max}(N)	E_{eq}^{Num}(MPa)	E_{eq}^{Exp}(MPa)	E_{eq}^{L} (MPa)	E_{eq}^{T}(MPa)	Error (%)
Benchmark	2.100	0.669	-	0.618	-	7.6
Real case	1.37	-	0.436	-	0.395	9.4

RESULTS & DISCUSSION

In this section results carried out by the benchmark and the real experimental are reported and discussed.

Benchmark case study

Data show that benchmark repeatability is very high. The MNMA provided in all cases the convergence to the solution identifying the global minimum and avoiding to fall in local minima by which the cost function Eq.(5) is characterized. Tab.(4) shows the values determined for both constants c_{10} and c_{01}, the mean and relative standard deviation, and the accuracy error respect to E_{eq}^{L} = 0.618 (MPa). Data demonstrate very different solutions from the percentages provided by Namani et al. [27] that reported absolute errors of 21.61% for c_{10} and 75.08% for c_{01} and a relative standard deviation of 13.6% and 22.5% respectively. This study demonstrates that the magnitude of the accuracy error for a benchmark study is very close to zero. Determination of the equivalent Young's modulus is much accurate with an error reaching an average value of 0.002%. Fig.(3) shows the searching path followed by MNMA to find the solution.

Tab. 4 Results for the benchmark (N=10): Mooney-Rivlin constatns c_{10} and c_{01}, equivalent Young's modulus $E_{eq(B)}^{MR}$ and its accuracy error respect to the equivalent Young's modulus $E_{eq.}^{L}$ determined by Eq.(4)

	Result (MPa)			Error (%)			Number of Iterations
	c_{10}	c_{01}	$E_{eq(B)}^{MR}$	c_{10}	c_{01}	$E_{eq(B)}^{MR}$	
$G_{B(0)}^{1}$	0.035045	0.067957	0.618012	0.13	0.06	0.002	83
$G_{B(0)}^{2}$	0.035024	0.067977	0.618006	0.07	0.03	0.001	113
$G_{B(0)}^{3}$	0.034935	0.068061	0.617976	0.18	0.09	0.004	106
Mean	0.035001	0.067998	0.617998	0.12	0.06	0.002	101
SD/Mean (%)	0.17	0.08	0.003				

The influence of the number of sampling points is evaluated considering N=10 and N=100 in Eq.(5) for the three sets $G_{R(0)}^{1}$, $G_{R(0)}^{2}$, and $G_{R(0)}^{3}$. A difference of less than 0.005% is noticeable for c_{10} and c_{01}. The increment in the calculation time is 10 time bigger for N=100.

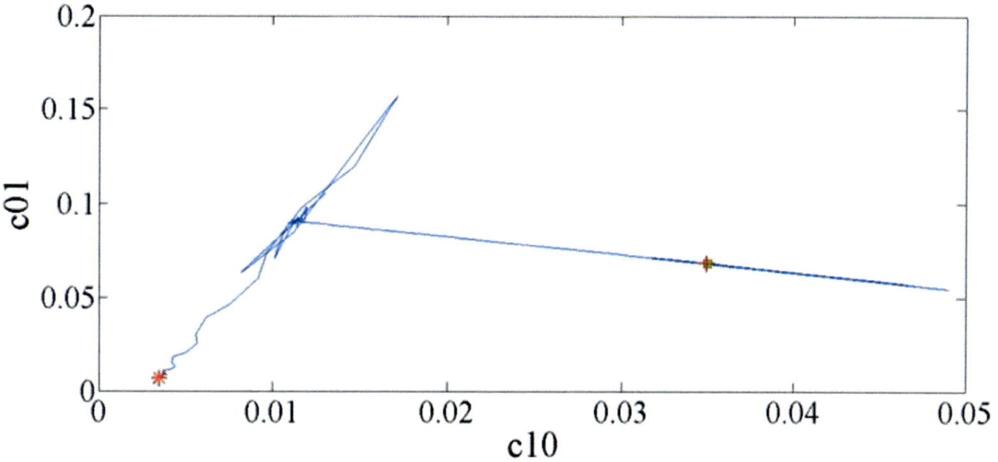

Fig. 3 MNMA search path: the red star indicates the starting point (initial guess), the red cross the final point (solution), and the green square the bechmark

Real experiment

The identified Mooney-Rivlin constants by the inverse technique are: c_{10}=0.022 (MPa) and c_{01}=0.045 (MPa). In this case study the reference value to determine the accuracy is $E_{eq.}^{T}$=0.395 (MPa). The equivalent Young's modulus calculated by Eq.(4) using the obtained Mooney-Rivlin material constants is $E_{eq(R)}^{MR}$=0.405 (MPa), 7.11% lower than E_{eq}^{Exp}. The accuracy average error is 2.6%. This demonstrates that the error margin in the real case is very low. The relative standard deviation is 0.19% for c_{10} and 0.09% for c_{01}. All the calculations are performed using N=10 sampling points for the experimental data. Results are showed in Tab.(5).

The MNMA showed the capability to distinguish the minima typology. Fig.(4) shows the local minima points for set $G_{R(0)}^{8}$. At iteration n. 43 the values for the two parameters are estimated as c_{10}^{l}= 0.001098 (MPa) and c_{01}^{l}=0.065547 (MPa). A standard Nelder-Mead termination test considers these as minimizing values and terminates giving a cost function value 4 times bigger. Iteration n. 44 represents the new initial guess adopted to find the global minimum.

According to Eq.(4) the equivalent Young's modulus referred to the local minima c_{10}^{l} and c_{01}^{l} is E_{eq}^{l}=0.400 (MPa): value 1.23% far from the solution $E_{eq(R)}^{MR}$=0.405 (MPa). This is the reason for which authors [26] stated that inverse techniques, based on indentation, cannot calculate more than one material parameter while can only "accurately" estimate the equivalent Young's modulus.

The MNMA, on the contrary of a classical Nelder-Mead implementation, recognizes the minima typology and re-starts the search, from a new guess, if the solution doesn't correspond to the global minimum.

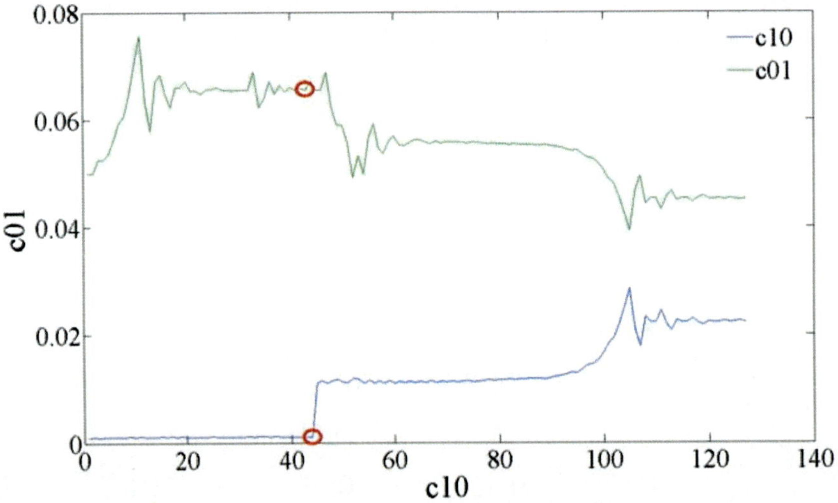

Fig. 4 Convergence trend for Mooney-Rivlin material constants c_{10} (blue) and c_{01} (green); the red circles indicate a local minimum

Benchmark V.S. Real experiment

It is possible to suppose that experimental data affect the measurement accuracy introducing an error of 2.6% referring to the equivalent Young's modulus. The error related to the benchmark values is almost zero. The relative standard deviations are almost the same. The average number of iterations is exactly the same. Reducing possible error sources deriving from standard tensile tests or for the micro-indenter it will be possible to assess the real accuracy error of the methodology which, however, cannot be lower than 0.002%.

Tab.5 Results for the real experiment: Mooney-Rivlin constatns c_{10} and c_{01}, equivalent Young's modulus $E_{eq(R)}^{MR}$ and its accuracy error respect to the equivalent Young's modulus $E_{eq.}^{T}$ determined by standard tensile tests

	Result (MPa)			Error (%)	Number of Iterations
	c_{10}	c_{01}	$E_{eq(R)}^{MR}$	$E_{eq(R)}^{MR}$	
$G_{R(0)}^{1}$	0.022328	0.045222	0.405299	2.607	120
$G_{R(0)}^{2}$	0.022381	0.045171	0.405316	2.612	124
$G_{R(0)}^{3}$	0.022327	0.045223	0.405298	2.607	154
$G_{R(0)}^{4}$	0.022368	0.045183	0.405308	2.610	87
$G_{R(0)}^{5}$	0.022295	0.045253	0.405287	2.604	110
$G_{R(0)}^{6}$	0.022335	0.045216	0.405301	2.608	125
$G_{R(0)}^{7}$	0.022422	0.045131	0.405318	2.612	158
$G_{R(0)}^{8}$	0.022341	0.045210	0.405301	2.608	127
$G_{R(0)}^{9}$	0.022354	0.045197	0.405302	2.608	82
$G_{R(0)}^{10}$	0.022329	0.045221	0.405298	2.607	86
$G_{R(0)}^{11}$	0.022401	0.045151	0.405314	2.611	81
$G_{R(0)}^{12}$	0.022421	0.045133	0.405323	2.613	117
$G_{R(0)}^{13}$	0.022424	0.045130	0.405325	2.614	64
$G_{R(0)}^{14}$	0.022322	0.045227	0.405296	2.607	73
$G_{R(0)}^{15}$	0.022313	0.045236	0.405292	2.606	113
$G_{R(0)}^{16}$	0.022378	0.045174	0.405315	2.611	80
$G_{R(0)}^{17}$	0.022348	0.045203	0.405303	2.608	85
$G_{R(0)}^{18}$	0.022429	0.045125	0.405325	2.614	70
$G_{R(0)}^{19}$	0.022307	0.045242	0.405294	2.606	74
$G_{R(0)}^{20}$	0.022375	0.045176	0.405308	2.610	90
Mean	0.022360	0.045191	0.405306	2.609	101
SD/Mean (%)	0.19	0.09	0.0028		

CONCLUSION

Results demonstrate that a mixed experimental/numerical method using non-destructive indentation is a reliable and accurate technique for characterization of Mooney-Rivlin incompressible material models. Assuming a priori, according to literature, to employ phenomenological hyperelastic constitutive laws to characterize soft biological tissues, this study demonstrate that the analyzed technique has the potential to be extended to in-vivo and ex-vivo tissues. Clinical studies, based on the collection of material parameters c_{10} and c_{01} of human soft tissues characterized by specific pathologies, age and living conditions, etc., could be carried on. Benchmark and experimental studies demonstrated that the methodology can be used to estimate more than one material parameter. The reported results allow to state firmly that the modified Nelder-Mead algorithm MNMA allowed to performing non-destructive material characterization of rubber-like materials using a single experimental procedure. A further study will evaluate the possible reduction of the accuracy error (2.6%) and the feasibility to simulate the mechanical behaviour of human breast tissue under quasi-static indentation using a hyperelastic Mooney-Rivlin incompressible material model.

REFERENCES

[1] Moerman K. M., Holt C. A., Evans S. L., and Simms C. K., Digital Image Correlation and Finite Element Modeling as a Method to Determine Mechanical Properties of Human Soft Tissue In Vivo, Journal of Biomechanics, 42, 1150–1153, 2009.

[2] Kauer M., Inverse Finite Element Characterization of Soft Tissues with Aspiration Experiments, PhD Thesis DISS. ETH No. 14233, Swiss Federal Institute of Technology Zürich, 2001.

[3] Samani A., Zubovits J., and Plewes D., Elastic Moduli of Normal and Pathological Human Breast Tissues: an Inversion Technique Based Investigation of 169 Samples, Phys. Med. Biol., 52, 1565–1576, 2007.

[4] Kellner A.L., Nelson T.R., Cervino L.I., and Boone J.M., Simulation of Mechanical Compression of Breast Tissue, IEEE Transactions On Biomedical Engineering, 54, 1885–1891, 2007.

[5] Lyyra T., Jurvelin J., Pitkänen P., Väätäinen U. and Kiviranta I., Indentation Instrument for The Measurement of Cartilage Stiffness Under Arthroscopic Control, Med. Eng. Phys., 17, 395-399, 1995.

[6] Vannah W. M., and Childress D. S., Indenter Tests and Finite Element Modeling of Bulk Muscolar Tissue in Vivo, Journal of Rehabilitation Research and Development, 33 (3), 239-252, 1996.

[7] Rome K., and Webb P., Development of a Clinical Instrument to Measure Heel Pad Indentation, Clinical Biomechanics, 15, 298-300, 2000.

[8] Miller K., Chinzei K., Orssengo G., Bednarz P., Mechanical Properties of Brain Tissue In-Vivo: Experiment and Computer Simulation, Journal of Biomechanics, 33, 1369-1376, 2000.

[9] Han L., Burcher M., and Noble J. A., Non-Invasive Measurement of Biomechanical Properties of In-Vivo Soft Tissues, MICCAI, 2488, 208-215, 2002.

[10] Korhonen R.K.,Saarakkala S., Töyräs J., Laasanen M. S., Kiviranta I., and Jurvelin J. S., Experimental and Numerical Validation of the Novel Configuration of an Arthroscopic Indentation Instrument, Phys. Med. Biol., 48, 1565-1576, 2003.

[11] Hu T., and Desai J.P., Characterization of Soft-Tissue Material Properties: Large Deformation Analysis, Proceedings of ISMS, 28-37, 2004.

[12] Kerdok A. M., Jordan P., Liu Y., Wellman P. S., Socrate S., and Howe R. D., Identification of Nonlinear Constitutive Law Parameters of Breast Tissue, Summer Bioengeneering Conference, 22-26, 2005.

[13] Liu Y., Kerdok A. E., and Howe R. D., A Nonlinear Finite Element Model of Soft Tissue Indentation, Proceedings of Medical Simulation: International Symposium - Lecture Notes in Computer Science, 3078, 67-76, 2004.

[14] Gales D. J., and Challis J. H., Validity of Hand-Held Heel Pad Indenter for Determining Heel Pad Properties, American Society of Biomechanics Blacksburg, VA, 2006.

[15] Kim J., Ahn B., De S., and Srinivasan M.A., An Efficient Soft Tissue Characterizaion Algorithm from In Vivo Indentation Experiments for Medical Simulation, Int. J. Med. Robotics and Comput. Assist. Surg., 4, 277-285, 2008.

[16] Jachowicz J., McMullen R., and Prettypaul D., Indentometric Analysis of In Vivo Skin and Comparison with Artificial Skin Models,Skin Research and Technology, 13, 299-309, 2007.

[17] Lu M. H., Yu W., Huang Q. W., Huang Y.P., and Zheng Y. P., A Hand-Held Indentation System

for the Assessment of Mechanical Properties of Soft Tissues In Vivo, IEEE Transactions on Instrumentation and Measurement, 58 (9), 3079-3085, 2009.
[18] Holzapfel G. A., and Austrell P. E., Similarities Between Soft Biological Tissues and Rubberlike Materials, Keri L. (eds.) ConstitutiveModels for Rubber IV, A.A. Balkema Publishers: Leiden, 607-6, 2005.
[19] Chung J.H., Rajagopal V., Nielsen P. M. F., and Nash M. P., Modeling Mammographic Compression of the Breast, MICCAI, 758–765, 2008.
[20] Azar F.S., Metaxa D.N., Schnall M.D., Methods for Modeling and Predicting Mechanical Deformations of the Breast Under External Perturbations, Medical Image Analysis 6, 1–27, 2002.
[21] Perez del Palomar A., Calvo B., Herrero J., Lopez J., and Doblar M., A Finite Element Model to Accurately Predict Real Deformations of the Breast, Med. Eng. Phys., 30, 1089–1097, 2008.
[22] Gao Z., Lister K., and Desai J. P., Annals of Biomedical Engineering, 38 (2), 505-516, 2010.
[23] Limtrakarn W., and Phakdeepinit W., Biomechanics Study of Knee Ligament, The First TSME International Conference on Mechanical Engineering 20-22 October, 2010.
[24] Costa K.D., and Yin F.C. P., Analysis of Indetation: Implications for Measuring Mechanical Properties with Atomi c Force Microscopy, Transactions of ASME, 121, 462-471, 1999.
[25] Wellman P. S., Tactile Imaging, PhD Thesis, Division of Engineering and Applied Sciences. Cambridge, Harvard University, 137, 1999.
[26] Namani R., and Simha N., Inverse Finite Element Analysis of Indentation Tests to Determine Hyperelastic Parameters of Soft-Tissue Layers, J. Strain Analysis, 44, 347-362, 2009.
[27] Timoshenko S., and Goodier J.N., Theory of Elasticity, McGRAW-HILL, BOOK COMPANY, Inc. 1951.
[28] Holzapfel G. A., JOHN WILEY & SONS, LTD eds., Nonlinear Solid Mechanics a Continuum Approach for Engineering, 2004.
[29] Abaqus 6.10 Online Documentation © Dassault Systèmes, 2010
[30] Lyyra-Laitinen T., Niinimäki M., Töyräs J., Lindgren R., Kiviranta I., and Jurvelin J.S., Optimization of the Arthroscopic Indentation Instrument for the Measurement of Thin Cartilage Stiffness, Phys. Med. Biol., 2511-2524, 1999.
[31] ASTM D412 - 06ae2 Standard Test Methods for Vulcanized Rubber and Thermoplastic Elastomers—Tension.

APPENDIX

Equivalent Young's modulus for a Mooney-Rivlin solid

The Mooney-Rivlin material model is characterized by the two material constants c_{10} and c_{01}. It is formulated, according to the strain energy formulation, as

$$\psi = c_{10}(\overline{I_1} - 3) + c_{01}(\overline{I_2} - 3) \tag{1}$$

where Ψ is the strain energy density function, $\bar{I}_1$ and $\bar{I}_2$ are the first and the second invariants of the deviatoric component of the left-Cauchy-Green deformation tensor, and c_{10}, c_{01} are material constants. In the case of simple tension, the constitutive law can be written considering the Cauchy stress in function of the stretches in the form,

$$\sigma = \left(2c_{10} + \frac{2c_{01}}{\lambda}\right) \cdot \left(\lambda^2 - \frac{1}{\lambda}\right) \tag{2}$$

where σ is the Cauchy stress in the applied tension direction, c_{10} and c_{01} are the material constants introduced in Eq.(1), $\lambda = \varepsilon + 1$ is the stretch, and ε is the true strain. For a complete calculation of Eq.(2) see Holzapefel (G. A., JOHN WILEY & SONS, LTD eds., Nonlinear Solid Mechanics a Continuum Approach for Engineering, 2004). This appendix gives the mathematical calculation which allow to state the relation

$$E_{eq.} = 6 \cdot (c_{10} + c_{01}) \tag{3}$$

Where $E_{eq.}$ is the linearized equivalent Young's modulus for a hyperelastic Mooney-Rivlin incompressible material model characterized by the constants c_{10} and c_{01}.

Starting from Eq.(2) and substituting the stretches λ as a cost function of the true strain in the tensile direction,

$$\sigma = \left(2c_{10} + \frac{2c_{01}}{\varepsilon + 1}\right) \cdot \left((\varepsilon - 1)^2 - \frac{1}{\varepsilon + 1}\right) = 2 \cdot \left(\frac{c_{10}(\varepsilon + 1) + c_{01}}{\varepsilon + 1}\right) \cdot \left(\frac{(\varepsilon + 1)^3 + 1}{\varepsilon + 1}\right) =$$

$$= 2 \cdot \left(\frac{c_{10}(\varepsilon + 1)^4 - c_{10}(\varepsilon + 1) + c_{01}(\varepsilon + 1)^3 - c_{01}}{(\varepsilon + 1)^2}\right) =$$

$$= 2 \cdot \left(\frac{c_{10}(\varepsilon^4 + 4\varepsilon^3 + 6\varepsilon^2 + 4\varepsilon + 1) - c_{10}\varepsilon - c_{10} + c_{01}(\varepsilon^3 + 3\varepsilon^2 + 3\varepsilon + 1) - c_{01}}{\varepsilon^2 + 2\varepsilon + 1}\right). \tag{4}$$

Considering a linear relationship between σ and ε, Eq.(4) can be formulated as,

$$\sigma = 2 \cdot \left(\frac{c_{10}(\varepsilon^4 + 4\varepsilon^3 + 6\varepsilon^2 + 4\varepsilon + 1) - c_{10}\varepsilon - c_{10} + c_{01}(\varepsilon^3 + 3\varepsilon^2 + 3\varepsilon + 1) - c_{01}}{(\varepsilon + 1)^2}\right). \approx$$

$$\approx 2 \cdot \left(\frac{c_{10}4\varepsilon + c_{10} - c_{10}\varepsilon - c_{10} + c_{01}3\varepsilon + c_{01} - c_{01}}{2\varepsilon + 1}\right) =$$

$$= 2 \cdot \left(\frac{c_{10}3\varepsilon + c_{01}3\varepsilon}{2\varepsilon + 1}\right). \tag{5}$$

Assuming $2\varepsilon + 1 = 1$ for small ε one can write,

$$E_{eq.} = \frac{\sigma}{\varepsilon} = 6 \cdot (c_{10} + c_{01}). \tag{6}$$

Eq.(6) is the equivalent Young's modulus for a hyperelastic Mooney-Rivlin incompressible material model in uniaxial extension.

Estimation of the 3D residual strain field in the arterial wall of a bovine aorta using optical full-field measurements and finite element reconstruction

Genovese Katia (a), PhD, Badel Pierre(b), PhD, Avril Stéphane(b), Pr

[(a)] Dept. of Engineering and Environmental Physics – Università degli Studi della Basilicata – Viale dell'Ateneo Lucano 10 – 85100 Potenza – Italy

[(b)] Center for Health Engineering – Ecole des Mines de Saint Etienne – 158, cours Fauriel – 42023 Saint Etienne – France

ABSTRACT

Arterial tissue consists of multiple types of structurally important constituents, each of which may have individual material properties and associated stress-free configurations that may evolve over time. This gives rise to residual stresses which contribute to the homeostatic state of stress in vivo as well as adaptations to perturbed loads, disease, or injury.

The existence of residual stress in an intact but traction-free excised arterial segment suggests compressive stresses in the inner wall and tensile stresses in the outer wall. Accordingly, an artery ring springs open into a sector after a radial cut. The measurement of the opening angle is used to deduce the residual stresses, which are actually the stresses required to close back the ring.

The open angle method provides an average estimate of circumferential residual stresses but it gives no information on either the local distribution through the thickness and along the axial direction. To address this lack, we propose to derive maps of residual stresses using a novel approach based on the contour method. Small pieces of freshly excised tissue are carefully cut into the specimen and the local distribution of residual stresses is determined from whole-body DIC measurements using a finite element model.

1. Introduction

Even removed from the body, the artery is not stress-free [1, 2]. Arterial tissue is similar to a laminated fibre-reinforced composite material consisting in multiple types of structurally important constituents, each of which may have individual material properties [3] and associated stress-free configurations that may evolve over time. This gives rise to residual stresses which contribute to the homeostatic state of stress in vivo as well as adaptations to perturbed loads, disease, or injury [2].

Understanding of residual stresses is crucial in consideration of vascular growth and remodelling. Several authors have shown that although residual stresses are small compared with in vivo wall stresses they have a strong influence on the in

T. Proulx (ed.), *Mechanics of Biological Systems and Materials, Volume 2*, Conference Proceedings of the Society for Experimental Mechanics Series 9999, DOI 10.1007/978-1-4614-0219-0_4, © The Society for Experimental Mechanics, Inc. 2011

vivo stress distribution. In particular, the residual stresses seem to have the effect of homogenizing the circumferential stress within each layer in the physiological load state [1, 2].

The existence of residual stress is manifest in the springing open of an arterial ring when it is cut in the radial direction. This opening relieves the residual stress, the circumferential component of which is compressive on the inner part of the ring and tensile on the outer part. The measurement of the opening angle is used to deduce the residual stresses, which are actually the stresses required to close back the ring.

The open angle method provides an average estimate of circumferential residual stresses but it gives no information on either the local distribution through the thickness and along the axial direction. Moreover, the procedure relies on the assumption that the residual stresses in the ring are completely released after one single radial cut. For this reason, the opening angle method is still to be improved. Zhang et al. [12] showed artefacts that may be induced by viscoelasticity. Recently, Ohayon et al. [13] used the opening angle method for an artery with an atheromatous plaque. This yielded an interesting study about the definition of plaque vulnerability. However, the approach is somehow questionable as the heterogeneity and localization of residual stresses in a plaque may not be fully revealed by the method. Actually, the opening angle method is well suited when the distribution of residual stresses is axisymmetric. There is still a lack of a more local method that may be used for identifying the residual stresses in an artery of any shape, especially with a plaque.

Several authors have tried to address this lack. Han and Fung [8] used markers to track local radial and circumferential deformations during the opening angle experiment. Sommer et al. [9] considered the existence of layer specific residual stretch by separating mechanically the different layers after performing the open angle experiment.

In order to have a new insight in this important question, we propose to derive the 3D local distribution of the residual stresses using a novel approach based on the most recent methods developed for the measurement of residual stresses in laminated fibre-reinforced composite materials.

Indeed, reconstruction of residual stresses has always been an important topic of research regarding engineering implications. Many techniques have been developed by materials engineers to measure residual stresses. For laminated composite materials for instance, at a ply or macroscopic scale, the hole-drilling method [4], the layer removal method and the compliance method may be used [5]. The compliance method applied to filament wound tubes consists in sectioning the tubes and measuring the change in strains on the external and internal surfaces with two biaxial strain gages, one bonded to each side of the tube wall. This leads to the estimation of the axial and circumferential internal bending moments of filament wound tubes. Using the hole-drilling or other removal approaches in combination with full displacement field optical techniques may provide some advantages with respect to the traditional use of strain gauges, in particular higher sensitivity and no-contact measurement [6].

Removal methods combined to the measurement of whole-body local deformations are referred to as the contour method [12, 13]. We propose here its application to arterial tissue, with the particularity that deformation measurement focuses on the removed part of the material. The approach combines whole-body digital image correlation (DIC) measurements and the resolution of an inverse problem through a finite element model updating (FEMU) method. Small pieces of freshly excised tissue are carefully cut into the specimen and the local 3D distribution of residual stresses is determined from their deformation after removal.

2. Materials and methods

2.1. Materials

A fairly straight segment of a fresh bovine descending aorta (about 40 mm length, 30 mm outer diameter, 4 mm thickness) was provided by a local slaughterhouse. The loose connective tissue was carefully removed in physiological solution and the artery was put vertically on a plastic cap (Fig. 1a). A ring of the artery was used to calculate the circumferential residual strains with the 'opening angle method' [1, 2]. An urethane expandable sealant foam was sprayed into the artery lumen while keeping it moist by means of cotton pads soaked in physiological solution (Fig. 1b). During the curing time (about 10 min) the foam gently expands and fits closely the artery inner geometry penetrating the branches. A preliminary test carried out by filling a latex glove finger with the foam assessed the absence of stress generation during its expansion and hardening. Once the foam hardened, 8 acupuncture needles (0.25 mm diameter, 20 mm length) were used to delimit fairly square areas on the arterial external surface (Fig. 1c). The needles penetrated through the arterial wall thickness and through the foam (for about 15 mm) allowing to fix the selected pieces in their unloaded configurations. A scalpel was then used to cut the specimens

(Fig. 1d), first removing the remaining part of the artery (Fig. 1e, note how the artery opened up in response to the radial cuts) and finally cutting the foam into the final shape (Fig. 1f). Two additional needles were used to handle the specimen during all the stages of the experimental protocol. The samples were checked carefully in order to verify the adhesion of the inner arterial surface to the foam and then were sprayed with black Indian ink by using a fine tipped airbrush. This allowed to create the speckle random pattern needed for the subsequent DIC-based measurement [14].

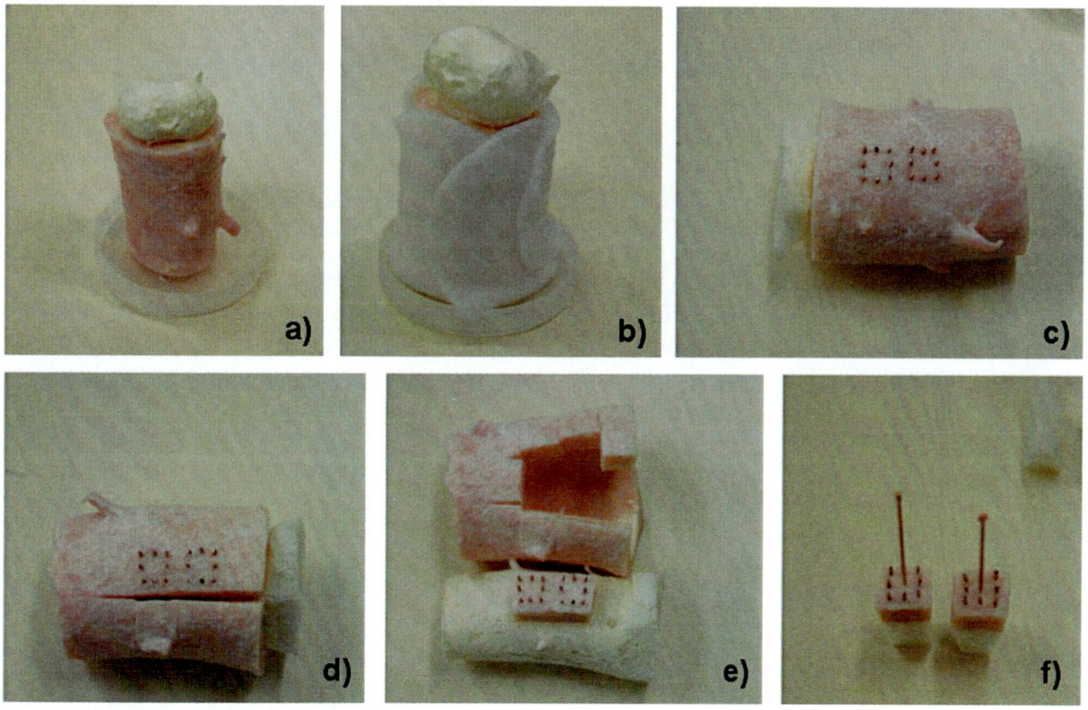

Fig. 1 Sample preparation: a) the artery was filled with expandable sealant foam, b) the foam was allowed to harden, c) the areas of interest were delimited by using acupuncture needles, d) the remaining part of the artery was cut and e) removed, f) the foam was cut into the final shape.

2.2. Experimental set-up and protocol for releasing residual stresses

Aim of the experimental test was to retrieve the shape of the sample before and after removing the needles that keep it into the unloaded configuration in order to get the displacement field due to the 'opening up' of the ring segment. A DIC-based 'whole body' measurement was obtained by positioning the sample on the top of a ½'' steel post fixed onto a rotational stage. A single camera was used to capture a 1 min video sequence of the sample throughout a full 360° rotation. In this way, each couple of frames of the video sequence (of about 200 frames) can be considered two views of a lateral DIC stereo-system and can be thus processed on the basis of the classic Stereo- Photogrammetry principles [15]. In particular, calibration was performed with the modified DLT method [16] that allows to retrieve the stereo-system parameters by using a single image (conversely, for example to the 'bundle-adjustement' method). To this scope, a square dot pattern was glued around the steel post allowing to have the calibration pattern always present in all the images of the sequence. Moreover, in this way it is possible to automatically reconstruct and merge the point clouds into a single global reference system. All the codes needed for calibration, DIC-matching, reconstruction and merging were compiled into the Matlab environment.

2.3. Deformation measurement

A first video sequence of the sample allowed to capture its shape in its unloaded configuration (i.e. with residual stresses, Fig. 2 left). Then, without moving the sample, the needles were carefully removed with the tweezers and the central needle was used to slightly lift the sample from the foam to ensure no contact between the foam and the inner surface of the artery. Once the stress-free configuration was obtained (Fig. 2 right), a second video sequence was captured.

Fig. 2 Two images of the same face of the sample with and without residual stresses. The square dot pattern glued on the post served for calibration purposes. The white arrows point out the deformation of the sample.

Due to the large amount of the frames captured it was possible to choose two closely corresponding views of the sample before and after the needles were removed and to match them via DIC. Image processing was performed by considering a measurement grid of about 2000 evenly spaced points for each face. DIC was performed by using template and analysis subset sizes of 21×21 pixels^2 and 41×41 pixels^2, respectively. Only points pairs matched with a correlation coefficient > 0.98 were considered for subsequent analysis. Eight couples of stereo-systems were considered for reconstructing the sample lateral surface all over 360° (4 for edges and 4 for the faces). The accuracy of the data overlap between the contiguous reconstructed surfaces (of the order of 10^{-2} mm^2) made not necessary to consider a larger number of stereo-views.

Figure 3 shows the data point cloud obtained for the reference (stress-free) configuration. In this case, it was possible also to reconstruct the top and the bottom of the arterial segment. Being these two surfaces load-free they don't need to be contoured also for the deformed configuration. Point data clouds so obtained were used as input to build the FE model.

Fig. 3 Point data obtained from the 'whole-body' DIC contouring of the stress-free configuration. The use of a single 'cylindrical' dot calibration pattern glued on the post (see Fig.2) allowed to automatically reconstruct and merge the point data of the 8 stereo-system in a single reference system.

2.4. Numerical reconstruction of residual strain fields

Based on the experimental shape and displacement measurements described above, a finite element (FE) model is built in order to obtain the three-dimensional residual strain field as well as an estimation of the residual stress field within the piece of artery. The principle of the procedure is to create a model of the stress-free specimen and to apply the displacement field which was previously measured experimentally between the stress-free and the residually stressed configurations. Thus, the deformed state of the FE model corresponds to the residually stressed state of the specimen.

The initial geometry of the FE model is created from the surface point cloud obtained after reconstruction of the specimen shape. The point cloud and its connectivity constitute the elements of the surface mesh. The volume mesh of the specimen is then created by filling the closed surface with tetrahedral elements. This operation is done using FE software Abaqus®. The obtained mesh is made of 293341 elements and 161652 degrees of freedom. The boundary conditions to be applied to the model are extracted from the DIC analysis performed experimentally. They consist of the displacement field on each individual node of the lateral faces of the specimen.

It is necessary to provide a constitutive equation for the FE calculation to be performed and then the strain field to be extracted. The constitutive model of the arterial material being unknown *a priori*, it is chosen to use a simple constitutive model. It is important to note that the behavior of the arterial material is expected to be little non-linear and little stiff. Indeed, the analysis is performed between the stress-free state and the residually stressed state, which means that the orders of magnitude of stresses, as well as that of strains, are very low with respect to the physiological conditions. For this reason, assuming the behavior is linear or quasi linear in such conditions seems to be a valuable assumption. Regarding anisotropy, it is neglected as a first approximation, assuming that stiffness is very low and does not vary much with orientation at such strains, where micro-structural and compositions effects are not fully activated.

A neo-Hookean model is implemented for its simplicity. This hyperelastic constitutive model is governed by a strain energy density function W of the following form:

$$W = \frac{G}{2}(I_1 - 3) + \frac{K}{2}(J - 1)^2 \quad (1)$$

where I_1 is the first deviatoric strain invariant, $J = \mathrm{tr}\left(\underline{\underline{\mathbf{F}}}.\underline{\underline{\mathbf{F}}}^T\right)$ is the volume ratio. The behavior of the material is driven by the two parameters G and K. For infinitesimal deformations, K denotes the bulk modulus (or compressibility modulus) and G denotes the shear modulus. The values of G and K being unknown, they were set according to the range of values reported in the literature for human and mammals aortas [17-21], also assuming the material to be quasi incompressible [22]. Most of the mentioned studies provide data in terms of incremental Young's modulus and/or other constitutive models, therefore the values of G and K were obtained by linearizing and identifying the parameters when needed. Quasi-incompressiblity was assumed to correspond to an equivalent Poisson's ratio of 0.49. Note, however, that the Poisson's ratio being set, the influence of the other parameter on the strain field is negligible. The magnitude of stresses may nevertheless be affected. In this study, G = 18 kPa and K = 800 kPa.

The problem is solved using an implicit scheme.

Thanks to the FE resolution, the actual surface strain field is then reconstructed. Remember that this strain field corresponds to the field of residual strains since the imposed deformation maps the stress-free specimen to its residually-stressed configuration. In order to evaluate the residual strain distribution through the specimen, average strains are computed within ten layers of the specimen. To this aim, the components of strain are averaged over each group i of elements with coordinates comprised between $r = r_i$ and $r = r_{i+1}$.

2.5. Derivation of the fields of residual stress

The field of residual stresses within the specimen is also an output of the FE simulation. It relies on the choice of the constitutive behavior of the material. In the previous section, the constitutive model is homogeneous throughout the specimen and it is governed by a neo-Hookean model.

3. Results

The circumferential Green strains calculated with the 'opening angle method' on the inner and outer surfaces are $\varepsilon_{\theta i}$ = -0.052 and $\varepsilon_{\theta i}$ = 0.068, respectively.

3.1. Residual strains and stresses

The map of axial and circumferential surface strains is shown in Fig. 4. No clear trend can be distinguished from this type of visual result. Therefore, the distribution of strains is also averaged at different radial coordinates and plotted as a function of the radial coordinate, see left graph in Fig. 5, which is more explicit. This graph emphasizes an almost linear increase of the circumferential strain component with increasing radius. Its value at the inner radius is about -1.11e-2, which corresponds to a compressive strain, and it increases up to values around 4.7e-2, corresponding to tensile strains. Note that the zero circumferential strain is obtained at 17% of the total thickness, not at mid-thickness. The axial strain is globally positive though not homogeneous. The influence of boundary effects may affect the results.

The distribution of residual stresses is plotted on the right graph in Fig. 5. When assuming the constitutive model of the material to be homogeneous, the residual strain components show the same evolution as the corresponding strains until mid-thickness. For larger radii, the stress distribution reaches a peak followed by a sharp decrease up to the outer radius. This response is related to the combination of axial and circumferential strain evolutions which induce strong three-dimensional effects on the stress field.

This graph raises, however, an important issue. The stress state of the specimen is not at equilibrium, otherwise the integral of the circumferential stress through the thickness would be zero. This issue is likely to be related to the application of foam inside the vessel segment during the experimental procedure. Though the ends of the segment are open, a slight pressure may remain in the section from which the specimen is extracted. For this reason, the curves are shifted towards the left by an amount guaranteeing a zero integral. Here, the offset is about 4 kPa which is reasonable and possibly resulting from the action of the foam onto the inner surface of the specimen. The corresponding graph is plotted in Fig. 6.

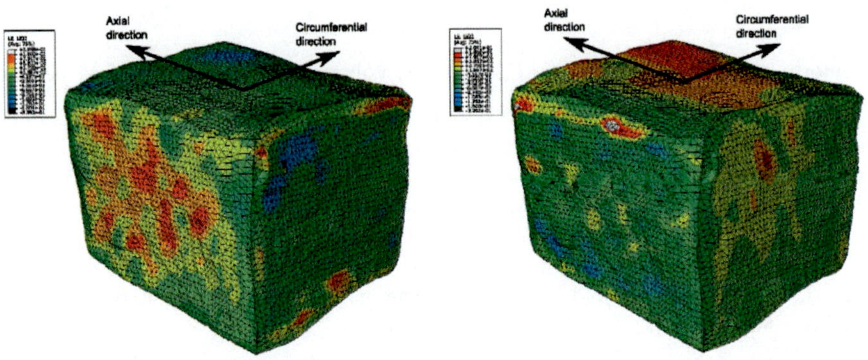

Figure 4. Surface logarithmic strain fields: on the right, axial strain and, on the left, circumferential strain.

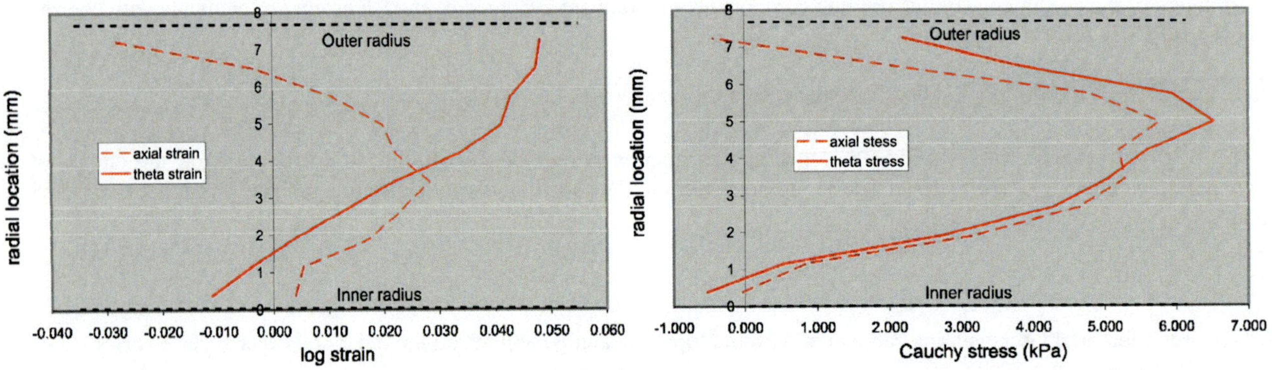

Figure 5. Logarithmic strain and Cauchy stress distribution as a function of the radial location.

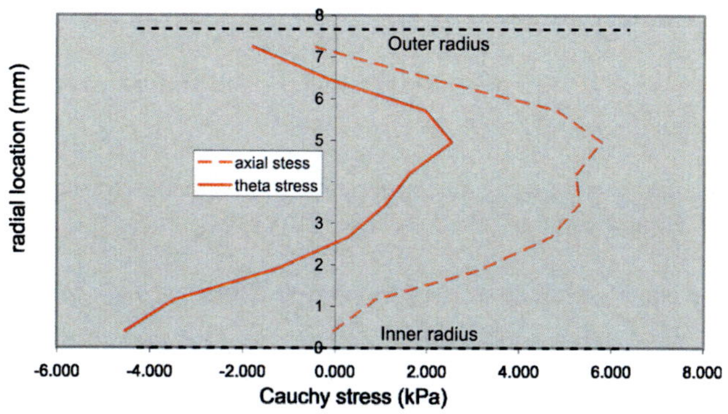

Figure 6. Cauchy stress (with a shift of -4 kPa corresponding to the action of the foam) distribution as a function of the radial location.

4. Discussion / conclusion

The current study was aimed at having a new experimental glance on the question of residual stresses in arteries. Only a few experiments have been carried out so far and no quantitative conclusion can be drawn for the moment, the experimental approach needing to be further validated.

However, the new protocol proposed in this study, based on imaging and finite element analyses, has revealed important subjects of discussion that could be investigated in the future regarding residual stresses in arteries.

More especially, the distribution of residual stresses through the thickness of an artery is not necessarily linear, even if the distribution of residual strains in the circumferential direction is linear (for instance in an open angle experiment). Two points that are generally not considered in the literature and may affect the distribution of residual stresses through the thickness of arteries are:

1. the constitutive behavior may be different in medial and adventitial layers, even at low strain. The rationale supporting this assumption relies on the nature of the composition of these layers [1]. The adventitia comprises mainly collagen fibers which are loose at the no-pressure state, whereas the media includes elastin sheets, smooth muscle cells and other components which provide an elastic stiffness even at low pressures. For this reason, different constitutive parameters may be affected to each of these layers, the adventitia being softer [9].

2. Axial strain is never considered in opening angle experiments (plane strain is generally assumed i.e. the axial stretch remains constant during an opening angle experiment [1]). We show here first that the axial strain is not zero when the residual stresses are released and second that it strongly affects the circumferential stress response by Poisson effect (due to the quasi-incompressibility of the medium).

Investigating these two points will require the development of a new theoretical background which is out of scope of this paper. It will also require applying the imaging and finite element approach developed in this study on a large number of specimens to generate sufficient database. To this aim, a certain number of points regarding tissue handling will be improved for rendering the approach more robust, in particular controlling better the fixing action of needles and also the foam influence that may induce some internal pressure and induce an offset on the residual stresses. Regarding the finite element model, it is envisaged to include more realistic constitutive equations based on local indentation tests for deriving the residual stresses in future experiments.

5. References

[1] Humphrey JD. Cardiovascular Solid Mechanics: Cells, Tissues, and Organs, Springer, 2002.

[2] Fung YC. Biomechanics. Mechanical Properties of Living Tissues. In: Springer, editor.; 1993.

[3] Matsumoto T., Gotob T., Furukawaa T., Satob M. Residual stress and strain in the lamellar unit of the porcine aorta: experiment and analysis. J Biomech 2004; 37: 807-815.

[4] Sicot O., Cherouat A., Gong X.L., and Lu J. Determination of residual stresses in composite laminates using the incremental hole-drilling method. J Compos Mater 2003; 27:575–598

[5] Ersoy N., Vardar O. Measurement of residual stresses in layered composites by compliance method. J Compos Mater 2000; 34:575–598.

[6] Baldi A. Full field methods and residual stress analysis in orthotropic material. I. Linear approach. Int J Sol Struct 2007; 44(25–26):8229–8243.

[7] Han HC, Fung YC. Species dependence of the zero stress state of aorta: pig versus rat. ASME J Biomech Eng 1991; 119:438:444.

[8] Cardamone L, Valentin A, Eberth JF, Humphrey JD. Origin of axial prestretch and residual stress in arteries. Biomech Model Mechanobiol 2008.

[9] Holzapfel GA, Sommer G, Auer M, Regitnig P, Ogden RW. Layer-specific 3D residual deformation of human aortas with non-atherosclerotic intimal thickening. Annals Biomed Eng 2007; 35(4):530-545.

[10] Murugan M, Naranayan R. Finite element simulation of residual stresses and their measurement by contour method. Mat Design 2009; 30:2067-2071.

[11] Prime MB. Residual stresses measured in quenched HSLA-100 steel plate. Proceedings of the 2005 SEM Annual Conference and Exposition on Experimental and Applied Mechanics, paper #52.

[12] Zhang W, Guo X, and Kassab GS. A Generalized Maxwell Model for Creep Behavior of Artery Opening Angle, J Biomech Eng 2008; 130(5): 054502.

[13] Ohayon J, Dubreuil O, Tracqui P, Le Floc'h S, Rioufol G, Chalabreysse L, Thivolet F, Pettigrew RI, Finet G. Influence of residual stress/strain on the biomechanical stability of vulnerable coronary plaques: potential impact for evaluating the risk of plaque rupture. Am J Physiol Heart Circ Physiol 2007;293(3):H1987-96.

[14] Sutton MA, Orteu J-J, Schreier H, Image Correlation for Shape, Motion and Deformation Measurements, Springer, 2009.

[15] Faugeras OD, Three-Dimensional Computer Vision: A geometric Viewpoint, MIT Press, 1993.

[16] Hatze H, High precision three-dimensional photogrammetric calibration and object space reconstruction using a modified DLT-approach, J Biomech, 21, 533-538, 1988.

[17] Cox RH, Passive mechanics and connective tissue composition of canine arteries, Am J Physiol, 234, H533-41, 1978.

[18] Dobrin PB, Mechanical properties of arteries, Physiol Rev, 58, 397-460, 1978.

[19] Armentano RL, Barra JG, Levenson J, Simon A, Pichel RH, Arterial wall mechanics in conscious dogs. Circ res, 76, 468-478, 1995.

[20] Thubrikar MJ, Labrosse M, Robicsek M, Al-Soudi J, Fowler B, Mechanical properties of abdominal aortic aneurysm wall, J med Eng Tech, 25, 133 – 142, 2001.

[21] Vorp DA, Biomechanics of abdominal aortic aneurysm, J Biomech, 40, 1887–1902, 2007.

[22] Carew TE, Vaishnav RN, Patel DJ, Compressibility of the arterial wall, Circ. Res, 23, 61-68, 1968.

Estimating Hydraulic Conductivity in Vivo Using Magnetic Resonance Elastography

A.J. Pattison[1], P.R. Perrinez[1], M.D.J. McGarry[1], J.B. Weaver[2], K.D. Paulsen[1,2]

[1]Dartmouth College, Hanover, NH 03755, [2]Dartmouth-Hitchcock Medical Center, Lebanon, NH 03756

1 ABSTRACT

Magnetic resonance elastography (MRE) is an imaging technique that estimates mechanical properties of *in vivo* tissue. Traditionally, a linear elastic or viscoelastic material model has been used to describe tissue behavior. Recently, a poroelastic algorithm has been developed to better estimate properties of biphasic tissues such as brain, which is 75-80% water. While shear modulus and pore-pressure are usually estimated from this model, hydraulic conductivity may also provide pertinent clinical information. Defined as the rate at which fluid penetrates through pores, estimates of this parameter would be an important biomarker in applications like tumor identification, detection of increased intracranial pressure, diagnosing traumatic brain injury, and drug delivery. Sensitivity analyses have been performed to demonstrate the feasibility of reconstructing this parameter, and hydraulic conductivity has been successfully estimated in simulations. The focus of current work is improving the reliability of this parameter for use in phantoms and *in vivo*. A robust hydraulic conductivity reconstruction would allow for a wide array of studies testing its applicability to diagnosing disease.

2 INTRODUCTION

Disease and injury often cause mechanical property changes in the affected tissue. This is apparent in cancer, where tumors are tangibly stiffer masses surrounded by healthy soft tissue, and liver fibrosis, where the entire liver tissue becomes stiff [1]. In many cases, the contrast between diseased and normal tissue is far greater than contrast provided by more common diagnostic tools like magnetic resonance imaging (MRI) or computed tomography (CT) [2]. Understanding the mechanical property distribution of tissues could allow for a stronger diagnostic tool than the current standard techniques.

One imaging modality that probes this contrast mechanism is known as 'elastography'. Here, the fundamental relationship between stress (σ) and strain (ϵ), known as Hooke's Law, is probed to estimate the Elastic modulus (E). In the earliest cases, a specified stress was applied to a tissue with an ultrasound probe and the resulting deformation (strain) was measured by correlating the two different radiofrequency fields, allowing for an estimate of E [3]. Drawbacks to this technique is that data is only taken in one dimension and it is highly user-dependent. Magnetic resonance elastography (MRE) is a version of elastography that characterizes a full 3-dimensional displacement field. In MRE, a low frequency, sinusoidal vibration is applied to the tissue of interest, either in one cycle (transient) or in multiple cycles (steady-state). The resulting tissue deformation is captured by a phase-contrast MRI sequence that has the gradients synchronized with the external vibration - allowing the MR to capture the phase accumulation at distinct time-points in the motion cycle. This displacement field can be fit to an equation that relates displacement and mechanical properties (ie. Navier's equation), with the final result being a mechanical property map known as an elastogram [4, 5].

This technique has been used to measure properties of muscle, liver, heart, breast, prostate, and brain [6, 7, 8, 9, 10, 11]. A major difference between these tissues that is not taken into account is their widely different mechanical responses. Equations to fit the tissue deformation to the mechanical property distribution can vary from tissue to tissue but, to date, have been limited to linear elasticity and viscoelasticity. Both of these models assume tissue to consist of only a solid elastic matrix. Linear elastic reconstructions have been shown to have limited use in *in vivo* tissue [12], while viscoelasticity, which describes an elastic and a damping viscous component, has shown to be more proficient at some *in vivo* work [13, 14]. However, neither of these models properly describe tissue that consists of a high amount of water, such as brain [15]. Due to a distinct solid and fluid phase, brain tissue is better characterized as a biphasic or poroelastic continuum [16, 17]. Poroelasticity is based on Biot's consolidation theory, which describes a solid porous matrix and a penetrating pore fluid [18].

This model has been recently developed for MRE and has been shown to better characterize porous materials than linear or viscoelastic material models [16]. Also, while linear elastic and viscoelastic models give a shear modulus estimate, the poroelastic algorithm gives an additional estimate of the pore-pressure. In studies done previously with the use of this algorithm, it

T. Proulx (ed.), *Mechanics of Biological Systems and Materials, Volume 2*, Conference Proceedings of the Society for Experimental Mechanics Series 9999, DOI 10.1007/978-1-4614-0219-0_5,

was shown that the pore-pressure estimate follows the elicited hydrostatic pressure of tofu [19]. However, the time-harmonic pore-pressure estimate may not fully characterize the hydrostatic pressure of the tissue. MRE experiments are run anywhere from 50-150*Hz*, and hydrostatic pressure is a constant (0*Hz*-like) parameter, therefore no hydrostatic pressure information should be present in the resulting displacements. This may indicate, instead, that keeping the estimates of the poroelastic parameters constant over different elicited hydrostatic pressures may be forcing a change in the estimated pore-pressure. Some of the parameters of interest in poroelasticity are hydraulic conductivity, porosity, and apparent mass density. Hydraulic conductivity (HC) is a measure of the ease at which fluid can pass through the pores in a material. Porosity is a volume fraction of the total pore space to the total volume. Apparent mass density is a measure of the work done by relative motion between the solid and fluid phases and is a function of the fluid density, porosity, excitation frequency, and geometry.

HC is an important parameter to consider because of its wide array of applications and the scope at which the parameter is estimated. In many brain pathologies, the perfusion levels change causing problems in overall function. However, perfusion is a macroscopic parameter and estimates higher levels of blood flow. Diffusion is another parameter that is altered drastically in many brain diseases. In this case, diffusion relies on microscopic alterations in molecular flow. HC may fall between these two parameters, allowing for a broad array of applications [20]. For instance, different regions of brain parenchyma have varied levels of matrix-to-water ratios, and thus, may result in different HCs for regions such as grey and white matter. While the solid tissue matrix may have similar mechanical structure in these regions, poroelastic parameters like HC may differentiate these regions better.

In pathology, it is widely known that normal and tumorous tissue have very different vasculatures. Even within tumor classes, perfusion properties change dramatically due to the type of tumor. For instance, benign tumors lack the tumor vasculature of a malignant tumor, which is described as irregular, tortuous, and heterogeneous. Furthermore, in a brain tumor that consists of very little brain tissue, blood flow is lower in the tumor than the surrounding tissue. On the other hand, if brain tumors consist of portions of brain tissue, flow can actually be higher than the surrounding tissue [21]. Estimating both stiffness and HC may allow for differentiation amongst healthy, benign, and malignant tumors. A benign tumor may show up as a stiff mass with very low HC, while a malignant tumor may also show up as a stiff mass, but with a much more variable level of HC.

Another application would be related to any disease in which noticeable increases in intracranial pressure are expected. Hydrocephalus is a disease characterized by a blockage in cerebrospinal fluid transport from the ventricles to the subarachnoid space. The fluid bulids up in the lateral ventricles, causing an increase in periventricular pressure. The current diagnostic procedure involves an MRI or CT scan that will depict the enlarged ventricles. However, this can be confused with any disease in which ventricles are larger due to atrophying of the tissue. The change that current diagnosis tools are not depicting is the increase in pressure. As stated prior, while pore-pressure may not be the best estimate of hydrostatic pressure, HC may depict this disease. With an increase in pressure on a porous matrix, the normal interstitial fluid flow is interrupted.

Traumatic brain injury is a current hot topic in medicine because of the prevalence of concussions in sports. Traumatic brain injury destroys neural connections and leaves large amounts of edematous tissue for fluid (blood or CSF) to flow freely. If the injury is mild enough, these neural connections begin to regenerate over time and improve the brain's overall function. HC estimates would be able to locate the impacted area by delineating the area of high flow and, also, would be able to track the recovery of the neural connections over time. One more motivation behind understanding HC is the interest in knowing these porous parameters for use in drug delivery. Convection-enhanced delivery (CED) is a current and popular option for drug delivery, but suffers from backflow, resulting from too great of a force of the drug infusion and is unsuccessful [22, 23]. In this case, the tissue has too low of a HC for the applied level of infusion. An understanding of the porous properties of brain tissue like HC and porosity would allow for more accurate delivery methods to neuropathologies like Alzheimer's and Parkinson's disease.

This work will attempt to reconstruct both shear modulus and hydraulic conductivity in a series of porous tissue phantom simulations. Noise tests will be used to check the robustness of the algorithm. The ability to reconstruct both a mechanical property and a poroelastic property will provide greater characterization of the material and will allow for extensions to *in vivo* brain tissue.

3 METHODS

3.1 Mathematical Model

The equations used to solve the time-harmonic poroelastic response in MRE extend from Biot's original work in 3D consolidation theory which described the settlement properties of soils under an applied load [18]. Cheng *et al.* [24] extended this work to the frequency domain, with the notation kept similar to Biot. More recent work by Perrinez *et al.* [15, 16] moved poroelasticity to the field of time-harmonic elastography. The resulting coupled set of equations can be written as

$$\nabla \cdot \mu \nabla \overline{\mathbf{u}} + \nabla \frac{\mu}{1-2\nu}(\nabla \cdot \overline{\mathbf{u}}) - (\alpha - \beta)\nabla p = -\omega^2(\rho - \beta\rho_f)\overline{\mathbf{u}} \tag{1}$$

$$\frac{\rho_f \omega^2(\alpha - \beta)}{\beta}(\nabla \cdot \overline{\mathbf{u}}) + \frac{\omega^2 \rho_f \phi^2}{\beta R}\overline{p} + \nabla^2 \overline{p} = 0 \tag{2}$$

where the vector $\overline{\mathbf{u}}$ is the three-dimensional displacement vector, $\overline{p}$ is the pore-pressure, μ is shear modulus, and ν is the Poisson's ratio. The overbar $(-)$ is the frequency-dependent part of the variable. Here, the assumption is made of fully saturated tissue (no air bubbles), therefore the parameters α and R can be simplified to 1 and ∞, respectively (For more information on meaning of parameters, please refer to Perrinez *et al.*). This simplification reduces the equations to

$$\nabla \cdot \mu \nabla \overline{\mathbf{u}} + \nabla \frac{\mu}{1-2\nu}(\nabla \cdot \overline{\mathbf{u}}) - (1 - \beta)\nabla \overline{p} = 0 \tag{3}$$

$$\frac{\rho_f \omega^2(1 - \beta)}{\beta}(\nabla \cdot \overline{\mathbf{u}}) + \nabla^2 \overline{p} = 0 \tag{4}$$

where β is given as

$$\beta = \frac{\omega \phi^2 \rho_f \kappa}{i\phi^2 + \omega\kappa(\rho_a + \phi\rho_f)} \tag{5}$$

The β term is a compilation of all of the poroelastic parameters that, to date, have been assumed based on literature research. κ is hydraulic conductivity - the ease at which fluid penetrates the pores. ϕ denotes porosity, which is the volume fraction of pore space to total volume. The other parameters are actuation frequency (ω), fluid density (ρ_f), and apparent mass density (ρ_a).

Shear modulus and pore pressure were the original parameters estimated in the poroelastic inversion algorithm. A sensitivity analysis has been performed by Perrinez *et al.* [17], that tested the sensitivity level of the other poroelastic parameters. This was done by slightly changing the parameter estimate and running a forward problem to calculate the estimated displacement field. The resulting displacement fields were compared to see the overall affect of the parameter on the material response. The parameters tested were hydraulic conductivity, porosity, and a C parameter that is a function of porosity and the fluid density. Figure 1 shows that hydraulic conductivity is the most sensitive parameter, thus being the most likely parameter to be accurately estimated using this inversion technique.

To estimate hydraulic conductivity, a series of modifications were made to the existing optimization code. In the original poroelasticity algorithm, β is placed outside of a divergence operator, thus assuming all terms in β are considered uniform. However, in most phantoms or tissue, HC varies drastically. A reformulation of the weak form was performed to find the updated version of the finite element model.

Reformulating the coupled set of equations while keeping β inside the divergence operator keeps Eqn. 3 as the same, but alters Eqn. 4 to give

$$\rho_f \omega^2 \nabla \cdot (\overline{\mathbf{u}}(1-\beta)) + \nabla \cdot (\beta \nabla \overline{p}) = 0 \tag{6}$$

Applying the divergence theorem to get the weak form becomes

$$\left\langle \beta \nabla \overline{p}, \nabla \phi_i \right\rangle - \left\langle \rho_f \omega^2 [\nabla \cdot (1-\beta)\overline{u}], \phi_i \right\rangle = \oint (\hat{n} \cdot \beta \nabla \overline{p})\phi_i ds \tag{7}$$

which changes the subelements of the stiffness matrix A to become

$$a_{41} = \left\langle -\rho_f \omega^2 (1-\beta)\frac{\partial \phi_j}{\partial x}\phi_i \right\rangle \tag{8}$$

$$a_{42} = \left\langle -\rho_f \omega^2 (1-\beta)\frac{\partial \phi_j}{\partial y}\phi_i \right\rangle \tag{9}$$

$$a_{43} = \left\langle -\rho_f \omega^2 (1-\beta)\frac{\partial \phi_j}{\partial z}\phi_i \right\rangle \tag{10}$$

$$a_{44} = \left\langle \beta \left(\frac{\partial \phi_j}{\partial x}\frac{\partial \phi_i}{\partial x} + \frac{\partial \phi_j}{\partial y}\frac{\partial \phi_i}{\partial y} + \frac{\partial \phi_j}{\partial z}\frac{\partial \phi_i}{\partial z} \right) \right\rangle \tag{11}$$

Once this update was performed, the stiffness matrix A and the Hessian and Jacobian matrices were updated. Lastly, addition

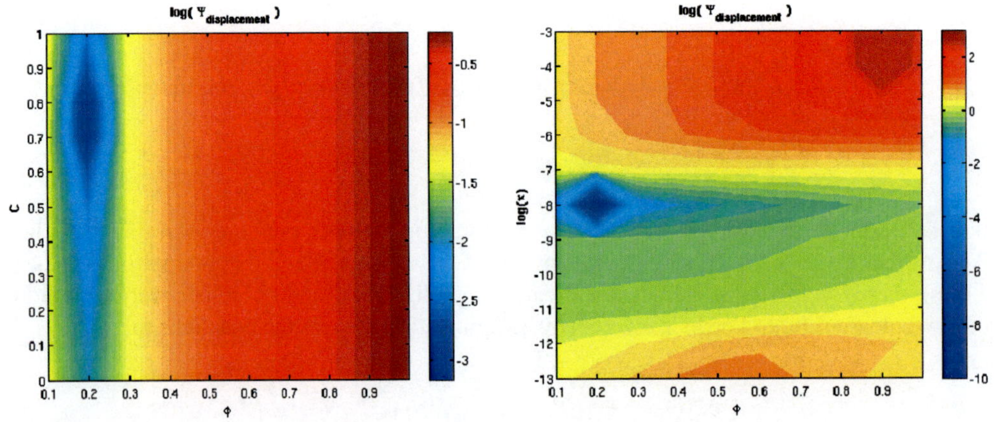

Figure 1: Contour plots of the sensitivity of three poroelastic parameters - hydraulic conductivity (κ), porosity (ϕ), and experimental parameter C. Images are comparing relative changes in the displacement when a change in a single parameter is made. Left image compares C and ϕ and the right image compares ϕ and κ. This figure is adapted from [17].

of more complex regularization techniques was needed due to the addition of a new, lower sensitivity parameter. Regularization was tailored towards the individual parameters and their inherent sensitivity levels and also to the current error level of the optimization scheme. Current regularization techniques include Marquadt, Joachimowicz, and Tikhonov, with future techniques such as Total Variation Minimization are expected to be used to increase stability and accuracy of the algorithm.

3.2 Simulated Phantom Studies

Two simulated phantom data sets were created to test the accuracy and the stability of the algorithm. The first test was a cubic porous phantom ($0.04m^3$) that had one large spherical inclusion. The background value was μ=3000*Pa* and κ=1e-9$\frac{m^3s}{kg}$. The inclusion was given contrast in both shear (3200*Pa*) and HC (2e-9$\frac{m^3s}{kg}$). The nodal distance was 2.0*mm*, which gave 17,261 nodes and 96,000 tetrahedral elements. Simulations were run over a range of frequencies (1-150*Hz*) and were supplied with a known displacement at the actuation surface (1*mm*) that is similar to what is expected during tofu and *in vivo* experiments. A noise experiment was conducted by adding Gaussian noise to the displacements that resembles the noise acquired during an MRI experiment. The level of displacement noise necessary to mock an MRI was calculated based on the error accrued in the phase and the number of measurements (phase offsets) taken in normal elastography experiments. Approximately 1.5% Gaussian noise was estimated for MR noise, as well as another similar level of noise accrued from MR elastography setup and post-processing. A total of 1.5% Gaussian noise was used for 1/2 MR noise and 3.0% noise was considered full MR noise.

To test resolution and robustness of the code, a second experiment was performed involving two conical inclusions in a brick porous phantom of size 50*mm* x 50*mm* x 100*mm*. The cones are meant to model theoretical hard masses in tissue (ie. tumors) where the tissue matrix is stiff and the fluid flow in the pores is lowered. With the same background values as the first experiment, one cone was given contrast in both shear modulus and HC (3200*Pa* and 9e-10$\frac{m^3s}{kg}$) while the second cone had both contrast as well, but in the opposite direction as the first cone (2800*Pa* and 2e-9$\frac{m^3s}{kg}$). A nodal distance of 2.5*mm* was used, which gave a total of 34,081 nodes and 192,000 elements. The same range of frequencies and actuation displacement was given to this simulated phantom.

4 RESULTS

The first simulated phantom result shows very accurate and precise estimation of shear modulus and HC. In both cases, a sharp outline of the inclusion is found and gives highly accurate estimates. As noise is added to the displacement fields, the shear modulus elastogram still locates the inclusion quite well, with only slightly more variation in the background estimate. The HC estimate suffers more when noise is added. At 1/2 MR noise, the image looks very similar, only at full MR noise does the image become more greatly distorted. The spherical inclusion loses the exact boundaries and the background has slightly higher variation. However, due to the magnitude of the values ($10^{-9}\frac{m^3s}{kg}$), the variations in the image are exaggerated.

When examining the regions of interest (background and inclusion), it is clear that the reconstruction algorithm estimates the values very well. Even with added noise, all of the relative errors between the estimated and true values stay below 1% and

Table 1: Estimates of the background and inclusion of the simulated phantom with one spherical inclusion. HC values are given in exponential value (ie. $10^{-9}\frac{m^3 s}{kg}$ is -9 on colorbar)

	Zero Noise	*1/2 MR Noise*	*MR Noise*
Background			
Shear Modulus (*Pa*)	3006.5 $\pm$ 22.7029	3006.9 $\pm$ 22.6093	3008.4 $\pm$ 24.631
Relative Error (%)	0.217	0.230	0.280
HC ($\frac{m^3 s}{kg}$)	-8.9869 $\pm$ 0.0325	-8.9798 $\pm$ 0.0342	-8.9551 $\pm$ 0.054
Relative Error (%)	0.145	0.224	0.498
Inclusion			
Shear Modulus (*Pa*)	3179.3 $\pm$ 22.1504	3173.8 $\pm$ 25.3778	3175.6 $\pm$ 24.659
Relative Error (%)	0.647	0.819	0.763
HC ($\frac{m^3 s}{kg}$)	-8.7649 $\pm$ 0.0215	-8.7792 $\pm$ 0.0219	-8.7438 $\pm$ 0.0311
Relative Error (%)	0.746	0.910	0.503

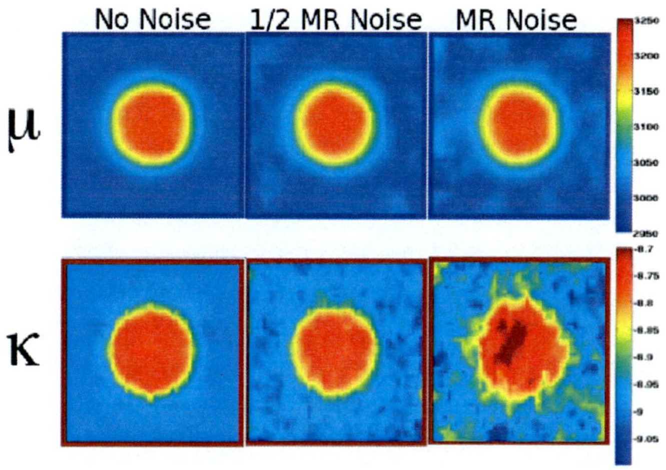

Figure 2: Elastograms of a simulated phantom with one spherical inclusion. Top row is shear modulus estimates (background=3000*Pa*, inclusion=3200*Pa*) with differing levels of Gaussian noise added to the displacements. Bottom row is hydraulic conductivity estimates (background=1e-9$\frac{m^3 s}{kg}$, inclusion=2e-9$\frac{m^3 s}{kg}$) with differing levels of noise. Colorbar for bottom images are the estimates' exponents (ie. $10^{-9}\frac{m^3 s}{kg}$ is -9 on colorbar)

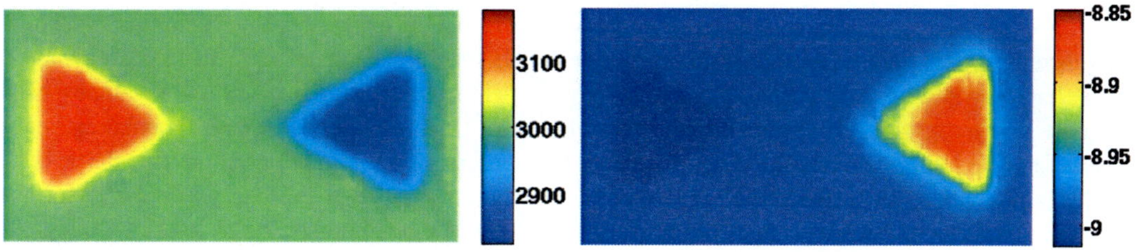

Figure 3: Elastogram of a simulated phantom with two conical inclusions. Top row is estimate of shear modulus (background=3000*Pa*, left inclusion=3200*Pa*, right inclusion=2800*Pa*) and bottom image is estimate of hydraulic conductivity (background=1e-9$\frac{m^3 s}{kg}$, left incl.=2e-9$\frac{m^3 s}{kg}$, right incl.=9e-10$\frac{m^3 s}{kg}$). Colorbar for bottom image is the estimate's exponent (ie. $10^{-9}\frac{m^3 s}{kg}$ is -9 on colorbar)

also the standard deviations of the regions were less than 1% of the true value. The influence of the level of noise was shown in the increasing standard deviation.

The second phantom involved two conical phantoms of different contrast - one was a stiffer inclusion with a higher HC and the second was softer than the background and had a lower HC. It, too, showed high accuracy and good resolution. While the point of the cone is slightly rounded, there is still good shear and HC definition for the both the inclusions.

5 DISCUSSION

To date, poroelastic MRE reconstructions have assumed that important mechanical parameters have a specific value and are homogeneous. The values assumed were based on literature or experimental results, but these 'normal' values can be different inter- or intra-subject. For instance, grey and white brain matter will have completely different cellular formations and thus could have different stiffnesses and different hydraulic conductivities.

A reformulation of the FE model allowed for spatially varying poroelastic parameters like hydraulic conductivity and porosity. The weak form kept β in the divergence operator and changed the resulting weak formulation. The resulting estimates were slightly more accurate the estimates taken using the original algorithm. While these simulations had homogeneous material properties in distinct regions, this reformulation will improve the reconstructions of phantoms or *in vivo* tissue more drastically.

As seen from Figure 2 and Table 1, the current algorithm very accurately estimates shear modulus and HC. Visually, the spherical inclusion is clearly delineated for both parameters and the integrity of the elastograms does not suffer much as higher levels of noise are added. Due to the lower sensitivity and the magnitude of HC, the noise affects it more than the robust shear modulus parameter. For instance, the background HC values fluctuate much more noticeably at 1/2 MR noise than the shear modulus values. And at full MR noise, the delineation of the inclusion is starting to distort, even if the values remain close to correct. The values shown in Table 1 give averaged values over the inclusion and background regions and give a relative error when compared to the expected value. While the error stays $< 1\%$ in all cases, it would be even smaller if it were not for the transition period that is seen between the inclusion and background. This results from the spatial filtering that is applied to the displacements to keep the derivatives stable. A partial volume type effect occurs, causing a higher level of variation in the regions of interest. Variations and errors in both regions would be less if a smaller region was analyzed that did not include any data near the region boundaries.

Figure 3 is a more complex simulated phantom that tests the robustness of the code as well as the resolution. Two cones of differing contrasts tests the sensitivity of the parameters and how specific a parameter can be when multiple values need to be estimated. The resolution of the parameter is tested by having small points at the ends of the cones. The algorithm, again, successfully estimates the parameters and also delineates the two cones to high accuracy.

Hydraulic conductivity could become an important biomarker in diseases of tissues that can be successfully modeled as poroelastic environments. The current focus is on neuropathologies like hydrocephalus, Alzheimer's, Parkinson's, and tumors. Each of these pathologies has a drastic and distinct influence on the macro- and micro-environment. From a macroscopic scale, global tissue properties change due to external problems like in hydrocephalus. From a microscopic scale, the interstitial environment is disrupted, like the loss of myelin sheath in Alzheimer's. Estimating a mechanical property and a biphasic property allow for a greater understanding of the true effect of the disease.

6 CONCLUSION

In elastography, the material models most commonly used in estimating mechanical properties of tissue are linear elasticity and viscoelasticity. However, it has been shown that these models are insufficient for tissues like brain that contain a high level of fluid infiltrating through pores and that poroelasticity better describes this biphasic environment [15, 25]. Magnetic resonance poroelastography has shown to be an improvement in the measurement of porous materials; however, many of the parameters in the algorithm are given constant and assumed values based on past literature values or experimental estimates. In most *in vivo* cases, parameters like hydraulic conductivity can change over orders of magnitude based on tissue type or disease. It was shown here that it is possible to reconstruct this parameter accurately and precisely. Further, while the effect of noise is more drastic in HC than in shear modulus, accurate estimates were gained at MR-level noise. Future work will make this parameter more robust and eventually be sensitive enough to accurately estimate *in vivo* brain tissue. This knowledge will be paramount in differentiating between healthy and diseased tissue in many types of neuropathologies.

References

[1] Y. C. Fung, Biomechanics: Mechanical Properties of Living Tissues, 2nd Edition, Springer-Verlag, New York City, NY, 1993.

[2] Y. K. Mariappan, K. J. Glaser, R. L. Ehman, Magnetic Resonance Elastography : A Review, Clinical Anatomy 23 (2010) 497–511. doi:10.1002/ca.21006.

[3] R. J. Dickinson, C. R. Hill, Measurement of soft tissue motion using correlation between A-scans, Ultrasound in Medicine and Biology 8 (1982) 263–271.

[4] R. Muthupillai, D. J. Lomas, P. J. Rossman, J. F. Greenleaf, A. Manduca, R. L. Ehman, Magnetic resonance elastography by direct visualization of propagating acoustic strain waves, Science 269 (1995) 1854–1857.

[5] R. Muthupillai, R. L. Ehman, Magnetic resonance elastography, Nature Medicine 2 (5) (1996) 601–603.

[6] G. Heers, T. R. Jenkyn, M. A. Dresner, J. R. Basford, M.-O. Klein, K. R. Kaufman, R. L. Ehman, K.-N. An, Measurement of muscle activity with magnetic resonance elastography, Clinical Biomechanics 18 (2003) 537–542. doi:10.1016/S0268-0033(03)00070-6.

[7] S. K. Venkatesh, M. Yin, J. F. Glockner, N. Takahashi, P. a. Araoz, J. A. Talwalkar, R. L. Ehman, MR Elastography of Liver Tumors: Preliminary Results, American Journal of Roentgenology 190 (2008) 1534–1540. doi:10.2214/AJR.07.3123.

[8] T. Elgeti, J. Rump, U. Hamhaber, S. Papazoglou, B. Hamm, J. Braun, I. Sack, Cardiac Magnetic Resonance Elastography, Investigative Radiology 43 (11) (2008) 762–772.

[9] A. L. McKnight, J. L. Kugel, P. J. Rossman, A. Manduca, L. C. Hartmann, MR Elastography of Breast Cancer: Preliminary Results, American Journal of Radiology 178 (2002) 1411–1417.

[10] R. Chopra, A. Arani, Y. Huang, M. Musquera, J. Wachsmuth, M. Bronskill, D. B. Plewes, In Vivo MR Elastography of the Prostate Gland Using a Transurethral Actuator, Magnetic Resonance in Medicine 62 (2009) 665– 671.

[11] A. J. Pattison, S. S. Lollis, P. R. Perriñez, I. M. Perreard, M. D. J. McGarry, J. B. Weaver, K. D. Paulsen, Time-harmonic magnetic resonance elastography of the normal feline brain, Journal of Biomechanics 43 (2010) 2747–2752. doi:10.1016/j.jbiomech.2010.06.008.
URL http://www.ncbi.nlm.nih.gov/pubmed/20655045

[12] I. M. Perreard, A. J. Pattison, M. M. Doyley, M. D. J. McGarry, Z. Barani, E. E. W. Van Houten, J. B. Weaver, K. D. Paulsen, Effects of frequency- and direction-dependent elastic materials on linearly elastic MRE image reconstructions., Physics in medicine and biology 55 (22) (2010) 6801–6815. doi:10.1088/0031-9155/55/22/013.
URL http://www.ncbi.nlm.nih.gov/pubmed/21030746

[13] R. Sinkus, M. Tanter, T. Xydeas, S. Catheline, J. Bercoff, M. Fink, Viscoelastic shear properties of in vivo breast lesions measured by MR elastography., Magnetic Resonance Imaging 23 (2005) 159–165. doi:10.1016/j.mri.2004.11.060.

[14] P. Asbach, D. Klatt, U. Hamhaber, J. Braun, R. Somasundaram, B. Hamm, I. Sack, Assessment of liver viscoelasticity using multifrequency MR elastography, Magnetic Resonance in Medicine 60 (2008) 373–379.

[15] P. R. Perriñez, F. E. Kennedy, E. E. W. Van Houten, J. B. Weaver, K. D. Paulsen, Modeling of Soft Poroelastic Tissue in Time-Harmonic MR Elastography, IEEE Transactions on Biomedical Engineering 56 (3) (2009) 598–608.

[16] P. R. Perriñez, F. E. Kennedy, E. E. W. Van Houten, J. B. Weaver, K. D. Paulsen, Magnetic resonance poroelastography: an algorithm for estimating the mechanical properties of fluid-saturated soft tissues., IEEE transactions on medical imaging 29 (3) (2010) 746–55. doi:10.1109/TMI.2009.2035309.
URL http://www.ncbi.nlm.nih.gov/pubmed/20199912

[17] P. R. Perriñez, A. J. Pattison, F. E. Kennedy, J. B. Weaver, K. D. Paulsen, Contrast detection in fluid-saturated media with magnetic resonance poroelastography, Journal of Medical Physics 37 (7) (2010) 3518–3526. doi:10.1118/1.3443563.

[18] M. A. Biot, General Theory of Three-Dimensional Consolidation, Journal of Applied Physics 12 (1941) 155–164.

[19] A. J. Pattison, P. R. Perriñez, M. D. J. McGarry, J. B. Weaver, K. D. Paulsen, Feasibility study to measure changes in intracranial pressure using magnetic resonance poroelastography, in: ISMRM, Vol. 608, 2010, p. 3406.

[20] D. Le Bihan, E. Breton, D. Lalleman, P. Grenier, E. Cabanis, M. Laval-Jeantet, MR Imaging of Intravoxel Incoherent Motions: Applicationo Diffusion and Perfusion in Neurologic Disorders, Radiology 161 (1986) 401–407.

[21] P. Vaupel, F. Kallinowski, P. Okunieff, Blood flow, oxygen and nutrient supply, and metabolic microenvironment of human tumors: a review., Cancer research 49 (23) (1989) 6449–65.
URL http://www.ncbi.nlm.nih.gov/pubmed/2684393

[22] A. A. Linninger, M. R. Somayaji, T. Erickson, X. Guo, R. D. Penn, Computational methods for predicting drug transport in anisotropic and heterogeneous brain tissue., Journal of biomechanics 41 (10) (2008) 2176–87. doi:10.1016/j.jbiomech.2008.04.025.
URL http://www.ncbi.nlm.nih.gov/pubmed/18550067

[23] O. Ivanchenko, N. Sindhwani, A. A. Linninger, Experimental techniques for studying poroelasticity in brain phantom gels under high flow microinfusion., Journal of biomechanical engineering 132 (5) (2010) 051008. doi:10.1115/1.4001164.
URL http://www.ncbi.nlm.nih.gov/pubmed/20459209

[24] A. H.-D. Cheng, T. Badmus, D. E. Beskos, Integral equation for dynamic poroelasticity in frequency domain with BEM solution, Journal of Engineering Mechanics 117 (5) (1991) 1136–1157.

[25] G. Franceschini, D. Bigoni, P. Regitnig, G. A. Holzapfel, Brain tissue deforms similarly to filled elastomers and follows consolidation theory, Journal of the Mechanics and Physics of Solids 54 (2006) 2592–2620. doi:10.1016/j.jmps.2006.05.004.
URL http://linkinghub.elsevier.com/retrieve/pii/S0022509606000883

Comparison of Iterative and Direct Inversion MR Elastography Algorithms

M.D.J. McGarry[1], E.E.W van Houten[2] A.J. Pattison[1], J.B. Weaver[3], K.D. Paulsen[1,3]

[1]Dartmouth College, Hanover, NH 03755, [2]University of Canterbury, Christchurch, New Zealand, [3]Dartmouth-Hitchcock Medical Center, Lebanon, NH 03756

1 ABSTRACT

Magnetic Resonance Elastography (MRE) is a medical imaging modality which aims to image the mechanical properties of tissue. Tissue stiffness maps can be calculated from measurements of the steady-state mechanical response of the tissue undegoing a low frequency (40-200Hz) mechanical excitation, which are taken using modified magnetic resonance imaging (MRI) sequences. There are two classes of methods to perform this stiffness calculation: Iterative inversion, and Direct inversion. Experiments using gelatin phantoms, consisting of a soft background and stiff inclusion, show that both methods accurately locate the inclusion. Iterative inversion provides better quantitative accuracy when compared to independent measurements using a dynamic mechanical analyzer, however, the computation time is significantly longer than direct inversions (iterative methods take hours, whereas direct inversion take seconds). The decision of which method to use for a given application must be made by weighing up the advantages of fast computation time for direct inversion against the better quantitative accuracy of iterative techniques.

2 INTRODUCTION

Magnetic Resonance Elastography (MRE) is a medical imaging modality that aims to produce images of the distribution of mechanical properties. The most common property to image is the shear modulus, due to its strong correlation with a range of diseases, including cancer [12, 14] cirrhosis of the liver due to hepatitis or alcoholism [4, 2], and several brain diseases [5, 11]. The most common approach to MRE is to excite the tissue of interest with some form of mechanical stimulus (usually a low frequency harmonic input), which sets up a steady state elastic wave field. An MRI imaging sequence with an additional motion sensitizing gradients [21] is then used to measure the amplitude and phase of the steady-state motion for each voxel of the tissue. This measurement can be repeated with the motion sensitizing gradient aligned in each of the 3 coordinate directions to fully describe the vector displacement of each voxel. Typical imaging times are on the order of 8 minutes per displacement direction.

The next step is to use an algorithm which estimates the mechanical properties using the displacement measurements. The mechanical properties and displacements are linked through a partial differential equation (PDE), for example, linear isotropic behavior is described by Navier's Equation

$$-\rho\omega^2\vec{u} = \nabla\cdot\left(\mu\left(\nabla\vec{u} + (\nabla\vec{u})^T\right)\right) + \nabla\left(\lambda\nabla\cdot\vec{u}\right), \tag{1}$$

where $\vec{u}$ is the vector displacement, ρ the density, ω the excitation frequency, μ the shear modulus, and λ is the second Lamẽ parameter (which is given in terms of the poisson's ratio, ν by $\lambda = \frac{2\nu\mu}{(1-2\nu)}$). The second RHS term of Navier's equation deserves special mention at this point due to the problems it can cause. This term governs the propagation of longitudinal waves in the medium. Tissue consists primarily of water, hence is close to incompressible (estimates of Poisson's ratio = 0.4999999 based on the speed of sound have been published [15]). As the incompressible limit is approached, $\lambda \longrightarrow \infty$, and the divergence, $\nabla\cdot\vec{u} \longrightarrow 0$. As a result, the second RHS term $\longrightarrow \infty \times 0$, which can cause numerical issues. Physically, there is a large imbalance between the propagation speed of the shear wave (governed by the first RHS term), and the longitudinal wave (governed by the second RHS term). The temporal frequency of each wave is fixed at the actuation frequency, therefore the wavelength of the longitudinal wave also becomes extremely large compared to the shear wavelength.

Navier's equation is commonly solved for the displacements, given the necessary boundary conditions and mechanical property distribution. Standard PDE solution techniques such as the Finite element (FE) method work well for this problem, which is known as the 'Forward Problem' in the MRE literature. At Poissons rations above 0.49, special care is required due to

T. Proulx (ed.), *Mechanics of Biological Systems and Materials, Volume 2*, Conference Proceedings of the Society for Experimental Mechanics Series 9999, DOI 10.1007/978-1-4614-0219-0_6, © The Society for Experimental Mechanics, Inc. 2011

numerical issues which cause standard finite element solutions to become numerically unstable. Using stabilized finite element forms, the equation can be accurately solved right up to the incompressible limit. [16].

Extracting the mechanical properties from Navier's Equation (known as the 'inverse problem') is a more difficult task. Simply solving the problem for the unknown properties using the Finite Element method, where the material properties are the unknown field variable, and the displacements are treated as known quantities and are used to build the Finite Element matrix produces solutions which are highly sensitive to measurement noise.

There are currently two major classes of algorithms which have been successful in accurately solving the MRE inverse problem in the presence of noise associated with MR measurements. Iterative methods use optimization techniques, where an estimate of the unknown property distribution is iteratively updated, such that the displacements predicted by solution of the forward problem closely match the measured displacements. Iterative techniques produce accurate stiffness images and deal with stiffness boundaries well, however a major issue is the computational cost, as hundreds of forward solutions are required for each inverse solution. The other approach is Direct Inversion (DI) strategies, which leverage the data rich nature of the MR measurements, and solve Navier's equation locally at each point in the volume. Computational cost is not an issue for these DI methods. Preprocessing of the data is required to allow the problematic second RHS term of Navier's equation to be neglected, and high order derivatives must be numerically evaluated, which can be difficult.

This paper deals with a direct comparison between Iterative and Direct Inversion MRE algorithms. Data collected from gelatin phantoms is used to test the spatial accuracy of the techniques, and comparison with independent Dynamic Mechanical Analyzer (DMA) stiffness measurements is used to evaluate the quantitative accuracy.

3 METHODS

3.1 Data Collection

Elastography phantoms were fabricated using porcine skin gelatin, which consisted of a soft background material (5% gelatin by weight), and stiffer inclusions (10% gelatin), of various shapes. Time-temperature superposition (TTS) [1] was performed on cylindrical samples of the same phantom materials using a TA Instruments Q800 Dynamic Mechanical Analyzer (DMA) to provide independent measurements of the shear modulus at 100 Hz. TTS is a technique originally designed for polymers which is used to estimate the mechanical response of a material at higher frequencies based on a relationship between the mechanical driving frequency and the temperature of the sample. TTS avoids the resonance behavior which often causes erratic results during standard dynamic measurements at high frequency (i.e. $F > 30$Hz) and in soft materials (i.e. $\mu < 20$kPa).

A pneumatic actuator which consists of an elastic membrane coupled to two subwoofers, driven at 100Hz by an amplified signal generator, was used to produce elastic waves in the phantoms (see Figure ??). A 3T Phillips Achieva MR system was then used with an EPI sequence [21] to measure the resultant steady state displacement field, on 2mm isotropic voxels.

3.2 Iterative Inversion

Iterative methods involve minimizing an objective function, which quantifies the difference between the measured displacements, $\vec{u}^m$, and a set of displacements generated by a computational model of the system, $\vec{u}^c$ with some estimate of the unknown material properties, $\vec{\theta}$. Once the minimum of the objective function is reached, the computational model is the closest possible match to the measured data, therefore, assuming the forward model is a good match to the tissue behavior, the current estimate of θ will be close to the true distribution. A common objective function is

$$\Phi = \frac{1}{2}\sum_{i=1}^{N_M}\left[|\vec{u}_i^c(\vec{\theta}) - \vec{u}_i^m|^2\right] \tag{2}$$

where N_M is the number of measurements, and $\vec{u}^c$ is the calculated displacement field (sampled at the measurement points), generated using the current estimate of the material properties, $\vec{\theta}$. In order to calculate $\vec{u^c}$, the necessary boundary conditions (BC's) for an elliptic PDE are required. These must be supplied on the entire boundary, and can be either prescribed displacements (Type I), or prescribed stresses (Type II). Fortunately with MRE, 3D displacements are measured at all points in the volume, therefore type I BC's are available for any arbitrary definition of the problem boundary.

The material properties are then iteratively updated to minimize Φ, using gradient descent based optimization techniques, such as the Gauss-Newton Method, Conjugate Gradient Method, and the Quasi-Newton Method. Derivatives of Φ required for these methods are calculated through modifications to the Finite element forward solution [20, 8].

The number of forward solutions required to minimize Φ and produce a solution can number in the hundreds, and a single MRE dataset can consist of 100,000+ measurement points. Simply solving one finite element forward problem on a mesh large

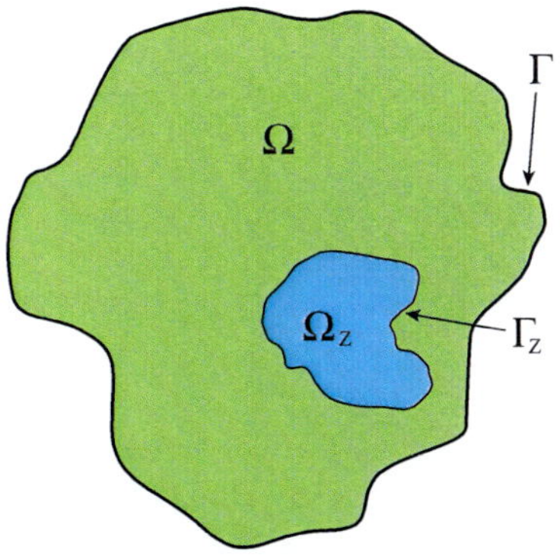

Figure 1: Schematic representation of subzone concept. The global problem domain is represented by Ω, with boundary Γ, and the domain of the subzone is represented by Ω_z, with boundary Γ_z. Sufficient boundary condition data is available around any arbitrarily chosen boundary, therefore, the global problem domain can be broken up into a number of smaller sub-domains for more efficient computation.

enough to accommodate the MRE data can exceed the memory capacity of modern computers, and the number of forward solutions required for the inverse problem can make the computation time unreasonably long. The extremely heavy computational load can be mitigated using the subzone method [20], which take advantage of the full-volume displacement data available in MRE. The boundary of the problem can be arbitrarily chosen, so the full problem domain is divided into a series of sub-regions, or 'Zones', which are processed independently, and the final material property solution over the full domain is made up of a assembly of the individual zone solutions. The subzone concept is illustrated by Figure 1. Each zone an be processed independently of the others, which creates a opportunity for efficient macro-parallelization, where computations for each zone are performed on separate processors. The subzone method is described in detail in these papers [20, 17, 19, 18].

3.3 Iterative Inversion Algorithm Used in this Study

The iterative algorithm for this study was subzone based, and used 27 node hexahedral finite elements, with a viscoelastic material model (incorporating a complex shear modulus). Nodally distributed real and imaginary shear moduli were reconstructed, using the conjugate gradient method. Typical runtime was 8 hours on 16 processors.

3.4 Direct Inversion

Direct inversion MRE algorithms perform local inversions of Navier's equation at each point in the dataset, using numerical evaluations of the required spatial displacement derivatives. Usually, locally homogeneous material properties are assumed, which simplifies Navier's equation to

$$-\rho\omega^2\vec{u} = \mu\nabla^2\vec{u} + (\lambda+\mu)\,\nabla\nabla\cdot\vec{u}. \quad (3)$$

(see [13] for a relaxing of the local homogeneity assumption). The divergence of the displacements, $\nabla\cdot\vec{u}$, describes the volume change of the material, which should be extremely small for nearly incompressible materials such as tissue. It is not possibly to accurately measure this very small value, divergences created by measurement noise and errors from numerical evaluation of the derivatives are much larger than the true value of $\nabla\cdot\vec{u}$, which gives a signal to noise ratio (SNR) $<< 1$. The $(\lambda+\mu)\,\nabla\nabla\cdot\vec{u}$ term in Eq. 5 is further complicated in tissue by the requirement that $\lambda\rightarrow\infty$ for a nearly incompressible material. It is possible to neglect the term by preprocessing of the data.

3.4.1 Local Frequency Estimation (LFE)

A simple method to allow the longitudinal wave term to be neglected is to apply a high pass filter, and recognize that the spatial wavelength of the longitudinal wave described by the problematic term is on the order of meters, whereas the shear wavelength described by the $\mu\nabla^2\vec{u}$ term is on the order of centimeters [7]. Most of the longitudinal waves should be removed from the data by the high pass filter, therefore the second RHS term of equation 5 can be neglected, leaving

$$-\rho\omega^2\vec{u} = \mu\nabla^2\vec{u}. \tag{4}$$

This represents three equations for one unknown (μ), and can be written as a matrix-vector system,

$$\begin{bmatrix} \nabla^2\vec{u})_1 \\ \nabla^2\vec{u})_2 \\ \nabla^2\vec{u})_3 \end{bmatrix} [\mu] = \begin{bmatrix} -\rho\omega^2\vec{u}_1 \\ -\rho\omega^2\vec{u}_2 \\ -\rho\omega^2\vec{u}_3 \end{bmatrix}, \tag{5}$$

which can be solved via linear least squares.

An alternative derivation of this technique is known as Local Frequency Estimation (LFE) [7, 6]. The inverse problem is posed as an image processing problem, where the goal is to extract the local spatial frequency, k, of the shear waves. If we assume the wave is traveling along the x direction, the shear modulus can be related to the spatial frequency through the relationship

$$\frac{1}{k} = \frac{1}{\omega}\sqrt{\frac{\mu}{\rho}}, \tag{6}$$

where ω is the actuation frequency. If we assume that the displacement pattern is locally sinusoidal, $u = A\sin(kx + P)$, which is valid in regions of homogeneous materials), k can be estimated by

$$k^2 = -\frac{\frac{d^2u}{dx^2}}{u}, \tag{7}$$

for a semi-infinite medium, with no boundary effects. Boundary conditions in non-infinite media can lead to wave guidance, where the spatial frequency is affected by boundary conditions, as well as the local stiffness of the material. A solution to this problem has been proposed using a series of spatio-temporal filters to remove reflected wave contributions [6]. Combining equations 6 and 7 leads to an expression which can be solved for μ,

$$\frac{d^2u}{dx^2}\mu = -\rho\omega^2 u. \tag{8}$$

Equation 8 is the one-dimensional equivalent of Eq. 6. This equation gives accurate results, provided the assumption of local homogeneity is met, and wave propagation is in the direction assumed. Performing a similar LFE analysis in 3D leads to the same form as Eq. 6.

3.4.2 Curl-Based formulation

There is another Direct inversion technique in use which involves a different preprocessing step, then a similar inversion. The curl operator is applied to Eq 5, which sets the term $(\lambda + \mu)\,\nabla\nabla \cdot \vec{u}$ to zero. This leaves the equation

$$-\rho\omega^2\vec{q} = \mu\nabla^2\vec{q}, \quad where\ \vec{q} = \nabla \times \vec{u}. \tag{9}$$

This set of 3 equations is solved for μ in an analogous way to equation 6. This approach has the advantage of not requiring a high pass filter to remove the effect of the longitudinal wave, the effect is removed from the governing equation mathematically. A disadvantage is the extra order of derivative required, 3rd order spatial derivatives must be numerically evaluated, which can be problematic in the presence of measurement noise.

3.4.3 Direct Inversion Algorithm used in this study

The Mayo clinic provides a high quality LFE package available for download on their website[3]. Options used for this program were 3 axis encoding, with a 3D complex LFE inversion, and a gaussian bandpass filter with cutoff limits 2.5 and 40 wavelengths/FOV. This package has been used in many of the LFE based MRE papers in the literature. Typical runtime is 5-10 seconds.

3.5 Comparison of Iterative and Direct Inversions

3 sets of data collected from gelatin phantoms were processed using both iterative and direct inversion MRE algorithms. Actuation amplitudes were modified to provide signal to noise ratios between 3.6 and 30. Individual slices and line profiles

near the center of the increased stiffness inclusion were compared to the MR magnitude inclusion to determine the spatial accuracy, and manual segmentation of the background and inclusion using the MR magnitude image allowed the quantitative accuracy of each inversion technique to be evaluated, by comparison with DMA measurements.

4 RESULTS

Figure 2 shows images from 3 phantom datasets. From left to right, images are MR magnitude, iterative MRE stiffness estimate, Direct Inversion MRE stiffness estimate images, and a line profile through the inclusion. Table 1 gives the median shear modulus of the background and inclusion for the 4 phantom datasets, along with SNR and DMA results.

Table 1: Modulus estimates of the background and inclusion shear modulus of the gelatin phantoms shown in figure 2, for iterative MRE, Direct inversion MRE, and independent dynamic mechanical analyzer measurements (units of kPa, given as median $\pm$ standard deviation). Manual segmentation of the inclusion using the MR magnitude image was used to produce the MRE values, which use data from the full 3D modulus estimate.

Background			Inclusion		
Iterative Inversion	Direct Inversion	DMA	Iterative Inversion	Direct Inversion	DMA
3.1 ± 0.5	3.7 ± 1.5	3.3 ± 0.8	8.3 ± 1.2	7.9 ± 2.7	8.8 ± 0.9
3.0 ± 0.4	2.9 ± 1.0	3.3 ± 0.8	8.2 ± 1.2	7.2 ± 2.3	8.8 ± 0.9
2.25 ± 0.35	2.9 ± 0.9	3.3 ± 0.8	6.0 ± 0.4	20 ± 5	8.8 ± 0.9

5 DISCUSSION

Results in gelatin phantoms presented in Fig. 2 show both iterative and direct inversion MRE achieve good spatial accuracy, boundaries of the stiff inclusions are clear in the shear modulus images for both techniques. Direct inversion appears to resolve the point of the conical inclusion better than iterative inversion, most likely because of the spatial continuity requirement enforced by the finite element support of the shear modulus in the iterative inversion algorithm, which is effectively a smoothing operation in areas of rapid spatial changes in shear modulus. Numerical accuracy of the iterative inversion is better, direct inversion shows both over and underestimation of the inclusion stiffness, however background stiffness estimates were accurate for both techniques.

The choice of which inversion technique to use is dependent on many factors. Obviously, accurate images are desirable so that artifacts and inaccuracies do not affect the diagnostic information provided by the images. If this was the only consideration, results form this study indicate that iterative inversion would be the most appropriate method. However, another important factor is the processing time, where Direct Inversion is orders of magnitude faster than iterative methods (seconds on a desktop computer compared to hours on a specialized multi-processor cluster). The iterative algorithm used in this study is not optimized for speed, development of robust convergence criteria could cut down on unnecessary iterations, and more efficient optimization strategies could also provide a speedup, however processing time would likely still be on the order of hours. Faster inversions fit much better into a clinicians workflow, MRE processing could be performed immediately after imaging, rather than having to wait hours for processing to be complete. Each MRE application must decide whether the better quantitative accuracy of iterative inversion is worth the increase in computational load and processing time.

Iterative methods have the advantage of being able to use different material models, such as poroelasticity [9, 10], which regarded as an appropriate model for fluid saturated materials such as brain tissue. There is currently no comparable direct inversion technique for poroelastic materials such as brain tissue, due to the increase in complexity of the governing constitutive equation.

A more detailed analysis of the advantages and disadvantages of iterative and direct inversion MRE, involving a larger collection of phantoms, and *in vivo* biological data is required to provide enough information to allow the important decision of which inversion strategy to use to be made. This work is currently underway.

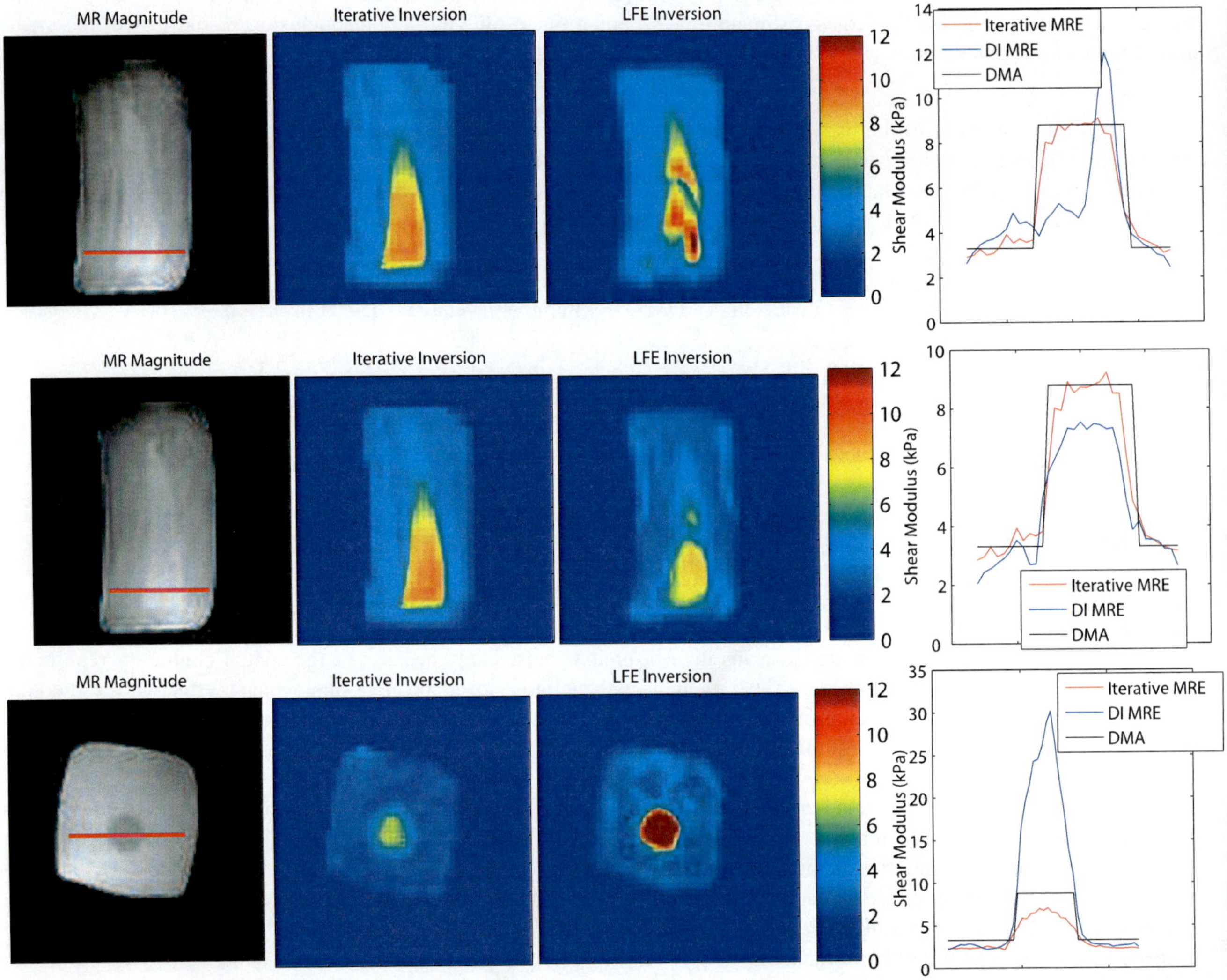

Figure 2: A collection of 3 separate gelatin phantom experiments, at a range of signal to noise ratios. The leftmost image shows the MR magnitude image, which show the true geometry of the phantoms. DMA measurements of the phantom material are 3.3 ± 0.8 kPa for the soft background (5% gelatin), and 8.8 ± 0.9 kPa for the stiff inclusions (10% gelatin). The second image from the left is the iterative MRE stiffness estimate, the 3rd image is the LFE estimate, and the rightmost plot is a profile through the inclusion, where the location is indicated by a red line on the MR magnitude image. The DMA estimate is shown as a black line.

References

[1] M.M. Doyley, I. Perreard, A.J. Patterson, J.B. Weaver, and K.M. Paulsen. The performance of steady-state harmonic magnetic resonance elastography when applied to viscoelastic materials. *Medical Physics*, 37:3970, 2010.

[2] R.L. Ehman. Science to Practice: Can MR Elastography Be Used to Detect Early Steatohepatitis in Fatty Liver Disease? *Radiology*, 253(1):1, 2009.

[3] R.C. Grimm, D.S. Lake, A. Manduca, and R.L. Ehman. Mre/wave. July 2006.

[4] L. Huwart, C. Sempoux, E. Vicaut, N. Salameh, L. Annet, E. Danse, F. Peeters, L.C. ter Beek, J. Rahier, R. Sinkus, et al. Magnetic resonance elastography for the noninvasive staging of liver fibrosis. *Gastroenterology*, 135(1):32–40, 2008.

[5] S.A. Kruse, G.H. Rose, K.J. Glaser, A. Manduca, J.P. Felmlee, C.R. Jack, and R.L. Ehman. Magnetic resonance elastography of the brain. *Neuroimage*, 39(1):231–237, 2008.

[6] A. Manduca, DS Lake, SA Kruse, and RL Ehman. Spatio-temporal directional filtering for improved inversion of MR elastography images. *Medical Image Analysis*, 7(4):465–473, 2003.

[7] A. Manduca, TE Oliphant, MA Dresner, JL Mahowald, SA Kruse, E. Amromin, JP Felmlee, JF Greenleaf, and RL Ehman. Magnetic resonance elastography: non-invasive mapping of tissue elasticity. *Medical Image Analysis*, 5(4):237–254, 2001.

[8] A.A. Oberai, N.H. Gokhale, and G.R. Feijoo. Solution of inverse problems in elasticity imaging using the adjoint method. *Inverse Problems*, 19(2):297–313, 2003.

[9] P.R. Perriñez, F.E. Kennedy, E.E.W. Van Houten, J.B. Weaver, and K.D. Paulsen. Modeling of Soft Poroelastic Tissue in Time-Harmonic MR Elastography. *IEEE transactions on biomedical engineering*, 56(3), 2009.

[10] P.R. Perriñez, A.J. Pattison, F.E. Kennedy, J.B. Weaver, and K.D. Paulsen. Contrast detection in fluid-saturated media with magnetic resonance poroelastography. *Medical Physics*, 37:3518, 2010.

[11] I. Sack, B. Beierbach, U. Hamhaber, D. Klatt, and J. Braun. Non-invasive measurement of brain viscoelasticity using magnetic resonance elastography. *NMR in Biomedicine*, 21(3):265–271, 2008.

[12] A. Samani, J. Zubovits, and D. Plewes. Elastic moduli of normal and pathological human breast tissues: an inversion-technique-based investigation of 169 samples. *Physics in Medicine and Biology*, 52(6):1565–1576, 2007.

[13] R. Sinkus, J.L. Daire, B.E. Van Beers, and V. Vilgrain. Elasticity reconstruction: Beyond the assumption of local homogeneity. *Comptes Rendus Mécanique*, 2010.

[14] R. Sinkus, K. Siegmann, T. Xydeas, M. Tanter, C. Claussen, and M. Fink. MR elastography of breast lesions: Understanding the solid/liquid duality can improve the specificity of contrast-enhanced MR mammography. *Magnetic Resonance in Medicine*, 58(6):1135–1144, 2007.

[15] R. Sinkus, M. Tanter, T. Xydeas, S. Catheline, J. Bercoff, and M. Fink. Viscoelastic shear properties of in vivo breast lesions measured by MR elastography. *Magnetic resonance imaging*, 23:159–65, 2005.

[16] R.L. Taylor. A mixed-enhanced formulation tetrahedral finite elements. *International Journal for Numerical Methods in Engineering*, 47(1-3):205–227, 2000.

[17] E. E. W. Van Houten. *Mechanical property Reconstruction from MR Detected Harmonic Displacement Data.* PhD thesis, Thayer School of Engineering at Dartmouth College, 2001.

[18] E. E. W. Van Houten, M. M. Doyley, F. E. Kennedy, K. D. Paulsen, and J. B. Weaver. A three-parameter mechanical property reconstruction method for MR-based elastic property imaging. *IEEE Trans. Medical Imaging*, 24:311–324, 2005.

[19] E. E. W. Van Houten, M. M. Doyley, F. E. Kennedy, J. B. Weaver, and K. D. Paulsen. Initial in-vivo experience with steady-state subzone-based MR elastography of the human breast. *J. Magn. Reson. Imaging*, 17:72–85, 2003.

[20] E. E. W. Van Houten, M. I. Miga, J. B. Weaver, F. E. Kennedy, and K. D. Paulsen. Three-dimensional subzone-based reconstruction algorithm for MR elastography. *Magnetic Resonance in Medicine*, 45:827–837, 2001.

[21] J.B. Weaver, E.E.W. Van Houten, M.I. Miga, F.E. Kennedy, and K.D. Paulsen. Magnetic resonance elastography using 3D gradient echo measurements of steady-state motion. *Medical Physics*, 28:1620, 2001.

Brain Response to Extracranial Pressure Excitation Imaged *in vivo* by MR Elastography

Erik H. Clayton and Philip V. Bayly
Mechanical Engineering & Materials Science
Washington University in St. Louis
1 Brookings Drive, Campus Box 1185
Saint Louis, MO 63130

ABSTRACT

Traumatic brain injuries (TBI) are common, and often lead to permanent cognitive impairment. Despite the prevalence and severity of TBI, the condition remains poorly understood. Computer simulations of injury mechanics offer enormous potential for the study of TBI; however, computer models require accurate descriptions of tissue constitutive behavior and brain-skull boundary conditions. Magnetic resonance elastography (MRE) is a non-invasive imaging modality that provides quantitative spatial maps of tissue stiffness *in vivo*. MRE is performed by inducing micron-amplitude propagating shear waves into tissue and imaging the resulting motion with a specialized "motion-sensitive" MRI pulse sequence. Invoking a restricted form of Navier's equation these data can be inverted to estimate material stiffness. As such, clinical interest in MRE has largely been driven by the direct empirical relationship between tissue stiffness and health. However, the so-called "raw" MRE data themselves (3-D displacement measurements) and calculated strains can elucidate loading paths, anatomic boundaries and the dynamic response of the intact human head. In this study, we use the MRE imaging technique to measure *in vivo* displacement fields of brain motion as the cranium is exposed to acoustic frequency pressure excitation (45, 60, 80 Hz) and calculate the resulting shear-strain fields (2-D).

INTRODUCTION

Traumatic Brain Injury, Diffuse Axonal Injury, and Shear Strain

The US Centers for Disease Control and Prevention estimate that each year 1.7 million Americans suffer a traumatic brain injury [1]. The prevalence of TBI is surely related to the variety of ways one can subject the brain to insult. Most TBI cases are caused by one or more of the following events: direct physical contact with an external rigid body, inertial loading due to linear or angular acceleration of the head, or external pressure loads (i.e. blast wave).

Traumatic brain injury (TBI) due to explosive blast may lead to permanent cognitive impairment, though the mechanism of injury remains poorly understood. In blast, the mechanical insult is an external pressure wave interacting with the outside of the skull; mechanical strain is likely the key kinematic parameter for injury. Despite the importance of brain deformation mechanics in TBI, little is known with confidence about strain patterns in injury. The relationship between extra-cranial forces and internal brain tissue response has been a topic of research since the early 1940's [2-7], yet little remains known about strain amplitudes and patterns that result from specific external forces. Computer simulations of injury mechanics offer enormous potential for diagnosis and prevention of TBI, but their validation requires direct comparison with experimental data.

In this study, we use MRE to measure human brain tissue motion *in vivo* as the cranium is exposed to acoustic frequency pressure loading of known amplitude. The unique features of this study are i.) knowledge of the external loading (pressure amplitude), ii.) quantification of the brain's response in terms of mechanical strain, and iii.) characterization of brain-skull filtering properties over multiple frequencies .

T. Proulx (ed.), *Mechanics of Biological Systems and Materials, Volume 2*, Conference Proceedings of the Society for Experimental Mechanics Series 9999, DOI 10.1007/978-1-4614-0219-0_7,

Magnetic Resonance Elastography

Magnetic resonance elastography (MRE) is a non-invasive imaging technique for quantitative measurement of the mechanical properties of biological tissue [8]. In essence, externally applied forces are used to excite propagating waves in tissue, and displacements are imaged by encoding spin phase.

To perform MRE the standard nuclear magnetic resonance imaging (MRI) method is modified to measure displacements due to wave propagation. The basic data in MR images are the radio-frequency (RF) signals produced by the precession of nuclear spins; ^1H is the predominant nucleus of interest in biological imaging. An MRE pulse sequence contains additional, temporally-varying, "motion-sensitive" magnetic-field gradients which produce changes in spin-emitted RF signal phase that are proportional to spin displacement. These changes in spin phase can be measured throughout the imaging volume, producing displacement fields of high spatial resolution. Components of displacement in any direction can be measured by applying the motion-encoding gradients along that direction. To date, clinical interest in MRE has largely been driven by the direct empirical relationship between tissue stiffness and health. However, the so-called "raw" MRE data themselves (3-D displacement measurements) and calculated strains can elucidate loading paths, anatomic boundaries and the dynamic response of the intact human head.

Insight into Brain Injury Biomechanics through MRE

Previous works by McCracken and colleagues [4, 9] identified transient wave propagation MRE as potential tool for studying TBI, but to date no study has analyzed raw MRE displacement data to examine the mechanical response and deformation mechanisms of the brain in terms of strain. As such, this study uses MRE, to illuminate the mechanical behavior of brain tissue while the skull is subjected to acoustic pressure waves. Results obtained will be applicable for characterization of mechanical features of both the brain and skull, as well as the validation of computer models. The strain amplitudes imposed for imaging in this study are much less than any known threshold for inducing TBI; however, result obtained provide valuable insight into the filtering properties of the skull as well as the ability of brain matter to dissipate mechanical energy.

METHODS

Acquisition of Spatiotemporal Wave Fields

Tissue displacements were acquired with a specialized motion-encoding, gradient-recalled echo (GRE) NMR imaging pulse sequence (Figure 1).

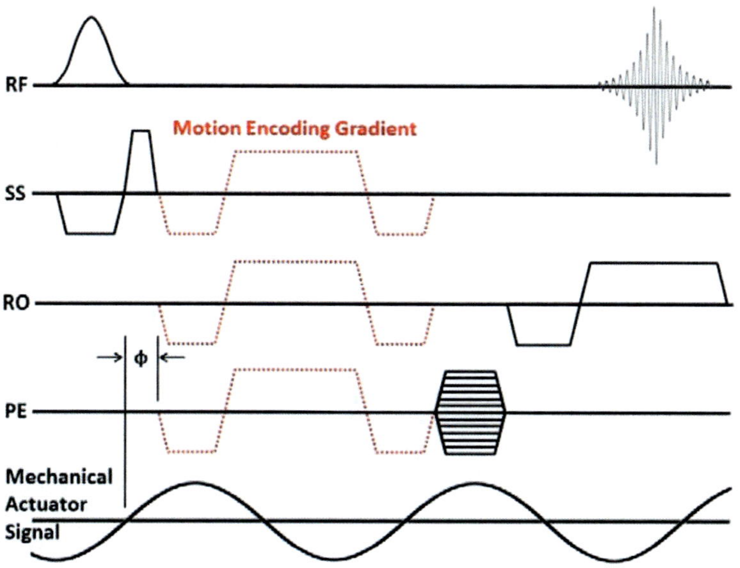

Figure 1: The specialized gradient-recalled echo (GRE) magnetic resonance pulse sequence used for data acquisition. In addition to the standard RF and magnetic-field gradient events required to create a GRE MRI, a trapezoidal motion-encoding gradient (MEG) is included. The MEG event and the mechanical actuation signal are temporally shifted by the parameter Φ. By varying the delay, Φ, temporal variations in the wave field are recorded (cf. Figure 2). In this example, a 1-cycle MEG (dashed) is shown on each gradient channel. As a result, motion-encoded phase images are acquired with contrast proportional to displacements perpendicular to an oblique plane in the imaged body.

The physics of measuring harmonically-varying displacement fields with NMR spin-phase accumulation is described in detail by Muthupillai et al. [10]; however the key points are summarized briefly in this section. Consider a single ^{1}H proton spin packet undergoing harmonic mechanical excitation in the presence of a harmonic, motion-encoding magnetic-field gradient, $G_i(t)$, parallel to an orthogonal reference coordinate *i*. The shift in NMR signal phase, θ_i, at position x_0, y_0, z_0, is governed by the following equation:

$$\theta_i(x_0, y_0, z_0, \Phi) = \gamma \int_0^T G_i(\tau) u_i\,(x_0, y_0, z_0) \sin(\omega\tau + \psi(x_0, y_0, z_0) + \Phi)\, d\tau \qquad \text{(no sum on } i\text{)} \qquad (1)$$

where, in our experiments $G_i(t)$ takes the following form:

$$G_i(t) = \begin{cases} -|G|, & \frac{2\pi(n-1)}{\omega} \le t \le \frac{\pi(4n-3)}{2\omega} \\ +|G|, & \frac{\pi(4n-3)}{2\omega} < t < \frac{\pi(4n-1)}{2\omega} \\ -|G|, & \frac{\pi(4n-1)}{2\omega} \le t \le \frac{2\pi n}{\omega} \end{cases} \qquad (2)$$

Here the duration of the applied gradient is $T = 2\pi n/\omega$, $u_i(x_0, y_0, z_0)$ and $\psi(x_0, y_0, z_0)$are the amplitude and phase of the displacement component of the spin packet at this location and ω is the mechanical actuator driving frequency; γ is the gyromagnetic ratio of ^{1}H nuclei (protons). The synchronization delay, Φ, is an additional temporal phase shift between the motion of a particular spin packet and the motion-encoding gradient. If multiple phase images are acquired, each with a different synchronization delay (corresponding to a fraction of the actuation period), a time history of spin phase is measured (Figure 2).

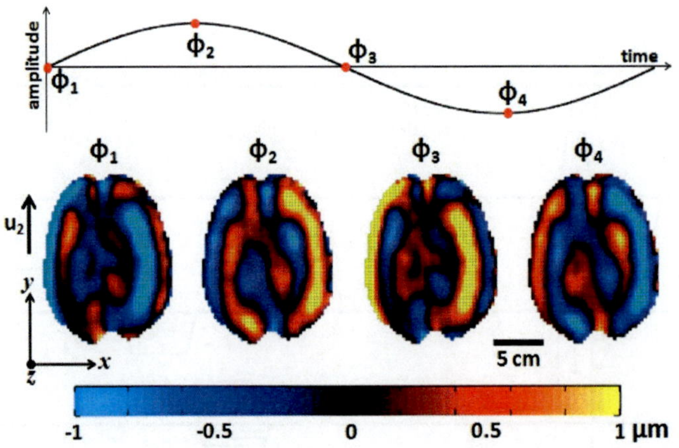

Figure 2: Imposed harmonic brain motion is acquired by the MRI scanner at four equally spaced phase offsets, Φ_i, through one actuation cycle.

Each displacement component, i.e., u_1, u_2, u_3, can be imaged if the experiment is repeated three times and the motion-encoding magnetic-field gradients are aligned to a different axis of an orthogonal reference coordinate system each time. The amplitude of each spin-packet displacement component, $u_i(x_0, y_0, z_0)$, can be determined directly from the amount of spin phase accrued, $\theta_i(x_0, y_0, z_0, \Phi)$, since the amplitude, frequency, and duration of the applied motion-encoding gradient, $G_i(t)$, are known.

Acquisition of Human Data

Human studies were conducted at 1.5 T on three healthy male subjects, aged 21-30 years-old (mean: 23.7 yr) at actuation frequencies of 45, 60, and 80 Hz. MRE data were collected using a MAGNETOM Avanto (Siemens) series whole-body clinical scanner equipped with a phased-array head coil. Imaging parameters: TR/TE: 138/27.5 ms, flip angle: 25°, NEX: 1, resolution: 3 mm^3.

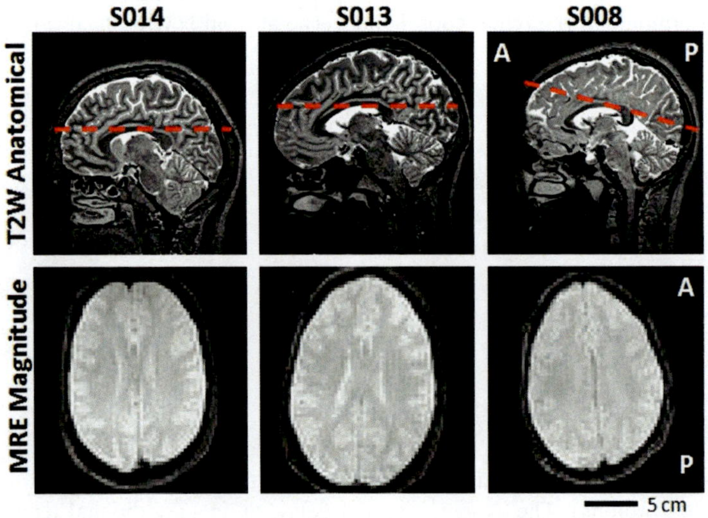

Figure 3: (Top row) T2-weighted sagittal images of each subject's brain. The dashed line indicates the approximate location and angle of the oblique imaging plane relative to anatomical features. (Bottom row) Elastography magnitude images of the oblique slice plane for each subject.

A single transverse-oblique slice of motion encoded data was acquired for each subject through the central (I/S) cerebrum, cf. Figure 3. All experimental protocols were approved by the institutional Human Research Protection Office Internal Review Board to ensure that the rights and welfare of each research participant were protected.

The imaging procedure was repeated three times with different motion-encoding, magnetic field gradient orientations to record three orthogonal displacement components (u_1, u_2, u_3) relative to the oblique imaging plane. The absolute peak motion-encoding gradient amplitude was 25 mT/m, nominally, for all test frequencies and spin-phase was accrued over a single gradient cycle. All data were acquired with a temporal resolution of four points per actuation cycle. Total scan time per subject was approximately 21 minutes. In this time, all motion components were acquired and for each actuation frequency.

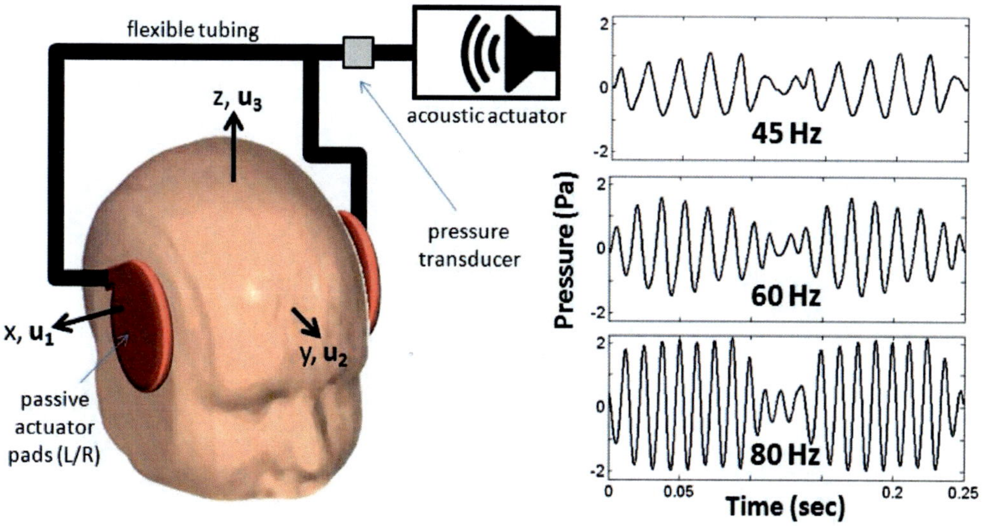

Figure 4: (Left) An active acoustic driver produced an oscillatory pressure wave which was transmitted to each passive actuator pad via flexible tubing. Passive actuator pads were positioned on the subject's skull near the left and right pterion bones and affixed via elastic bandage (not shown). (Right) Acoustic pressure was measured via transducer embedded in the tubing. A typical waveform is shown.

Motion was induced in the brain using an acoustic actuation system (Resoundant™, Resoundant Inc.) modified slightly so that a single active driver could power two passive drivers with equal amplitude and phase. Each passive actuator was positioned on the subject's skull near the left and right pterion bones and affixed with an elastic bandage (not shown), cf. Figure 4. The actuation system was configured to transmit a 4, 6, and 8 cycle pressure-wave train synchronized with the MRE sequence at 45, 60 and 80 Hz, respectively. Imposed acoustic pressure loads were measured with the PCB Piezotronics model 103B01 dynamic pressure sensor.

Data processing

Motion encoded MR data were obtained using switched-polarity encoding acquisition scheme to remove systematic phase errors and enhance displacement contrast. This scheme increases the displacement contrast by a factor of two since two motion-encoded images, each with opposite polarity, are required. Phase-contrast images were obtained by complex division of positive and negative polarity phase images, and converted into units of length. Sensitivity factors of 5.63 (45 Hz), 7.66 (60 Hz), and 10.6 μm/rad (80 Hz) data were calculated by numerical integration of Equation 1 – taking into consideration actual gradient performance specifications.

Data were filtered with a 3 x 3 kernel median filter and circular, 4th-order Butterworth bandpass filter (high cut: 2.14 cm, low cut: 40.0 cm). Displacement gradients and shear angles were observed to be small for all experiments performed; therefore, the 2-D infinitesimal strain tensor was calculated,

$$\varepsilon_{ij} = \frac{1}{2}\left(u_{i,j} + u_{j,i}\right) \quad (i,j = 1,2). \qquad (3)$$

Shear-strains were evaluated in terms of the root-mean-squared (RMS) scalar quantity and normalized by the RMS applied pressure load. Pressure-normalized RMS shear-strains were evaluated in two fashions for each subject: i.) globally – across all brain matter, and ii.) regionally – in three concententric ring-like zones, in hopes of illuminating the dispative properties of brain tissue.

RESULTS

The RMS acoustic pressure supplied to each actuator pad was calculated over the duration of each scan to be 0.54, 0.76, and 1.2 Pa at 45, 60, and 80 Hz. Example displacement and shear strain fields are shown for one subject in Figure 5.

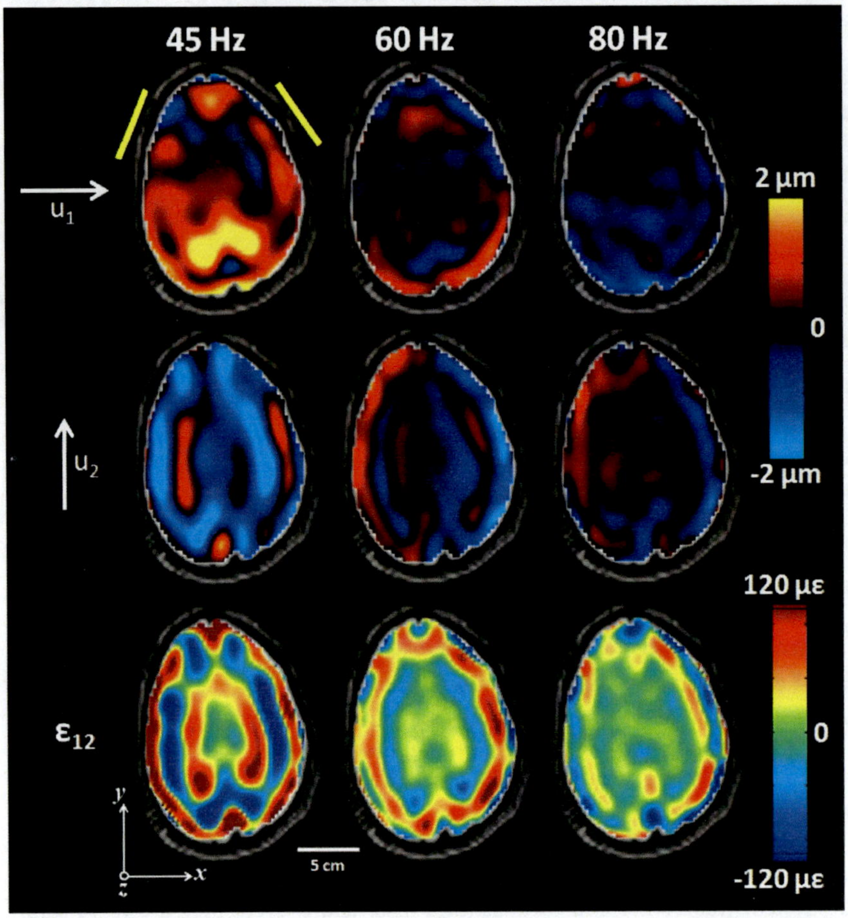

Figure 5: Brain deformation and strain maps for subject S008. (Row 1-2) The in-plane displacement components are shown for each actuation frequency column-wise. Yellow bars denote the approximate location of the L/R acoustic pads. (Row 3) The in-plane shear-strain component of the infinitesimal strain tensor is presented. The ε_{12} shear-strain clearly reveals the emergence of a propagating shear wave due to extracranial acoustic loading.

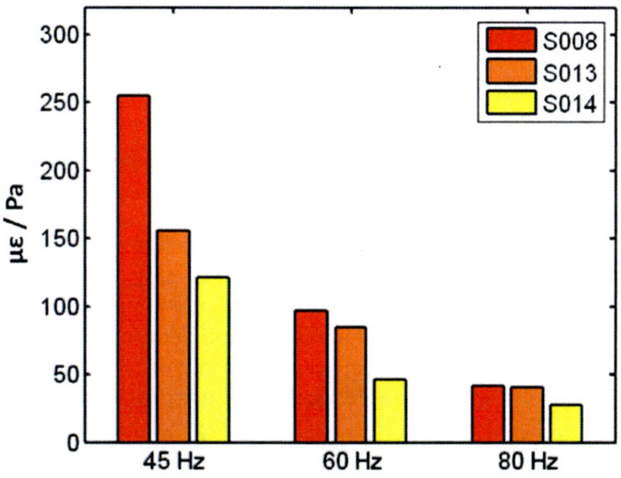

Figure 6: Mean global RMS shear strain normalized by RMS excitation pressure for each subject.

Mean RMS shear strains for each subject, normalized by input RMS pressure, are shown in Figure 6. At 45 Hz, the global strain amplitude in subject S008 is significantly higher than subjects S013 and S014. On the other hand, at 80 Hz the global strain amplitude is approximately equal across all subjects.

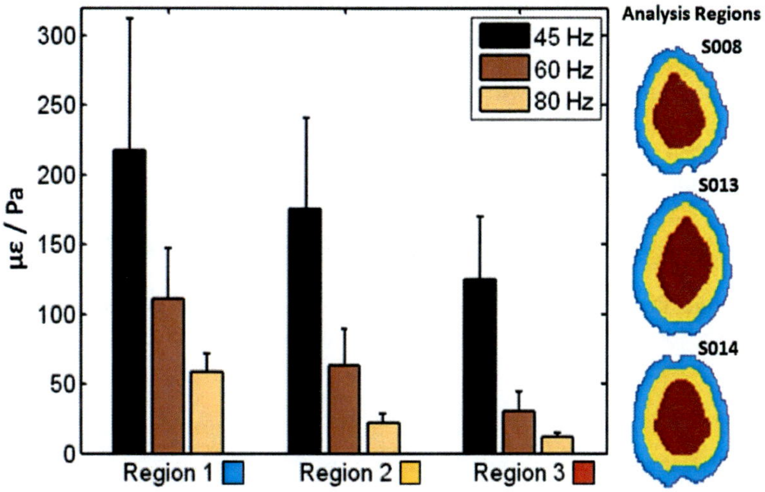

Figure 7: Mean regional RMS shear strain normalized by RMS excitation pressure, all subjects (n=3). Error bars signify one standard deviation. (Outset) Regions 1-3 are color-coded for each subject.

Mean regional RMS shear strains, normalized by input RMS pressure, are shown in Figure 7. At 45 Hz, the strain amplitude reaches 217 με per Pa of applied pressure in the outer ring of this imaging plane; at 80 Hz the response is approximately 58 με per Pa, reflecting reduced transmission.

DISCUSSION

When the skull exterior is subjected to symmetric acoustic pressure loading, concentric elliptical shear strain (ε_{12}) bands are produced in brain tissue, cf. Figure 5. Shearing strains that propagate inward from the skull boundary towards the center of the brain predominate; however, we also observe motion emanating outward from the anterior and posterior falx. We believe this behavior can be attributed to reflections from interior structures (the falx cerebri between the brain hemispheres, and the tentorium between cerebrum and cerebellum).

The frequency response of each subject exhibits a consistent pattern: with increasing frequency the amplitude of strain in the brain interior decreases. These data provide basic insight into the filtering properties of the skull, as well as the strain-rate dependent (energy-absorbing) mechanics of brain tissue. The data in Figure 7 are direct measures of strain attenuation. For all driving frequencies, the average normalized strain response in each region decreased with increasing distance from the brain-skull boundary. However, it is clear that the observed shear waves do not simply propagate radially inward from the skull. The mechanisms of transmission and attenuation of waves by specific anatomical structures will be a topic of future studies.

There is subject-to-subject variation in the measured internal brain response, cf. Figure 6. Further studies should explore the possible effects of differences in the age and anatomy of the subjects, but image plane orientation, actuator-skull contact, and brain volume between subjects may also contribute. The quantitative and qualitative features of the response offer insight into brain attachment and the filtering effects of the skull to acoustic pressure loading. These results highlight the importance of understanding skull transmissibility and brain boundary conditions in the response to blast.

CONCLUSION

MRE-based displacement measurements and input pressure data, provide new insight into the study of blast biomechanics. Symmetric acoustic pressure excitation of the human skull leads to propagation of shear waves in the brain. The magnitude and phase of these oscillations relative to known pressure excitation can be used to validate computer models and to illuminate fundamental mechanical properties of the skull, brain, and associated intracranial anatomy *in vivo.* Future work will investigate the response to alternative loading locations, and will include acquisition of data from multiple contiguous slices of brain. This will permit through-image-plane derivatives to be calculated, allowing all components of the 3-D strain tensor to be calculated.

ACKNOWLEDGEMENTS
Authors' gratefully acknowledge Dr. Bradley Bolster and Dr. Agus Priatna at Siemens Healthcare for supporting MR elastography pulse sequence development. Financial support was provided by NIH RO1 NS055951 (Bayly).

REFERENCES

1. Faul, M., et al., *Traumatic Brain Injury in the United States: Emergency Department Visits, Hospitalizations and Deaths 2002-2006*, C.f.D.C.a.P. U.S. Department of Health and Human Services, National Center of Injury Prevention and Control, 2010: Atlanta, GA.
2. Holbourn, A.H.S., *Mechanics of Head Injuries.* The Lancet, 1943. **242**(6267): p. 438-441.
3. Smith, D.H., D.F. Meaney, and W.H. Shull, *Diffuse axonal injury in head trauma.* The Journal of head trauma rehabilitation, 2003. **18**(4): p. 307-16.
4. McCracken, P.J., et al. *Transient MR Elastography: Modeling Traumatic Brain Injury.* in *MICCAI.* 2004.
5. Green, M.A., L.E. Bilston, and R. Sinkus, *In vivobrain viscoelastic properties measured by magnetic resonance elastography.* NMR in Biomedicine, 2008. **21**(7): p. 755-764.
6. Sack, I., et al., *Non-invasive measurement of brain viscoelasticity using magnetic resonance elastography.* NMR in Biomedicine, 2008. **21**(3): p. 265-271.
7. Meaney, D.F., et al., *Biomechanical Analysis of Experimental Diffuse Axonal Injury.* Journal of Neurotrauma, 1995. **12**(4): p. 689-694.
8. Muthupillai, R., et al., *Magnetic resonance elastography by direct visualization of propagating acoustic strain waves.* Science (New York, N Y), 1995. **269**(5232): p. 1854-7.
9. McCracken, P.J., A. Manduca, and R.L. Ehman, *MR Elastography for Studying the Biomechanics of Traumatic Brain Injury*, in *International Society for Magnetic Resonance in Medicine.* 2003. p. 799.
10. Muthupillai, R., et al., *Magnetic Resonance Imaging of Transverse Acoustic Strain Waves.* Magnetic Resonance in Medicine, 1996. **36**(2).

Mechanical Properties of Abnormal Human Aortic and Mitral Valves

K. Paranjothi[1], U. Saravanan[2], R. KrishnaKumar[3], K.R. Balakrishnan[4]
[1]Research scholar, Department of Engineering Design, IIT Madras, Email: kpjothi@iitm.ac.in
[2]Assistant Professor, Department of Civil Engineering, IIT Madras, Email: saran@iitm.ac.in
[3]Professor, Department of Engineering Design, IIT Madras, Email: rkkumar@iitm.ac.in
[4]Director, Cardiovascular Surgery, Malar Hospitals, Email: cvskrb@gmail.com

ABSTRACT
Details about custom built experimental set up to perform uniaxial and biaxial tests on planar soft tissues are presented. This displacement controlled set up can apply and measure loads ranging up to 100 N. The surface deformation is determined from tracking markers in 3D space using 2 CCD cameras. Discarded valve tissue from patients undergoing valve replacement surgery were collected and stored at 4°C in normal saline and experiments were completed within 24 hours from harvest. Uniaxial tests on aortic valve leaflets and biaxial tests on mitral valve leaflets were conducted. Results show that the deformation of these tissues is not homogeneous. The principal stretches in the plane orthogonal to the direction of stretching in the uniaxial stretch experiment are not the same but the principal directions do not change much with loading. Further, the results show that both the valve leaflets are compressible. The loading and unloading path is nearly the same and is not sensitive to the rate of displacement when it is varied between 200μm/s and 800μm/s. These results have implications in the development of constitutive models for these tissues.

1. Introduction

Cardiovascular diseases claim around 17.1 million lives a year worldwide. Valve disease which requires replacement of valves occurs in approximately 0.25 million people worldwide [1]. The knowledge on the etiology of the valve diseases is very limited, although valve replacements have increased the life expectancy of patients with valve disease [1]. Though there are two semilunar and atrio-ventricular valves each, much of the studies on valves have been limited to aortic valve and mitral valve [2] because they are the most problematic valves that require surgical intervention or replacement more regularly. Recently, Sacks et al. [2] has reviewed the literature concerning the biomechanics of heart valve function.

In this study, we report results of experiments that we performed to arrive at the constitutive relation for the diseased aortic and mitral valve. There have been studies devoted to the mechanical response of normal aortic [3-6] and mitral [7-11] valves, but to the knowledge of the authors there are none on diseased valves. Knowing the constitutive relation for diseased valves would be useful for two reasons. One, it would help answer the question whether the diseased valve can be repaired. Two, as and when the growth and remodeling framework becomes available, knowledge of both the initial and final states would allow one to conjure the causes for this final state and possible preventive strategies against the valve getting damaged.

While the aortic valve leaflet is subjected to uniaxial stretch, mitral valve leaflet is subjected to both uniaixal and biaxial stretch experiments. From these we find that the deformation of both the valve leaflets is not homogeneous. Contrary to what is assumed in the literature [9] we find that these valves are compressible. The loading and unloading paths are nearly the same and the rate of stretching when increased by 4 times produced nearly the same response, suggesting that these leaflets are undergoing a non-dissipative process. These observations have implications on the development of constitutive relations for these valve leaflets.

2. Material and Methods

2.1 Details of the Experimental Setup

This experimental set-up to perform multi-axial mechanical tests consists of three subsystems. First, a video based system that allows 3D tracking of surface markers. By tracking the movement of the surface markers during the experiment the deformation field is found from which the principal stretch ratios are obtained. This system consists of two charged couple device (CCD) camera(AVT GUPPY F-146 B) with 1390 by 1040 pixel resolution, each connected to a PCI card (AVT 1394a adapter) and a micro-lens (NAVITAR zoom 6000 lens with 3mm fine focus and internal coaxial illuminator attached to a 1x

T. Proulx (ed.), *Mechanics of Biological Systems and Materials, Volume 2*, Conference Proceedings of the Society for Experimental Mechanics Series 9999, DOI 10.1007/978-1-4614-0219-0_8,

adapter) whose magnification can be varied from 0.7x to 4x. The lens with a working distance of 92mm, has a field of view of 10mm at 0.7x magnification and 2mm at 4x magnification. The depth of field also varies from 1mm at low magnification to 0.1mm at high magnification. The camera can capture at most 17 frames per second. While one of the camera is vertical, the other is inclined at 45 degrees in the vertical plane (see figure - 1). The camera and the lens are mounted on damped mounting rods using a custom designed mounting system.

Second subsystem consists of a suit of computer controlled linear translation stages to apply the required multi-axial loads. Two stages are used along each of the two mutually orthogonal directions (see figure 1) to perform displacement controlled bi-axial tests on planar specimens. As seen in figure 1, on top of each of these stages, manually controlled precision stages (25mm travel, 0.01mm resolution) is provided for each of the two directions to allow positioning of the specimen accurately in the field of view of the cameras. The computer controlled linear translation stages (Holmarc Opto-Mechatronics) has a 100mm travel with 0.5 micron precision and its velocity can be set between $2\mu m/s$ to $6mm/s$. The stepper motor used in the translation stage can apply a maximum force of 100 N at its peak velocity. The motor controller communicates with the computer using a RS-232 port. By appropriately actuating these stages uniaxial extension/compression, biaxial extension/compression, rigid body translations can be achieved.

The final subsystem is the data acquisition system. It consists of two load cells (Honeywell-Sensotec Model 31E) and a 18 bit analog to digital converter (UEI model PowerDNA with DNA-AI-208 as input layer), which can take 8 voltage input signals and whose sampling rate can be varied between 1 to 8000 samples per second is used for sensing the applied load . The data acquisition system communicates with the computer through the ethernet interface. After, a preliminary study on the expected load that needs to be applied to engender a stretch ratio of 1.2 in the tissue one of the two types of load cells one that can measure loads up to 200N in both tension and compression and the other measures 2.5 N is used. The reported deviation from linearity for the load cells is less than 0.15 percent over the full range and resolution limited by only the analog to digital convertor. All these subsystems are connected to HP Workstation XW4600 with Intel Core 2 Duo processor running at 2.66 GHz and having 4GB RAM.

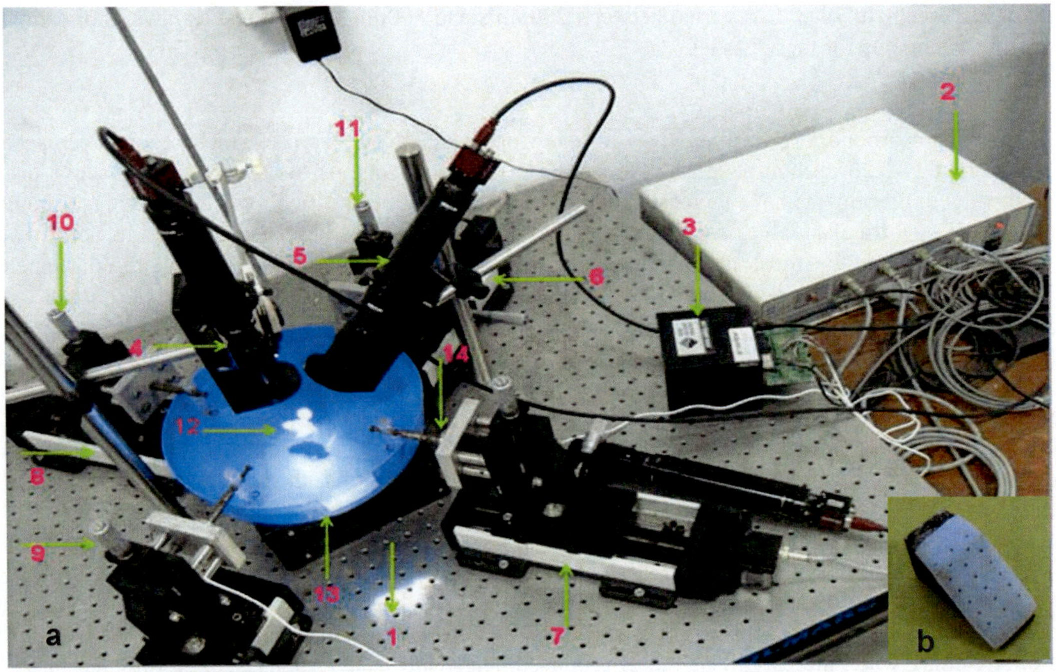

Figure 1 (a) Photograph of the experimental set up (1.Bread board, 2. Stepper Motor Controller (W,X,Y & Z axes) 3. Data acquisition system, 4. Vertical camera and lens, 5. Inclined camera and lens, 6. Z-axis motorized stage, 7. W-axis motorized stage, 8. Y-axis motorized stage, 9. X-translational stage, 10. Y-translational stage, 11. Z-translational stage, 12. Sample, 13. Saline Container, 14. Load cell) (b) Calibration wedge

2.2 Details of the Software Modules

Software for simultaneous operation (video, motor, DAQ) and synchronous acquisition of data (video, load) from this hardware has been developed in LABVIEW. The experimental protocols are completely automated by separate software modules. Each of these modules has many sub-modules whose details are given below in the order of their execution.

2.2.1 Load Cell Calibration

The main functionality of this module is to find the relationship between applied load and the voltage output from the load cells. This relationship between the applied load and developed voltage being linear, the two parameters (slope and intercept) that characterize the linear relationship, is alone determined through calibration and stored in a text file for later use.

2.2.2 Reference Configuration

This module controls the video system and the DAQ system. It acquires images containing the surface markers from the vertical and inclined cameras and stores their centroid locations. Also it acquires data from DAQ and stores the initial readings of the load cell. All these data namely the centroid positions of the surface markers from the cameras, the computed thickness values and the load cell readings are stored in a text file at the end of the recording.

2.2.3 Motor Actuation

This model controls the stepper motors and actuates them to induce uniaxial/biaxial extension or rigid body translation of the specimens. The strain rate is set by controlling the speed of the motors. While for uniaxial test only two of the stepper motors are actuated, for the biaxial test all the four motors are actuated. Using this module cyclic test with specified number of cycles can be performed. Apart from this, the module has all the functionalities of the reference configuration module and so as the motors load/unload the specimen it also acquires data from the cameras and load cells simultaneously and stores the load and the location of the centroid of the markers in a text file at the end of the protocol.

2.2.4 Video Calibration

Here a calibration wedge (see figure 1) containing markers separated by 2mm along the horizontal direction and by 1.5mm in the vertical direction is imaged in both the vertical as well as the inclined camera. Then, the relationship between the 3D coordinates of these markers and the two projections of these markers onto the imaging plane is established off-line. Since, here we need to just record the centroid of stationary markers, which is what we do while recording the reference configuration too, the same reference configuration recording module is used for video calibration.

2.3 3D Reconstruction from Two Perspective Views

The video subsystem has to be calibrated so as to obtain the 3D coordinates of the markers from two perspective views. This is achieved using global reconstruction techniques. The relationship between 3D coordinates of the centroid of the markers referenced with respect to a laboratory coordinate system (X, Y, Z) and its 2D coordinates obtained by projecting it on to the imaging surface and referenced with respect to the camera coordinate system $((u^v, v^v)$ or (u^i, v^i) depending on which camera is being referred to) is assumed to be linear, i.e.,

$$\mathbf{P}^k\mathbf{X} = \mathbf{u}^k - \mathbf{o}^k, \tag{1}$$

where k can be v or i, referring to the vertical camera and the inclined camera respectively, $\mathbf{X}$ denotes the 3D coordinates of the centroid of the marker, $\mathbf{u}^k$ denotes the 2D projected coordinates of the centroid of the marker on the kth camera, $\mathbf{o}^k$ denotes the 2D projected coordinates of the origin of the laboratory coordinate system on the plane of the imaging surface of the kth camera, $\mathbf{P}^k$ denotes the projection matrix for the kth camera. The projection matrix is a 2 by 3 matrix. Thus, for each camera we have to find 8 constants (6 in the projection matrix(P_{lm}) and 2 corresponding to the projected coordinates of the origin of the laboratory coordinate system). For this we require the coordinates of the centroid of at least four non-coplanar markers both in the laboratory coordinate ($\mathbf{X}$) and the projected camera coordinate system ($\mathbf{u}^k$). Usually more than 4 markers are used to find the eight constants and hence these calibration constants are found in a least square sense.

During the experiment we need to find the 3D coordinates of the centroid of markers, $\mathbf{X}$, given the two 2D coordinates of the projections of them onto the imaging surface. For this the relation (1) has to be inverted. However, now there are four equations with just three variables. Hence, these equations are also solved in least square sense to obtain the 3D coordinates of the centroid of the markers.

The assumption that the relationship between the laboratory coordinates and the projected camera coordinates is linear holds only if the camera lens provides uniform magnification over the depth of view and there is no distortion in the field of view. The validity of these assumptions is verified later by examining if the stretch ratios become 1 when the body is subjected to a rigid body translation along any axis. It is found that the stretch ratios deviate from the expected value by $\pm 10^{-3}$ and the invariants, to be defined next, by $\pm 10^{-2}$ due to these rigid body translations, irrespective of the axis, as long as one is in the depth and field of view. Hence, it is concluded that the relationship between the laboratory coordinates and the projected camera coordinates is linear and that the accuracy of the measured stretch ratios is $\pm 10^{-3}$.

2.4 Determination of Deformation Gradient

In this work, 9 to 15 markers are placed on the surface of the valve tissue and are tracked as the specimen is loaded. In general, it is not possible to experimentally determine the exact deformation field by tracking 12 markers (or for that matter even a larger number of markers), we can at most get a reasonable approximation of the deformation field and its gradient locally. However, we can verify if the observed deformation field is consistent with that theoretically predicted. Since, we are going to stretch the valve along two axis; we expect that the deformation field be homogeneous. This prompts us to assume a deformation of the form:

$$x = a_x^0 + a_x^1 X + a_x^2 Y + a_x^3 Z, \qquad y = a_y^0 + a_y^1 X + a_y^2 Y + a_y^3 Z, \qquad z = a_z^0 + a_z^1 X + a_z^2 Y + a_z^3 Z, \tag{2}$$

where (X,Y,Z) are the Cartesian coordinates of the centroid of a marker in the reference configuration and (x,y,z) are the Cartesian coordinates of the centroid of the same marker in the current configuration, a_x^i, a_y^i and a_z^i (for $i = \{0,1,2,3\}$) are constants. Knowing the location of five markers in the current and reference configurations, we obtain a_x^i, a_y^i and a_z^i, by solving equation (2) in the least square sense.

Then, the matrix components of the deformation gradient, **F** is computed as:

$$\mathbf{F} = \begin{pmatrix} a_x^1 & a_x^2 & a_x^3 \\ a_y^1 & a_y^2 & a_y^3 \\ a_z^1 & a_z^2 & a_z^3 \end{pmatrix}. \tag{3}$$

It is now straightforward to compute the right Cauchy-Green deformation tensor, $\mathbf{C} = \mathbf{F}^t\mathbf{F}$ and its invariants:

$$I_1 = \mathrm{tr}(\mathbf{C}), \qquad I_2 = \mathrm{tr}(\mathbf{C}^{-1}), \qquad I_3 = \det(\mathbf{C}). \tag{4}$$

Sometimes, we find it illustrative to quantify the deformation in terms of the following stretch ratios:

$$\lambda_1 = \sqrt{p_1}, \qquad \lambda_2 = \sqrt{p_2}, \qquad \lambda_3 = \sqrt{p_3} \tag{5}$$

where p_1, p_2, p_3 are the principal values of the right Cauchy-Green deformation tensor, **C**.

Before proceeding further a few comments have to be made. Since, we are tracking markers in 3D space, and the markers are not on the same plane, we are able to estimate all the nine components of the deformation gradient. However, since all the markers are on the surface of the valve tissue, the determined deformation gradient is for the surface of the valve tissue. Also, though the constants a_x^i, a_y^i and a_z^i could be found using just four markers, it resulted in larger errors because the volume of the tetrahedron with these four markers at its corners is small that the noises get magnified. Consequently, we use five markers to find the unknown constants in the deformation field (2). We observe that comparing the invariants of the Cauchy-Green deformation tensor is necessitated because one cannot ensure that the laboratory coordinates is the same between the specimens that are tested and therefore one cannot compare the components of the deformation gradient. One cannot ensure that the laboratory coordinates is the same between the experiments because the calibrating wedge dimension is small (12 mm x 7.5 mm x 10 mm), it is difficult to orient it accurately in the same direction each time. Of course any invariants could be used.

2.5 Materials

Table – 1: Details of Aortic Valve Samples tested

Sample No	Age (Years)	Sex	Length (mm)	Width (mm)
1	31	F	15	10

Table – 2: Details of Mitral Valve Samples tested

Sample No	Age (Years)	Sex	Length (mm)	Width (mm)
1	20	F	30	30
2	27	F	30	30
3	64	M	30	25
4	36	F	30	20
5	36	M	33	20
6	24	F	31	15

Figure 2: Close up view of the mitral valve tissue in uniaxial test set up with markers in the middle

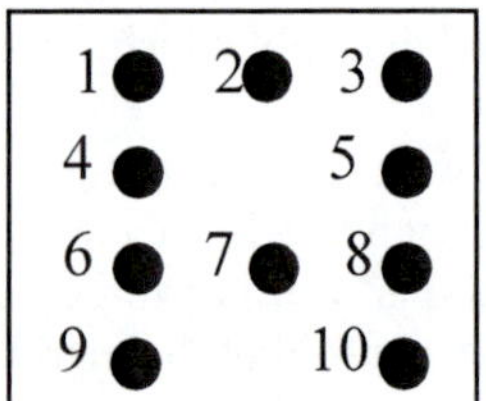

Figure 3: Arrangement of viable markers for uniaxial test of aortic valve

Diseased Aortic and Mitral Valves from patients undergoing heart valve replacement surgery were collected from Fortis Malar Hospital Chennai and stored at 4°C in normal saline and experiments were completed within 24 hours from harvest. The details of the samples tested are given in tables 1 and 2. The sample is tied using a thin metallic wire of diameter 0.3 mm to the hooks provided in the attachments made on the motor stage (see figure 2). Then, more than 9 markers are placed in the center of the tissue within the area of 4mm^2. After the marker has dried the tissue is immersed in normal saline and all the experiments are conducted in the immersed condition.

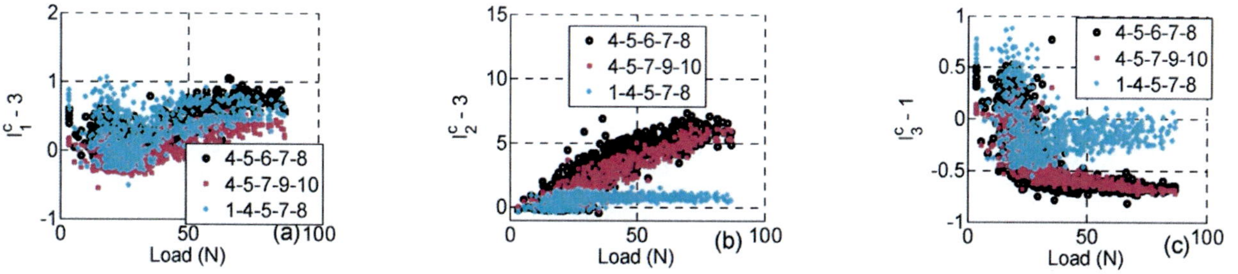

Figure 4: Plot of (a) I_1^C-3 (b) I_2^C-3 (c) I_3^C-1 vs. uniaxial load for aortic valve when stretched at the rate of 200 μm/s

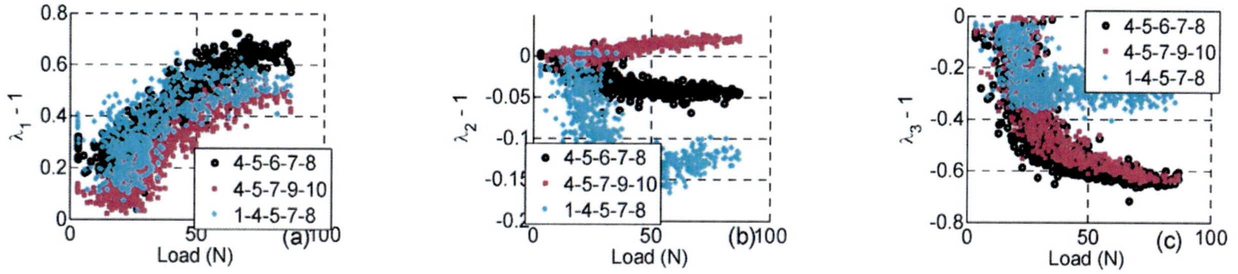

Figure 5: Plot of (a) λ_1-1 (b) λ_2-1 (c) λ_3-1 vs. uniaxial load for aortic valve when stretched at the rate of 200 μm/s

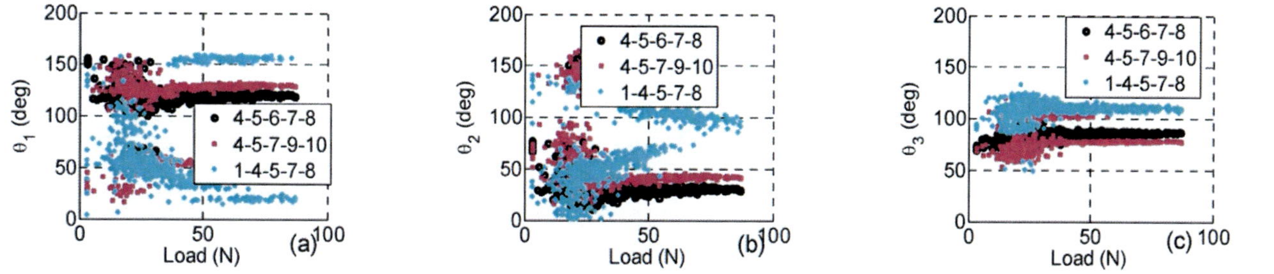

Figure 6: Plot of the orientation of principal directions (a) $\mathbf{n}_1$ (b) $\mathbf{n}_2$ (c) $\mathbf{n}_3$ with the laboratory x – axis vs. uniaxial load for aortic valve when stretched at the rate of 200 μm/s

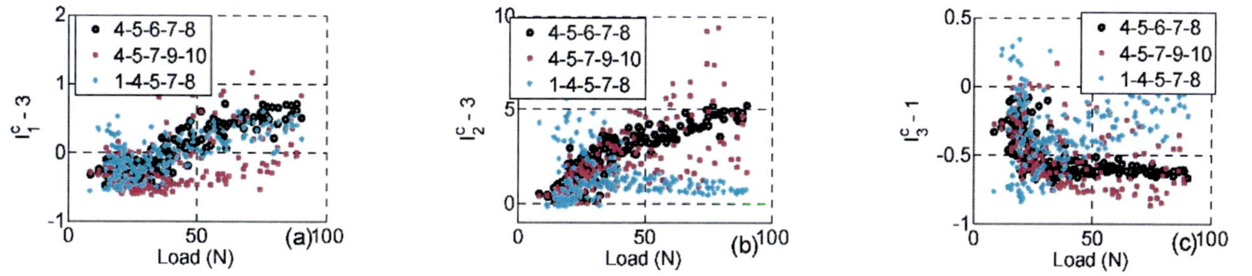

Figure 7: Plot of (a) I_1^C-3 (b) I_2^C-3 (c) I_3^C-1 vs. uniaxial load for aortic valve when stretched at the rate of 800 μm/s

3. Results and Discussion

While on each aortic leaflet only uniaxial tests were done on mitral valve leaflets both the uniaxial and biaxial tests were conducted. For biaxial tests the stretch along one direction is held fixed and that in the other orthogonal direction is cyclically stretched between 1 and 1.2 times its original length along the stretching direction. In uniaxial tests reported here the specimen is stretched to 1.15 times its original length along the stretching direction.

3.1 Aortic Valve Leaflet

While figure 4 plots the variation of the invariants when the aortic valve is stretched uniaxially, figure 5 plots the variation of the stretch ratios along three principal directions of **C** for the same experiment. Figure 6 plots the variation of the principal direction for this experiment. We do these for various selections of markers whose configuration is given in figure 3. It can be seen from these graphs that the deformation is not homogeneous, since different marker selections lead to different values for the invariants. This may be because the leaflets are inhomogeneous. Further it is evident from figure 4c that the aortic leaflet is compressible, since det(**C**) is not equal to 1, even approximately. Also, from figures 5b and 5c we infer that λ_2 and λ_3 are not same. Moreover, even though the stretch applied is 1.15, we observe stretches up to 1.7. This could be an artifact of the deformation being inhomogeneous. Moreover, from figure 6 we find that the principal directions do not change appreciably with applied load, implying the material under observation is isotropic or anisotropic with fiber orientation along the loading direction. These observations suggest that modeling the heterogeneous structure of the leaflet is important. Comparing figure 4 with figure 7 where the only difference is the rate of stretching, we find that there is no visible difference when the rate of stretching increased by 4 times. Consistently, we find that the loading and unloading paths are not widely different. These suggest that the valve is undergoing a non-dissipative process.

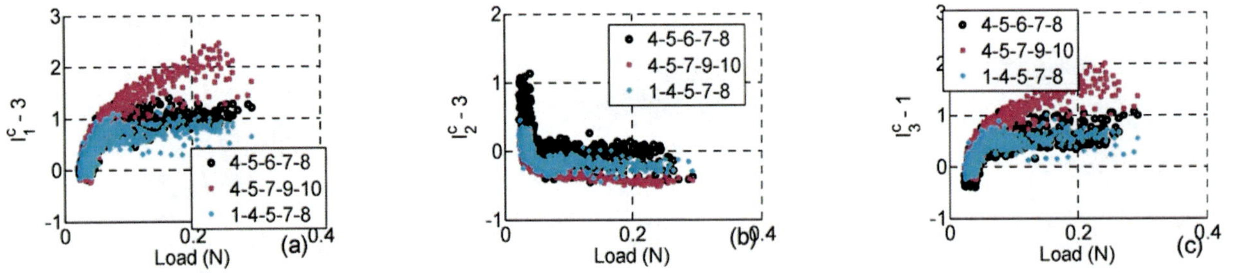

Figure 8: Plot of (a) I_1^C-3 (b) I_2^C-3 (c) I_3^C-1 vs. uniaxial load for mitral valve when stretched at the rate of 200 μm/s

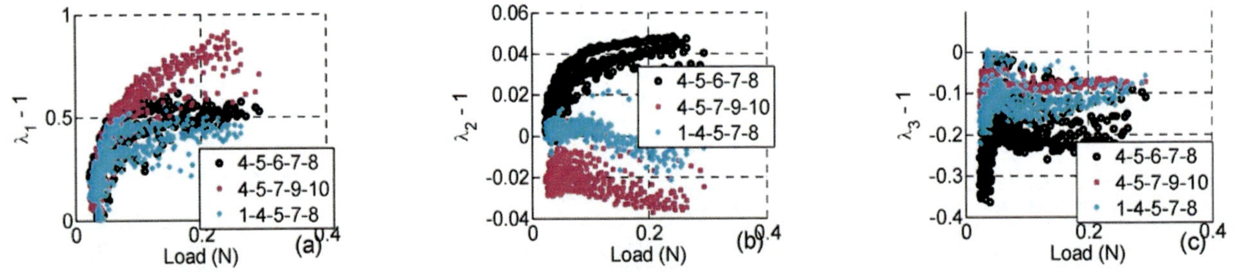

Figure 9: Plot of (a) λ_1-1 (b) λ_2-1 (c) λ_3-1 vs. uniaxial load for mitral valve when stretched at the rate of 200 μm/s

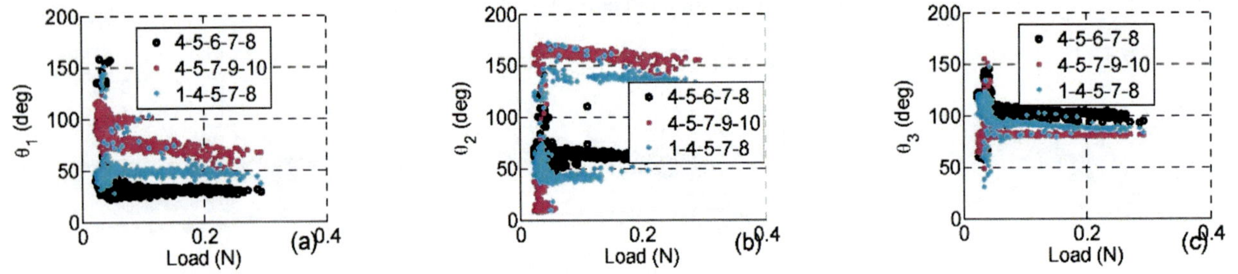

Figure 10: Plot of the orientation of principal directions (a) $\mathbf{n}_1$ (b) $\mathbf{n}_2$ (c) $\mathbf{n}_3$ with the laboratory x – axis vs. uniaxial load for mitral valve when stretched at the rate of 200 μm/s

3.2 Mitral Valve Leaflet

While figure 8 plots the variation of the invariants when the mitral valve is stretched uniaxially, figure 9 plots the variation of the stretch ratios along three principal directions of **C**, for the same experiment. Figure 10 plots the variation of the principal directions for this experiment. We do these for various selections of markers whose configuration is given in figure 15a. It can be seen from these graphs that the deformation is not homogeneous, since different marker selections lead to different values for the invariants. This may be because the leaflets are inhomogeneous. Further it is evident from figure 8c that the mitral valve leaflet is compressible, since det(**C**) is not equal to 1, even approximately.

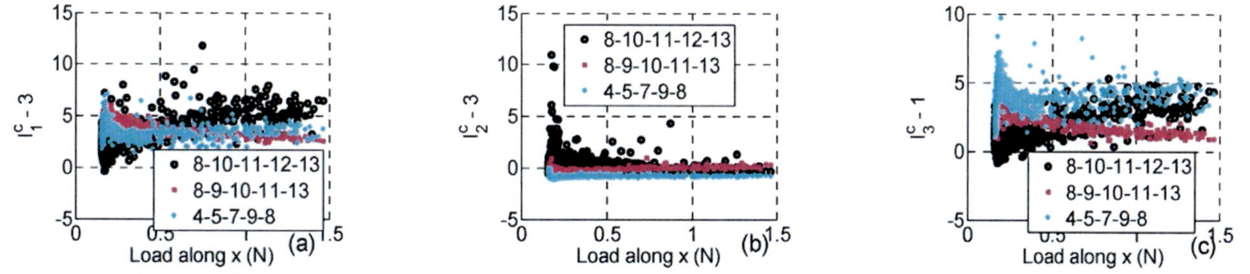

Figure 11: Plot of (a) I_1^C-3 (b) I_2^C-3 (c) I_3^C-1 vs. load in the x direction for mitral valve when biaxially stretched such that $\lambda = 1$

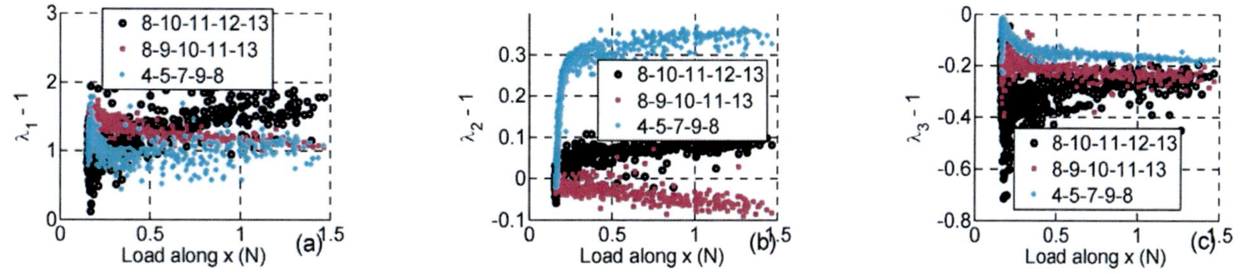

Figure 12: Plot of (a) λ_1-1 (b) λ_2-1 (c) λ_3-1 vs. load in the x direction for mitral valve when biaxially stretched such that $\lambda_y = 1$

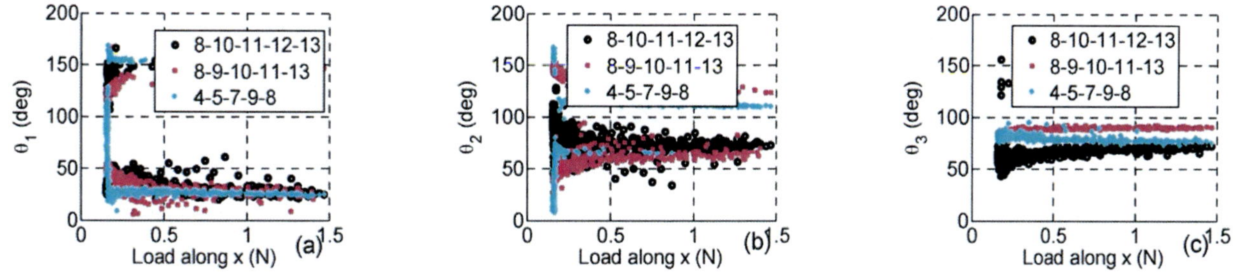

Figure 13: Plot of the orientation of principal directions (a) $\mathbf{n}_1$ (b) $\mathbf{n}_2$ (c) $\mathbf{n}_3$ with the laboratory x – axis vs. load in the x direction for mitral valve when biaxially stretched such that $\lambda_y = 1$

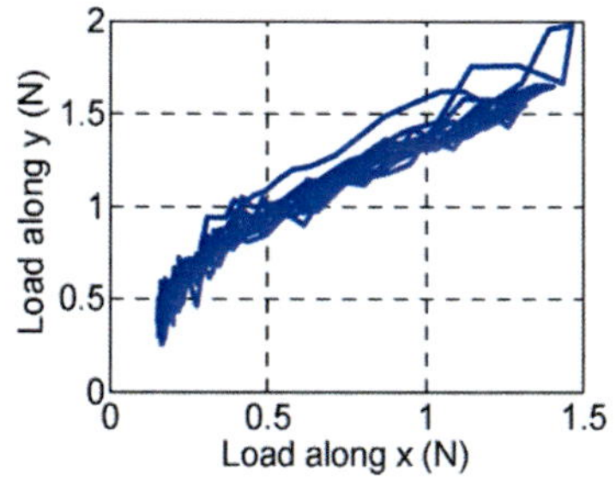

Figure 14: Load along y direction vs. load along x direction for mitral valve when biaxially stretched along x direction such that $\lambda_y = 1$.

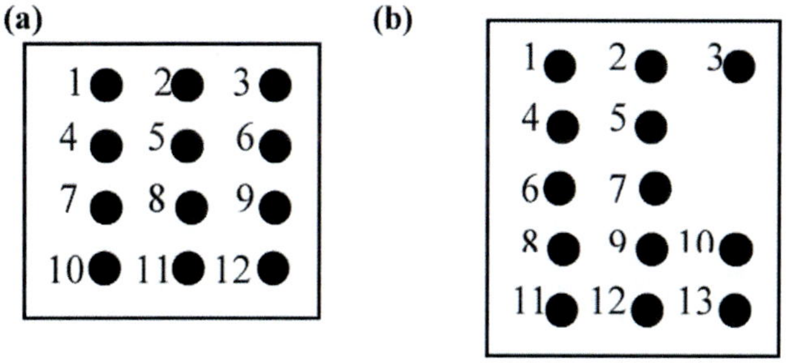

Figure 15: Arrangement of viable markers for (a) uniaxial test (b) biaxial test of mitral valve

Also, from figures 9b and 9c we infer that λ_2 and λ_3 are not same. Moreover, even though the stretch applied is 1.15, we observe stretches up to 1.7. This could be an artifact of the deformation being inhomogeneous. Moreover, from figure 10 we find that the principal directions do not change appreciably with applied load, implying the material under observation is isotropic or anisotropic with fiber orientation along the loading direction. These observations suggest that modeling the heterogeneous structure of the leaflet is important.

Thus, qualitatively the response of mitral valve to uniaxial stretch is similar to that of the aortic valve leaflet. However, quantitatively their responses are different. Though all the tested mitral valve leaflets show qualitatively similar response as the sample 6, whose response is presented here, quantitatively they are different.

Figure 11 plots the variation of the invariants when the mitral valve is stretched biaxially holding the stretch along one direction fixed and cyclically stretching along a direction perpendicular to the direction along which the stretch ratio is held fixed. Figure 12 plots the variation of the stretch ratios along three principal directions of **C** and figure 13 plots the variation of the principal directions for this experiment. We do these for various selections of markers whose configuration is given in figure 15b. It can be seen from these graphs that the deformation is not homogeneous, since different marker selections lead to different values for the invariants. This may be because the leaflets are inhomogeneous. Further it is evident from figure 11c that the mitral valve leaflet is compressible, since det(**C**) is not equal to 1, even approximately. Also, from figures 12 we infer that none of the principle stretch ratios is constant. Moreover, even though we stretched the leaflet only up to a stretch ratio of 1.2, locally the tissue seems to have been stretched even up to a stretch ratio of 2. This may be because of the heterogeneous nature of the tested specimen. Further, from figure 13 we find that the principal directions do not change significantly with applied load, implying the material under observation is isotropic or anisotropic with fiber orientation along the loading direction.

In this work, we presented the results in terms of the load rather than the stress because the stress distribution is non-uniform. The thickness of the specimen varies, thus making the stress distribution non-uniform. Hence, we are exploring on ways to characterize the non-uniform stress distribution.

These results suggest that contrary to what is assumed in the literature [9], these valve leaflets are compressible. Further, it seems that it is important to consider the heterogeneous nature of these tissues. These have implications with respect to the development of the constitutive relation for these leaflets.

Acknowledgement
We thank Department of Biotechnology, Government of India for funding this work.

References

[1.] Stella, J.A., Sacks, M.S., On the biaxial mechanical properties of the layers of the aortic valve leaflet, Journal of Biomechanical Engineering, 129(5), 757–766, 2007.

[2.] Sacks, M.S., Merryman, D.W., Schmidt, D.E., On the biomechanics of heart valve function, Journal of Biomechanics, 42 (12), 1804–1824, 2009.

[3.] Adamczyk, M.M., Vesely, I., Characteristics of compressive strains in porcine aortic valves cusps, Journal of Heart Valve Disease 11 (1), 75–83, 2002.

[4.] Billiar, K.L., Sacks, M.S., Biaxial mechanical properties of the natural and glutaraldehyde treated aortic valve cusp. Part I: experimental results, Journal of Biomechanical Engineering, 122 (1), 23–30, 2000.

[5.] Billiar, K.L., Sacks, M.S., Biaxial mechanical properties of the native and glutaraldehyde-treated aortic valve cusp: part II-A structural constitutive model, Journal of Biomechanical Engineering, 122 (4), 327–335, 2000.

[6.] Christie, G.W., Barratt-Boyes, B.G., Age-dependent changes in the radial stretch of human aortic valve leaflets determined by biaxial stretching, Annals of Thoracic Surgery, 60, S156–S159, 1995.

[7.] May-Newman, K., Yin, F.C.P., Biaxial mechanical behavior of excised porcine mitral valve leaflets, American Journal of physics, 269, H1319-H1327, 1995.

[8.] Gorman 3rd, J.H., Gupta, K.B., Streicher, J.T., Gorman, R.C., Jackson, B.M., Ratcliffe, M.B., Bogen, D.K., Edmunds Jr., L.H., Dynamic three-dimensional imaging of the mitral valve and left ventricle by rapid sonomicrometry array localization, Journal of Thoracic and Cardiovascular Surgery, 112 (3), 712–726, 1996.

[9.] May-Newman, K., Yin, F.C.P., A constitutive law for mitral valve tissue, Journal of Biomechanical Engineering, 120, 38-47, 1998.

[10.] Gorman 3rd, J.H., Jackson, B.M., Enomoto, Y., Gorman, R.C., The effect of regional ischemia on mitral valve annular saddle shape, Annals of Thoracic Surgery, 77 (2), 544–548, 2004.

[11.] Sacks, M.S., He, Z., Baijens, L., Wanant, S., Shah, P., Sugimoto, H., Yoganathan, A.P., Surface strains in the anterior leaflet of the functioning mitral valve, Annals of Biomedical Engineering 30 (10), 1281–1290, 2002.

Proceedings of the SEM Annual Conference
Mohegan Sun, Uncasville, Connecticut, USA, June 13 - 16, 2011

Bio-prosthetic Heart Valve Stress Analysis: Impacts of Leaflet Properties and Stent Tip Deflection

Caitlin Martin and Wei Sun

Tissue Mechanics Laboratory
Biomedical Engineering and Mechanical Engineering
University of Connecticut, Storrs, USA

ABSTRACT

Bio-prosthetic heart valves (BHV) have complex geometry, and their normal function involves interactions between the leaflets and the supporting stent, thus, the alteration of tissue mechanical properties can have a confounding influence on the stress distribution throughout the valve. Recently, structural models have been proposed to capture the nonlinear, anisotropic mechanical properties of native and engineered tissues. This anisotropic behavior arises as a consequence of inherent structural heterogeneity present in tissue arising from regional variations in fiber orientation and density. In this study, we utilize a nonlinear fiber-based structural model to evaluate the impact of the leaflet properties on the stress distribution of a pericardial BHV. Valve deformations under 120 mmHg hydrostatic pressure are simulated with various perturbations of the tissue properties including the collagenous fiber orientation, as well as the fiber and matrix stiffness, in addition to perturbations of the stent elastic modulus. The maximum principal stress of the valve leaflets under each condition was normalized with respect to that of a 45 degree fiber orientation and the original un-perturbed material constants. The results indicated that the valve leaflet peak stress can be reduced, provided the proper manipulation of the structure and property of tissue constituents. These results can provide guidelines to engineer and design leaflet tissues to improve valve durability.

INTRODUCTION

Artificial heart valves have been used for over five decades as replacements for diseased heart valves and have saved millions of lives. Over 50 different valve designs have been developed, yet despite their widespread use, there remains no optimal design. Currently, bioprosthetic heart valves (BHVs) made from biologically-derived, glutaraldehyde-treated bovine pericardium or porcine aortic leaflets, continue to be the dominant replacement valve modality due to their superior hemodynamic function. However, BHVs are only recommended for patients who are 65 years or older, due to concerns of long-term durability [1].

The main factors contributing to the limited durability of BHV have been identified as: [2,3] 1) cuspal mineralization and calcification, causing cuspal stiffening, and 2) non-calcific cuspal damage, including mechanical fatigue and possible proteolytic degradation of the collagenous extracellular matrix. Studies have shown that mechanical fatigue damage alone can be a major clinical cause of BHV failure [4]. The regions of tearing and calcification in BHVs correlate with the regions of high tensile and bending stresses [5,6]. Therefore, stress concentrations within the leaflet can either directly accelerate tissue structural fatigue damage, or initiate calcification by causing structural disintegration, thus enabling multiple calcification pathways that lead to valve failure [3,7].

T. Proulx (ed.), *Mechanics of Biological Systems and Materials, Volume 2*, Conference Proceedings of the Society for Experimental Mechanics Series 9999, DOI 10.1007/978-1-4614-0219-0_9, © The Society for Experimental Mechanics, Inc. 2011

Studies have shown that changing the valve design can affect the magnitude and distribution of stress acting on the valve leaflets. For example, results by Cacciola et al.[8] indicate that for a polymer valve, a stentless design could reduce peak leaflet stresses by up to 80% for a sinusoidal fiber reinforcement layout with respect to a stented valve with the same fiber reinforcement. Similarly, studies on surgical BHVs have shown that a flexible leaflet attachment stent, with radial stent-tip deflection, reduces the peak stress on valve leaflets [8-11]. The effects of using different leaflet material models have also been investigated by Patterson et al. [12], who compared the effects of using linear and nonlinear isotropic models of leaflets. Burriesci et al. [13] studied the effects of orthotropy of a pericardial heart valve and found that even a small amount of orthotropy can significantly affect the mechanical behavior of the valve.

In our previous study [14], we developed a computational pericardial BHV model and nominal valve deformations under 120 mmHg hydrostatic pressure were simulated and rigorously validated by experimental data. Built upon the aforementioned model, in this study, we utilized a nonlinear fiber-based structural model to evaluate the impact of leaflet properties on the stress distribution of the BHV. Various perturbations of tissue properties, including various fiber orientations, as well as fiber and matrix stiffnesses were performed. In addition, we also studied the impacts of valve stent-tip deflections on the valve leaflet stress distribution.

METHODS

BHV Leaflet Material Model

A hyperelastic constitutive law, based on the anisotropic fiber-based formulation by Holzapfel et al. [15], was developed to define a strain energy function that incorporates both matrix and fiber components, with independent contributions of two or more fiber families of distinct orientations. Briefly, following the concept of pseudo-elasticity, the symmetric 2nd Piola-Kirchhoff stress, **S**, can be derived from a strain energy function, **W**, through:

$$\boldsymbol{S} = \frac{\partial \boldsymbol{W}}{\partial \boldsymbol{C}} \tag{1}$$

where **C** is the right Cauchy-Green strain tensor. For computational convenience, a multiplicative decomposition of the strain energy function is performed into a volumetric part, **U**, and a deviatoric part, $\overline{\boldsymbol{W}}$. The deviatoric part, $\overline{\boldsymbol{W}}$, is further decomposed into a contribution from the elastin matrix of the composite, $\overline{\boldsymbol{W}}_m$, and a contribution from the collagen fibers, $\overline{\boldsymbol{W}}_f$. Therefore, the total strain energy function can be described as,

$$\boldsymbol{W} = (1 - c_f)\overline{\boldsymbol{W}}_m + c_f\overline{\boldsymbol{W}}_f + \boldsymbol{U} \tag{2}$$

where, c_f is the fiber volume fraction. As detailed elsewhere [16], the overall constitutive law is given as:

$$\boldsymbol{W} = (1 - c_f)C_1\left(\exp\left(C_2(\bar{I}_1 - 3)\right) - 1\right) + c_f\frac{k_1}{k_2}\left((\exp(k_2(\bar{I}_4 - 1)^2) - 1) + (\exp(k_2(\bar{I}_6 - 1)^2) - 1)\right) + \frac{1}{D}(J - 1)^2 \tag{3}$$

where c_f is fiber volume fraction ratio, C_1, C_2, k_1, k_2, and D are material constants. $\bar{I}_1$, $\bar{I}_4$ and $\bar{I}_6$ are deviatoric strain invariants, and J is the volume ratio before and after deformation. The material constants of the model were obtained by fitting the model to biaxial testing data of glutaraldehyde treated bovine pericardium (GLBP) that was previously reported by Sun et al. [17] using the Marquardt-Levenberg nonlinear regression algorithm.

Simulation of Biaxial Tissue Responses

To evaluate the effects of varying the fiber orientation, as well as the fiber and matrix stiffness on the mechanical response of the tissue, load-control biaxial testing simulations were performed in ABAQUS [18]. The biaxial testing simulation model consisted of 400 plane stress CPS8R elements giving a specimen geometry of 25 mm x 25 mm x 0.4 mm (Fig. 1). Four evenly spaced node forces, with 5 mm between two adjacent nodes and 2.5 mm inside the specimen edge, were imposed on each side (Fig. 1). Each node force was 2.5 N, imposing a net total of 1 MPa Lagrangian stress on each edge. Similar to the actual experimental biaxial testing setup [19], only the central region of the specimen was considered for stress–strain measurements. This was accomplished by averaging the stress and strain tensor components for sixteen elements located in the center of the FE model, delimiting a 5 mm x 5 mm region.

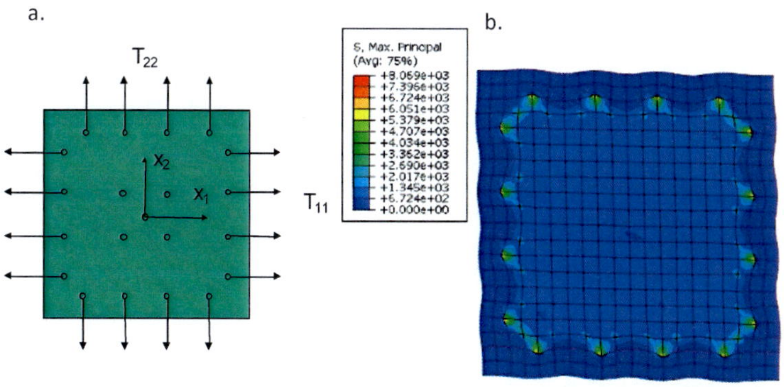

Figure 1 - (a) Diagram of a biaxial test specimen and (b) simulated biaxial test specimen stress distribution at peak load.

BHV Finite Element Model

The BHV finite element model was developed previously by Sun et al. [14] (Fig. 2). Briefly, tri-leaflet BHVs are fabricated from GLBP sheets that are die-cut to form leaflets, which are then mounted onto a metal stent. In the finite element model the stent was constructed using beam elements. The wireframe was assigned a Young's modulus of 2.33×10^7 kPa and a Poisson ratio of 0.3. The leaflets were attached to the valve stent and modeled using large strain brick elements. Each leaflet had its own local coordinate system for material property definitions that are fully defined by the constitutive law, Eq. (3), and incorporated into the user subroutine UMAT. Details of the constitutive model implementation into ABAQUS have been previously presented [18,20]. It was assumed that each leaflet can be modeled with one set of material parameters, which was supported by experimental findings [14]. The contact between two leaflets was modeled using a master-slave contact pair (an option in ABAQUS). The leaflet that was stiffer in the radial direction was assigned as the master surface while the other was specified as the slave surface. A quasi-static approach was used to analyze the deformation of the model from a stress-free position to a fully loaded configuration by applying a uniform pressure to the aortic side of the leaflet (Fig. 2).

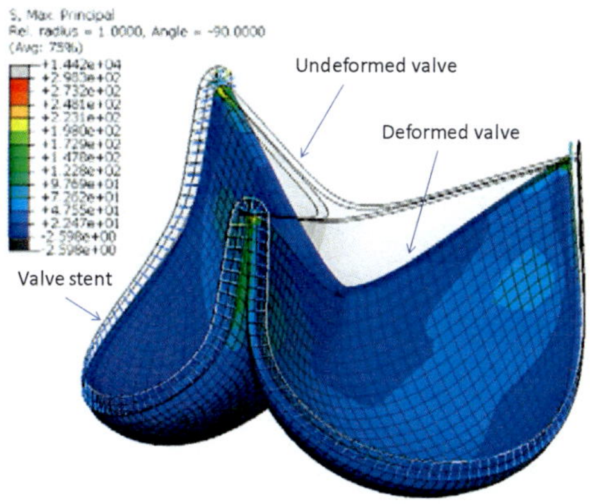

Figure 2 – An overlay plot of undeformed and deformed valve finite element models, showing stress distribution of Maximum Principal Stress under a valve closing pressure of 120 mmHg.

FE Parametric Analysis

In this study, we first examined the effects of fiber orientations, thus, the effects of fiber orientations of 15, 30, 45 and 60 degrees with respect to the leaflet circumferential direction were simulated. Afterwards, we fixed the fiber orientation at 45 degrees and perturbed the tissue matrix and fiber stiffness parameters, i.e., matrix stiffness c_1, fiber stiffness k_1 and k_2. In addition, we also perturbed the Young's modulus of the stent wireframe to induce varying degrees of stent-tip deflection. The complete list of parameter perturbations are given in Table 1.

Table 1.

C1	*C2*	*D*	*K1*	*K2*	*Fiber Orientation*	*Stent Modulus (E)*
1.73E-01	**1.73E+01**	**1.38E-03**	**2.96E+00**	**1.70E+01**	**4.50E+01**	**2.33E+07**
0.2C1			0.1k1	0.1k2	15 deg	0.6E
0.5C1			10k1	10k2	30 deg	0.8E
5C1					60 deg	1.2E
						1.4E

RESULTS/DISCUSSION

As shown in Fig. 3, substantial alterations of tissue anisotropy and compliance could be induced by changing fiber orientation and fiber/matrix mechanical properties. When tissue fibers are aligned at a 45 degree angle with respect to the 11 axis (the 11 axis is equivalent to the circumferential direction in the valve leaflet), the tissue responses from the two loading axes of the biaxial simulation (Fig. 1) are identical. Thus in Fig. 3a, only one 45 degree stress-strain curve is illustrated. It can also be seen from Fig. 3a that the fiber orientation determines the tissue compliance in a certain direction, for example, the 15 degree fiber orientation would induce a stiffer tissue response in the 11 axis than that of the 30 degree fiber orientation. In addition, it is apparent from Fig. 3a that a fiber orientation of a smaller angle would induce a higher degree of tissue anisotropy, e.g., the 15 degree angle fiber orientation would give a higher degree of tissue anisotropy than would the 30 and 45 degree fiber orientations.

Tissue matrix and fiber stiffness also have a significant influence on the overall tissue stiffness. Interestingly, the overall tissue compliance is predominately determined by the tissue matrix stiffness, whereas changing the fiber stiffness does not appear to induce substantial changes in the tissue compliance.

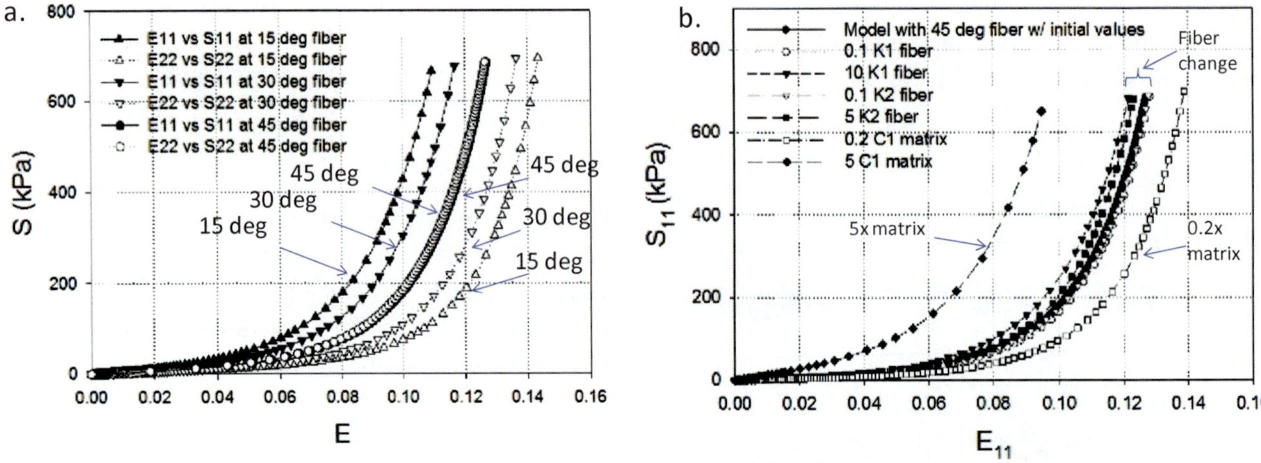

Figure 3 – Predicted tissue mechanical responses when varying the (a) fiber orientation and the (b) fiber and matrix stiffness in the model of Eqn (3) in a simulated equi-biaxial loading condition.

Figure 4a-c shows that changes in the BHV design parameters can significantly alter the stress distribution throughout the valve. The maximum principal stress on the valve leaflets under each perturbation condition, as illustrated in Fig. 4d, was normalized with respect to that of the BHV with 45 degree fiber orientation and the original un-perturbed material constants. The maximum principal stress in each scenario was in proximity of the leaflet commissures (Fig. 4 a-c). Interestingly, as shown in Fig. 4d, alterations of the K1 and K2 fiber stiffness had the most dramatic effect on the maximum principal stress in the BHV leaflets, yet the biaxial response of GLBP with the unperturbed parameters is nearly identical to those with altered fiber stiffness. Similarly, increasing the matrix stiffness by 5x made the GLBP biaxial specimen significantly stiffer in the 11 direction, however it did not have a significant impact on the maximum principal stress in the BHV leaflets. These discrepancies demonstrate the confounding relationships between the design parameters in the complex BHV environment as compared to the biaxial testing conditions. Reduction of the stent elastic modulus by 40% resulted in the greatest reduction of the max principal stress in the leaflets (-6.88%). Overall, the results indicated that the max principal stress in the leaflets may be reduced by increasing the matrix stiffness, decreasing the fiber stiffness, or reducing the stent modulus.

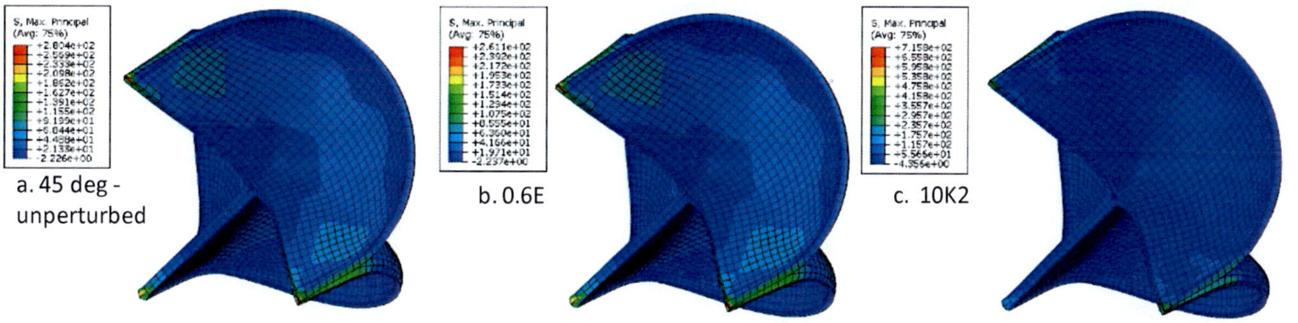

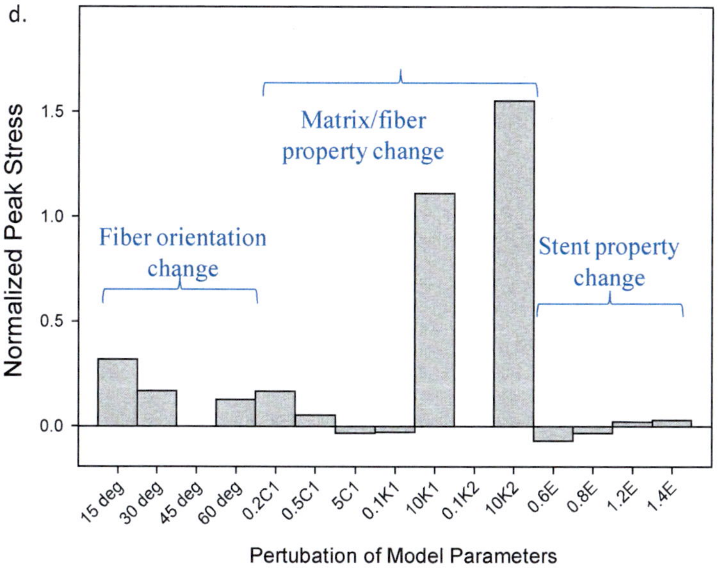

Figure 4 – Contour plot of the max principal stress in the BHV with (a) 45 deg fiber orientation and unperturbed material parameters, (b) 0.6x elastic modulus of valve stent, and (c) 10x K2 fiber family stiffness. (d) Comparison of predicted valve peak stresses at a fiber orientation of 45° with other perturbations of tissue mechanical properties, including fiber orientation and fiber/matrix properties, as well as stent properties.

It should be noted that a reduction in the leaflet stresses does not necessarily mean that the valve will be more durable. For example, increasing the fiber stiffness may increase the stress in the leaflets, but as the fibers act to reinforce the GLBP material, this may increase the leaflet durability regardless of the increase in stress. The stress distribution on the leaflets with respect to the fiber orientation is also critical in evaluating the BHV performance.

In the future we will conduct a probabilistic study to allow for the systematic analysis of multiple design uncertainties and the coupling effects between them. Probabilistic distribution functions for these design variables will be utilized as random variable inputs, in contrast to the deterministic approach adopted in this study, to accurately quantify the sensitivity of each design parameter, as well as the coupling effects between GLBP tissue properties, BHV geometry, and stent properties for BHV design optimization.

ACKNOWLEDGEMENTS

Research for this project was funded in part by the CT DPH grant #2010-0085 and an AHA SDG grant #0930319N.

REFERENCES

1. AHA. American Heart Association: Heart Disease and Stroke Statistics — 2006 Update; 2006.
2. Turina J, Hess OM, Turina M, Krayenbuehl HP. Cardiac bioprostheses in the 1990s. Circulation 1993;88(2):775-81.
3. Schoen FJ. Future directions in tissue heart valves: impact of recent insights from biology and pathology. J Heart Valve Dis 1999;8(4):350-8.
4. Sacks MS, Schoen FJ. Collagen fiber disruption occurs independent of calcification in clinically explanted bioprosthetic heart valves. J Biomed Mater Res 2002;62(3):359-71.
5. Sacks MS, Mirnajafi A, Sun W, Schmidt P. Bioprosthetic heart valve heterograft biomaterials: structure, mechanical behavior and computational simulation. Expert Review of Medical Devices 2006;3(6):817-834.
6. Sacks MS, Schoen FJ, Mayer JE. Bioengineering Challenges for Heart Valve Tissue Engineering. Annual Review of Biomedical Engineering 2009;11(1):289-313.
7. Gross JM. Calcification of bioprosthetic heart valves and its assessment. J Thorac Cardiovasc Surg 2003;125(90030):6S-8.
8. Cacciola G, Peters GW, Baaijens FP. A synthetic fiber-reinforced stentless heart valve. J Biomech 2000;33(6):653-8.
9. Christie GW. Computer modelling of bioprosthetic heart valves. Eur J Cardiothorac Surg 1992;6(suppl_1):S95-101.
10. Krucinski S, Vesely I, Dokainish MA, Campbell G. Numerical simulation of leaflet flexure in bioprosthetic valves mounted on rigid and expansile stents. J Biomech. 1993;26(8):929-43.
11. Li K, Sun W. Simulated thin pericardial bioprosthetic valve leaflet deformation under static pressure-only loading conditions: Implications for percutaneous valves. Annals of Biomedical Engineering 2010;38(8):2690-2701.
12. Patterson EA, Howard IC, Thornton MA. A comparative study of linear and nonlinear simulations of the leaflets in a bioprosthetic heart valve during the cardiac cycle. J Med Eng Technol 1996;20(3):95-108.
13. Burriesci G, Howard IC, Patterson EA. Influence of anisotropy on the mechanical behaviour of bioprosthetic heart valves. J Med Eng Technol. 1999;23(6):203-15.
14. Sun W, Abad A, Sacks MS. Simulated bioprosthetic heart valve deformation under quasi-static loading. J Biomech Eng 2005;127(6):905-14.
15. Holzapfel GA, Gasser TC, Ogden RW. A new constitutive framework for arterial wall mechanics and a comparative study of material models. J Elasticity 2000;61:1-48.
16. Sun W, Chaikof E, Levenston M. Development And Finite Element Implementation Of A Nearly Incompressible Structural Constitutive Model For Artery Substitute Design. ASME Summer Bioengineering Conference. Marco Island, FL; 2008.
17. Sun W, Sacks MS, Sellaro TL, Slaughter WS, Scott MJ. Biaxial mechanical response of bioprosthetic heart valve biomaterials to high in-plane shear. Journal Biomechanical Engineering 2003;125:372-380.
18. Sun W, Sacks MS. Finite element implementation of a generalized Fung-elastic constitutive model for planar tissues. Biomechanics and Modeling in Mechanobiology 2005;4(Nov.,(2-3)):190-9.
19. Sacks MS, Sun W. Multiaxial Mechanical Behavior of Biological Materials. Annu Rev Biomed Eng 2003.
20. Sun W, Chaikof E, Levenston M. Numerical approximation of tangent moduli for finite element implementations of nonlinear hyperelastic material models. Journal of Biomechanial Engineering 2008 130(6):20-27.
21. Grunkemeier G, Wu Y, Jin R. Statistical analysis of heart valve outcomes. J Heart Valve Dis. 2002;11(Suppl 1):S2-7.

Mechanical Properties of Human Saphenous Vein

K. Paranjothi[1], U. Saravanan[2], R. KrishnaKumar[3], K.R. Balakrishnan[4]
[1]Research scholar, Department of Engineering Design, IIT Madras. Email: kpjothi@iitm.ac.in
[2]Assistant Professor, Department of Civil Engineering, IIT Madras. Email: saran@iitm.ac.in
[3]Professor, Department of Engineering Design, IIT Madras. Email: rkkumar@iitm.ac.in
[4]Director, Cardiovascular Surgery, Malar Hospitals. Email: cvskrb@gmail.com

ABSTRACT
Details about custom built experimental set up to perform inflation tests at constant length on blood vessels are presented. Using this displacement controlled set up we can apply and measure pressures up to 100 kPa and axial loads ranging up to 100 N. The surface deformation is determined from tracking twelve markers in 3D space using 2 CCD cameras. Discarded vein tissue from patients undergoing Coronary bypass surgery were collected and stored at 4°C in normal saline and experiments were completed within 24 hours from harvest. Inflation tests at different axial stretch ratios on saphenous vein were conducted. Results show that the deformation of the vein is not axially symmetric. These suggest that the vein is inhomogeneous and/or residually stressed not only in the radial direction, but also in the circumferential and/or axial direction. The loading and unloading path is not different, suggesting that the vein is being subjected to non-dissipative process. Checking for the incompressibility condition, results show that the vein is compressible. These results have implications in the development of constitutive models for the vein.

Introduction

The great saphenous vein (GSV) is the large (subcutaneous) superficial vein of the leg and thigh and is used widely in coronary artery bypass grafting (CABG). In order to understand how it adopts to changes in its mechanical environment it is important to know the stress and deformation field experienced by this vein due to the altered mechanical environment. For this one requires to develop a constitutive relation for these veins [1, 2]. These constitutive relations would help determine how the graft material has to be handled by the surgeons so that the vein does not get damaged [3]. Because of these needs mechanical tests on GSV has been performed [3-7] but to the knowledge of the authors a constitutive relation for the same has not been reported. This article details experiments conducted on GSV towards arriving at a constitutive relation. Though salient conclusions pertaining to the development of a constitutive relation is arrived at, no such relation is proposed. It is found that the GSV is compressible and is heterogeneous and residually stressed not only along the radial direction but also along the circumferential and/or axial direction. These have implications in the development of constitutive relation for the vein.

2 Material and Methods
2.1 Details of the Experimental Setup

The experimental set-up to perform inflation tests at constant length consists of three subsystems. First, a video based system that allows 3D tracking of up to 12 surface markers. By tracking the movement of the surface markers during the experiment the deformation field is found from which the principal invariants of the right Cauchy-Green deformation tensor is obtained. This system consists of two charged couple device (CCD) camera(AVT GUPPY F-146 B) with 1390 by 1040 pixel resolution, each connected to a PCI card (AVT 1394a adapter) and a micro-lens (NAVITAR zoom 6000 lens with 3mm fine focus and internal coaxial illuminator attached to a 1x adapter) whose magnification can be varied from 0.7x to 4x. The lens with a working distance of 92mm, has a field of view of 10mm at 0.7x magnification and 2mm at 4x magnification. The depth of field also varies from 1mm at low magnification to 0.1mm at high magnification. The camera can capture at most 17 frames per second. While one of the camera is vertical, the other is inclined at 45 degrees in the vertical plane (see figure - 1). The camera and the lens are mounted on damped mounting rods using a custom designed mounting system. Second subsystem consists of two computer controlled linear translation stages to apply axial displacement and a syringe pump to inflate the arteries. As seen in figure 1, on top of each of these stages, manually controlled precision stages (25mm travel, 0:01mm resolution) is provided for each of the two directions to allow positioning of the specimen accurately in the field of

T. Proulx (ed.), *Mechanics of Biological Systems and Materials, Volume 2*, Conference Proceedings of the Society for Experimental Mechanics Series 9999, DOI 10.1007/978-1-4614-0219-0_10, © The Society for Experimental Mechanics, Inc. 2011

view of the cameras. The computer controlled linear translation stages (Holmarc Opto-Mechatronics) has a 100mm travel with 0.5 micron precision and its velocity can be set between $2\mu m/s$ to $6mm/s$. The stepper motor used in the translation stage can apply a maximum force of 100 N at its peak velocity. The motor controller communicates with the computer using a RS-232 port. By appropriately actuating the motors uniaxial extension/compression can be achieved. The syringe pump (Cole Parmer) can deliver $60ml$ of fluid at flow rates ranging between 0.73 $\mu l/hr$ to 2100 ml/hr. The motor running the pump can apply a maximum linear force of $135N$ at the minimum velocity. The pump communicates with the computer through RS-232 port.

The final subsystem is the data acquisition system (DAQ). It consists of a load cell (Honeywell-Sensotec Model 31E), a pressure transducer (Honeywell-Sensotec, Model LM) and a 18 bit analog to digital converter (UEI model PowerDNA with DNA-AI-208 as input layer), which can take 8 voltage input signals and whose sampling rate can be varied between 1 to 8000 samples per second is used for sensing the applied loads. The data acquisition system (DAQ) communicates with the computer through the ethernet interface. The load cell can measure loads up to 200N in both tension and compression. The reported deviation from linearity for the load cell is less than 0.15 percent over the full range and resolution limited by only the analog to digital convertor. The pressure transducer is rated to measure gage pressures up to $100kPa$ with an accuracy of $50Pa$ (0.5 percent over the full scale). All these subsystems are connected to HP Workstation XW4600 with Intel Core 2 Duo processor running at 2.66 GHz and having 4GB RAM.

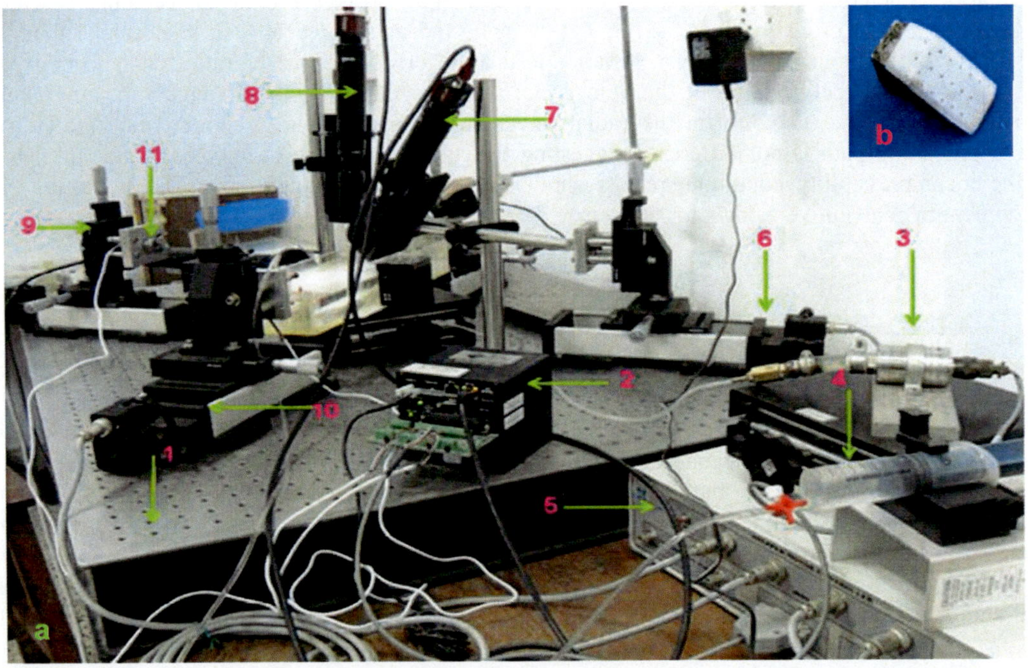

Figure 1. (a) Photograph of the experimental set up (1.Bread board, 2.Data acquisition system, 3.Pressure sensor, 4.Syringe pump, 5.Stepper Motor Controller (W,X,Y&Z-axes), 6.Z-axis Motorized Stage, 7. Inclined camera and lens, 8.Vertical camera and lens, 9.Manual X-translation stage, 10.W- axis Motorized Stage 11. Load cell). (b). Calibration Wedge

2.2 Details of Software Modules

Software for simultaneous operation (video, motor, DAQ) and synchronous acquisition of data (video, load) from this hardware has been developed in LABVIEW. The experimental protocols are completely automated by separate software modules. Each of these modules has many sub-modules whose details are given below in the order of their execution.

2.2.1 Load Cell Calibration

The main functionality of this module is to find the relationship between applied load (pressure) and the voltage output from the load cells (pressure transducer). This relationship between the applied load and developed voltage being linear, the two parameters (slope and intercept) that characterize the linear relationship, is determined through calibration and stored in a text file for later use. The load cell is calibrated using a 400 N digital spring balance with a resolution of 0.1 N. A sphygmomanometer with a resolution of 266 Pa is used to calibrate the pressure transducer

2.2.2 Reference Configuration
This module controls the video system and the DAQ system. It acquires images containing the surface markers from the vertical and inclined cameras and stores their centroid locations. Also it acquires data from DAQ and stores the initial readings of the load cell. All these data namely the centroid positions of the surface markers from both the cameras and the load cell readings are stored in a text file at the end of the recording.

2.2.3 Syringe Pump Module
This module controls two stepper motors and the syringe pump. The stepper motors are actuated to axially stretch the blood vessels to a predetermined value then the syringe pump is actuated to inflate the blood vessel. By controlling the speed at which the motors are actuated and the infusion/withdrawal rate of the syringe pump the rate at which the blood vessel is deformed can be altered. Using this module cyclic inflation tests at constant length or cyclic axial extension test at constant pressure with user specified number of cycles can be performed. Apart from this, the module has all the functionalities of the reference configuration module. So as the pump (or motors) load/unload the specimen it also acquires data from the cameras and load cells simultaneously. It then stores the load and pressure values and the location of the centroid of the markers in a text file at the end of the protocol.

2.2.4 Video Calibration
Here a calibration wedge (see figure 1b) containing markers separated by 2mm in the horizontal and 1.5mm in vertical direction is imaged in both the vertical as well as the inclined camera. Since, here we need to just record the centroid of stationary markers, which is what we do while recording the reference configuration too, the same reference configuration recording module is used for video calibration. Then, the relationship between the 3D coordinates of these markers and the two projections of these markers onto the imaging plane is established off-line.

2.3 3D Reconstruction from Two Perspective Views
The video subsystem has to be calibrated so as to obtain the 3D coordinates of the markers from two perspective views. This is achieved using global reconstruction techniques. The relationship between 3D coordinates of the centroid of the markers referenced with respect to a laboratory coordinate system (X, Y, Z) and its 2D coordinates obtained by projecting it on to the imaging surface and referenced with respect to the camera coordinate system $((u^v, v^v)$ or (u^i, v^i) depending on which camera is being referred to) is assumed to be linear, i.e.,

$$\mathbf{P}^k\mathbf{X} = \mathbf{u}^k - \mathbf{o}^k, \tag{1}$$

where k can be v or i, referring to the vertical camera and the inclined camera respectively, $\mathbf{X}$ denotes the 3D coordinates of the centroid of the marker, $\mathbf{u}^k$ denotes the 2D projected coordinates of the centroid of the marker on the kth camera, $\mathbf{o}^k$ denotes the 2D projected coordinates of the origin of the laboratory coordinate system on the plane of the imaging surface of the kth camera, $\mathbf{P}^k$ denotes the projection matrix for the kth camera. The projection matrix is a 2 by 3 matrix. Thus, for each camera we have to find 8 constants (6 in the projection matrix(P_{lm}) and 2 corresponding to the projected coordinates of the origin of the laboratory coordinate system). For this we require the coordinates of the centroid of at least four non-coplanar markers both in the laboratory coordinate ($\mathbf{X}$) and the projected camera coordinate system ($\mathbf{u}^k$). Usually more than 4 markers is used to find the eight constants and hence these calibration constants are found in a least square sense.

During the experiment we need to find the 3D coordinates of the centroid of markers, $\mathbf{X}$, given the two 2D coordinates of the projections of them onto the imaging surface. For this the relation (1) has to be inverted. However, now there are four equations with just three variables. Hence, these equations are also solved in least square sense to obtain the 3D coordinates of the centroid of the markers.

The assumption that the relationship between the laboratory coordinates and the projected camera coordinates is linear holds only if the camera lens provides uniform magnification over the depth of view and there is no distortion in the field of view. The validity of these assumptions is verified by examining if no strain is developed due to rigid body translation along any axis. It is found that the invariants deviate from the expected value by $\pm 10^{-2}$ due to these rigid body translations, irrespective of the axis, as long as one is in the depth and field of view. Hence, it is concluded that the relationship between the laboratory coordinates and the projected camera coordinates is linear and that the accuracy of the measured invariants is $\pm 10^{-2}$.

2.4 Determination of Deformation Gradient
In this work, 9 to 12 markers are placed on the surface of the vein and are tracked as the specimen is loaded. In general, it is not possible to experimentally determine the exact deformation field by tracking 12 markers (or for that matter even a larger number of markers), we can at most get a reasonable approximation of the deformation field and its gradient locally. However, we can verify if the observed deformation field is consistent with that theoretically predicted. Since, we are going

to inflate the vein; we expect that the deformation field be axially symmetric and hence uniform on the surface. This prompts us to assume a deformation of the form:

$$x = a_x^{\,0} + a_x^{\,1}X + a_x^{\,2}Y + a_x^{\,3}Z, \qquad y = a_y^{\,0} + a_y^{\,1}X + a_y^{\,2}Y + a_y^{\,3}Z, \qquad z = a_z^{\,0} + a_z^{\,1}X + a_z^{\,2}Y + a_z^{\,3}Z, \tag{2}$$

where (X,Y,Z) are the Cartesian coordinates of the centroid of a marker in the reference configuration and (x,y,z) are the Cartesian coordinates of the centroid of the same marker in the current configuration, $a_x^{\,i}$, $a_y^{\,i}$ and $a_z^{\,i}$ (for $i = \{0,1,2,3\}$) are constants. Knowing the location of five markers in the current and reference configurations, we obtain $a_x^{\,i}$, $a_y^{\,i}$ and $a_z^{\,i}$, by solving equation (2) in the least square sense.

Then, the matrix components of the deformation gradient, **F** is computed as:

$$\mathbf{F} = \begin{pmatrix} a_x^1 & a_x^2 & a_x^3 \\ a_y^1 & a_y^2 & a_y^3 \\ a_z^1 & a_z^2 & a_z^3 \end{pmatrix}. \tag{3}$$

It is now straightforward to compute the right Cauchy-Green deformation tensor, $\mathbf{C} = \mathbf{F}^t\mathbf{F}$ and its invariants:

$$I_1^{\,C} = \mathrm{tr}(\mathbf{C}), \qquad I_2^{\,C} = \mathrm{tr}(\mathbf{C}^{-1}), \qquad I_3^{\,C} = \det(\mathbf{C}). \tag{4}$$

Before proceeding further a few comments have to be made. Since, we are tracking markers in 3D space, and the markers are not on the same plane, we are able to estimate all the nine components of the deformation gradient. However, since all the markers are on the surface of the vein, the determined deformation gradient is for the surface of the vein. Also, though the constants $a_x^{\,i}$, $a_y^{\,i}$ and $a_z^{\,i}$ could be found using just four markers, it resulted in larger errors because the volume of the tetrahedron with these four markers at their corners is small that the noises get magnified. Consequently, we use five markers to find the unknown constants in the deformation field (2). We observe that comparing the invariants of the Cauchy-Green deformation tensor is necessitated because one cannot ensure that the laboratory coordinates is the same between the specimens that are tested and therefore one cannot compare the components of the deformation gradient. One cannot ensure that the laboratory coordinates is the same between the experiments because the calibrating wedge dimension is small (12 mm x 7.5 mm x 10 mm), it is difficult to orient it accurately in the same direction each time. Of course any invariants could also be used. But there is some merit in using the invariants defined in (4). Any deviation of $I_3^{\,C}$ from 1 indicates that the material is compressible. Further, if the deformation gradient, **F** is singular then $I_3^{\,C}$ would be close to 0 and $I_2^{\,C}$ be a large number. The deformation gradient would become singular if the markers selected to find the deformation field happened to lie nearly on a plane. Thus, the validity of the determined deformation field could also be ascertained by using the invariants defined in equation (4).

2.5 Materials

Discarded Saphenous veins from patients undergoing coronary artery bypass surgery were collected from Fortis Malar Hospital Chennai and stored at 4°C in normal saline. The experiments were completed within 24 hours from harvest. Details of the specimens tested are tabulated in table 1. Test samples were connected using plexiglass ferrules on both the ends and then connected to 2-way stopcock. This arrangement again connected to the 3-way stopcock. The vein is inflated by pumping saline through the syringe pump at one end and the other end of the sample is connected to the pressure transducer. Markers are placed in the middle portion of the vein using water insoluble eyeliner (see figure 2).

Table 1: Details of vein samples tested

Sample No	Age (Years)	Sex	Length (mm)	Diameter (mm)	Wall thickness (mm)
1	67	F	40	4.0	0.5
2	63	M	25	5.0	0.5

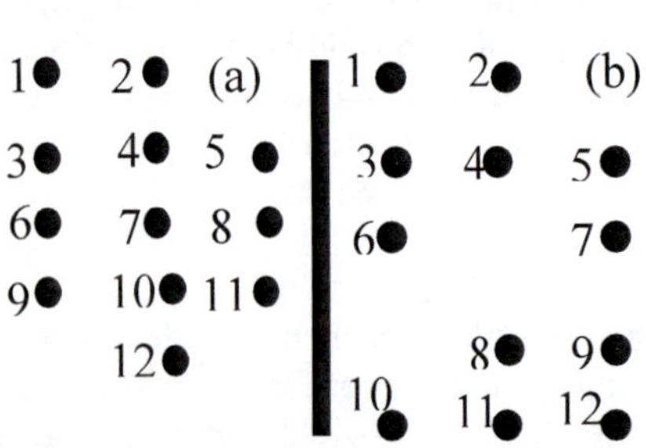

Figure 2: Arrangement of markers for (a) sample 1 (b) sample 2

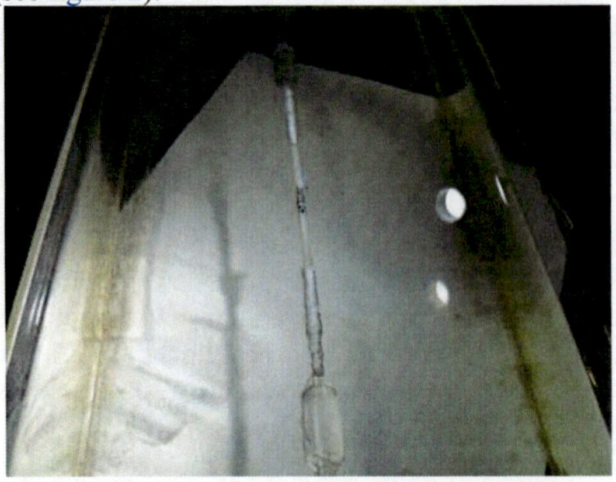

Figure 3: Close up view of the vein connected to the plexiglass ferrules with markers in the middle.

3. Results and Discussion

On each of the specimen, inflation tests at constant but five different axial stretch ratios, λ, were performed. The inflation pressure was cycled between 0 to 200 mm Hg (approximately 26 kPa) eight times while the vein being held at constant length. In each of this test the deformation field, the applied negative of the radial component of the normal stress and the axial load applied to ensure that the length of the vein is constant are measured simultaneously for all the eight cycles of loading and unloading. The negative of the radial component of the normal stress at the inner surface of the vein is hence forth called as pressure here. The pressure was increased by infusing normal saline at a constant rate of 10 ml/hr. The pressure was decreased by withdrawing normal saline at the same constant rate. The five different stretch ratios at which the vein was held are 1, 1.05, 1.10, 1.15 and 1.20. Some of the results are shown in figures 4 through 7 and in figure 9.

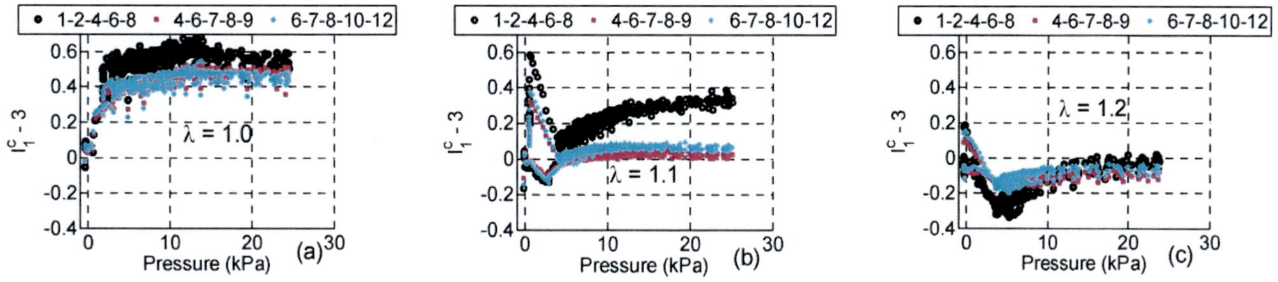

Figure 4: Plot of $I_1^C - 3$ vs inflation pressure for GSV at constant axial stretch ratio, (a) λ=1.0 (b) λ=1.1 (c) λ=1.2

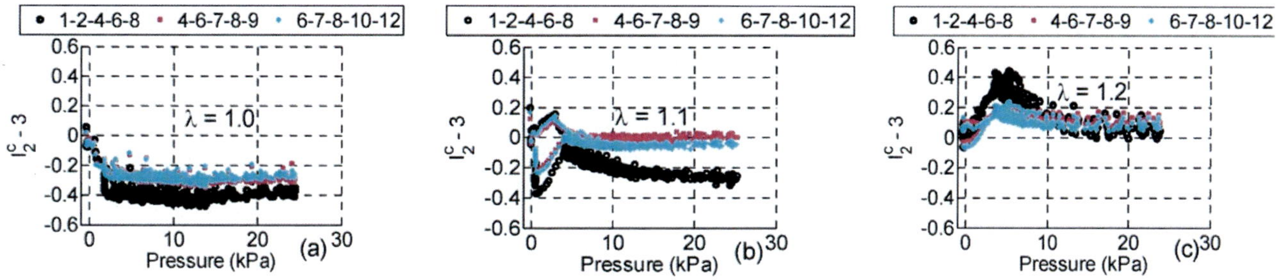

Figure 5: Plot of $I_2^C - 3$ vs inflation pressure for GSV at constant axial stretch ratio, (a) λ=1.0 (b) λ=1.1 (c) λ=1.2

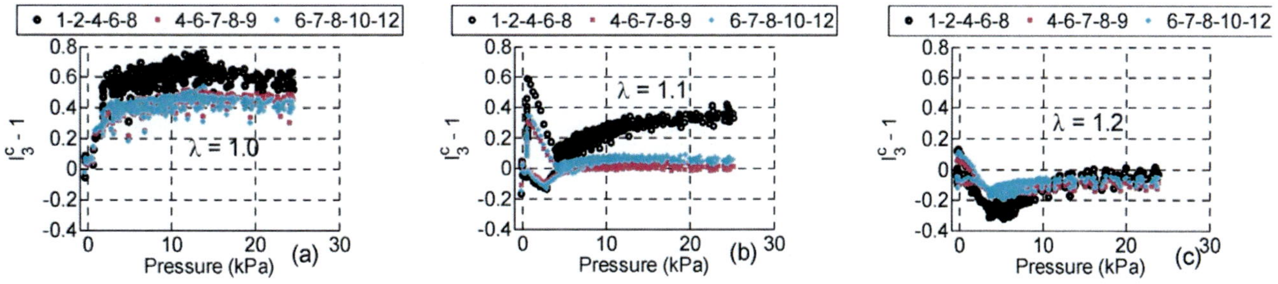

Figure 6: Plot of $I_3^C - 3$ vs inflation pressure for GSV at constant axial stretch ratio, (a) λ=1.0 (b) λ=1.1 (c) λ=1.2

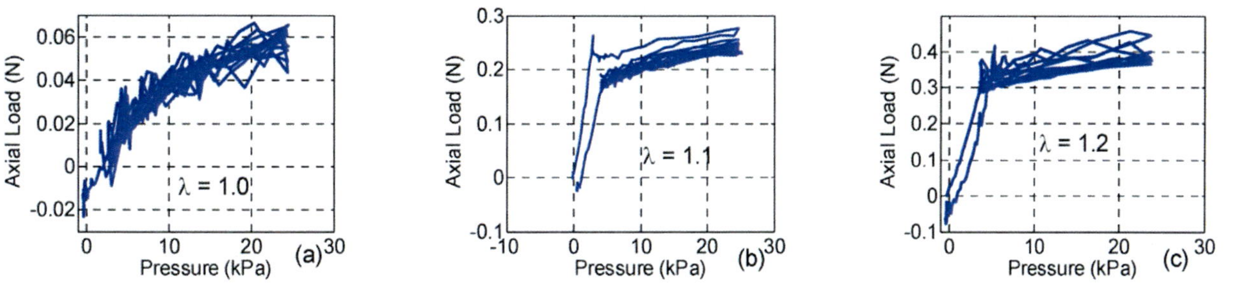

Figure 7: Plot of Axial load required to maintain a constant axial stretch ratio, (a) λ=1.0 (b) λ=1.1 (c) λ=1.2 vs inflation pressure for GSV (Note that the scale of the y-axis is different in these figures)

In figures 4 through 6 we plot the invariants as a function of pressure for sample 2. Also, in these figures we plot the variation of these invariants as determined using three different sets of five markers. The configuration of the markers is given in figure 2. If the deformation is axially symmetric and does not vary along the axis, then the value of the invariants would not depend on the choice of markers used to estimate them. However, it is clear from the figures 4 through 6 that the value of the invariants depends upon the choice of markers. Here it is pertinent to observe that the error in the estimated invariants is ±0.02 as is evident from figure 8. This error in the estimated invariants is arrived at by translating a balloon along the horizontal direction as a rigid body and computing the invariants for this motion. Ideally, the invariant I_1^C and I_2^C should be 3 and I_3^C should be 1. Deviation from this value is deemed as error in the estimated invariants.

The observation that the invariants of the Cauchy-Green deformation tensor are not a constant on the surface of the vein is significant for the following reason. Saravanan [8] showed that if this happens then the vein should be heterogeneous and/or residually stressed in the axial and/or circumferential direction.

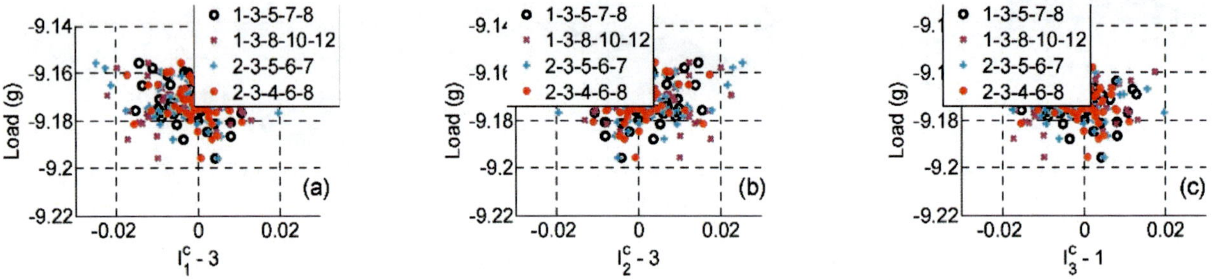

Figure 8: Plot of invariants (a) $I_1^C - 3$ (b) $I_2^C - 3$ (c) $I_3^C - 1$ vs. axial load as a rubber balloon is translated as a rigid body horizontally along its axis

It can be seen from figure 6 that I_3^C is not 1, as it should be if the vein is incompressible. Thus, we conclude that the vein is compressible. It should be noted that we are checking the compressibility of a small chunk of the solid material that constitute the wall of the vein, namely the region occupied between the markers used to estimate the deformation field and not the overall volume change in the vein.

From the figures 4 through 6 and figure 9 it is evident that the loading and unloading paths are not different in the GSV. However, if we look at figure 7, where the axial load required for maintaining the vein at constant length while inflating and deflating is plotted, we find that the axial load required for maintaining the axial length decreases with increasing number of cycles of inflating and deflating pressure. Moreover, this being a displacement controlled set up when the vein is brought back to its original length after eight cycles of pressurization, the axial load becomes negative (see figure 7b and 7c). These suggest the artery has stress relaxed along the axial direction, a viscoelastic phenomenon. Hence, it appears that the vein is a special kind of viscoelastic material that stress relaxes but does not dissipate in cyclic loading or an anisotropic viscoelastic material with differing viscoelastic properties along the radial and axial directions.

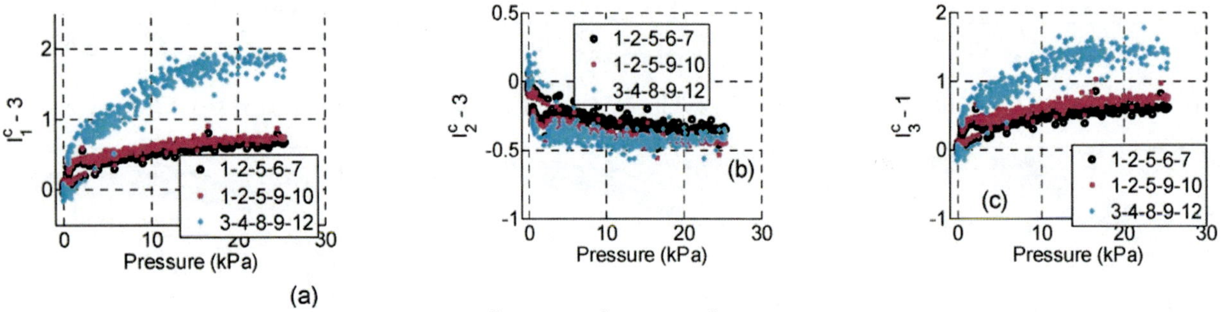

Figure 9: Plot of invariants (a) $I_1^C - 3$ (b) $I_2^C - 3$ (c) $I_3^C - 1$ vs. inflation pressure for sample 1

Figure 4 through 6 also capture the variation in the invariants with increasing axial stretch ratio. Here we find that as the axial stretch ratio increases, the value of I_1^C decreases, that of I_2^C increases and I_3^C decreases. However, unlike some of the other blood vessels [2], the axial load increases with increasing pressure for all stretch ratios at which the vein is held fixed.

While figures 4 through 7 depict the response of sample 2, figure 9 shows the response of sample 1. Comparing these figures we observe that sample 1 has been stretched more for the same pressure since the value of I_1^C and I_3^C realized are greater for this sample. Also, the variation in the invariants with spatial location is more pronounced in sample 1. These markedly different responses is at the least surprising given the work of Donovan et al. [4] because the factors that are known to affect the mechanical property of these saphenous veins is different between these two samples.

Following the work of Saravanan [8] and the observation here that the residual stresses could vary even along the circumferential and axial directions we did not measure the opening angles of these veins. This is because if the residual stresses vary even along the circumferential and axial direction a radial cut will not relieve the residual stresses and hence making opening angle measurement is of little use. On the other hand if the residual stresses vary only along the radial direction, Saravanan [8] showed that inflation test itself can be used to infer the residual stresses.

Further we observe that if the vein is anisotropic, then reporting the three invariants as done here alone would not suffice to characterize the vein. Ways to address this lacuna is being explored.

Conclusion

Our studies seem to suggest that the saphenous vein is compressible. Further, we observe that the deformation gradient evaluated at the surface of the vein varies circumferentially and axially, suggesting that the material response functions and/or residual stresses can depend on circumferential direction or axial direction or both. These have implications with respect to the development of the constitutive relation for these veins. Thus, at Indian Institute of Technology Madras, a systematic study of soft tissues has just begun with the necessary infrastructure being created. It is hoped that on testing more samples and development of constitutive relations for these tissues, the factors that influence the mechanical response of the soft tissues like age, sex, food habits, etc. can be quantified and correlated.

Acknowledgement

We thank Department of Biotechnology, Government of India for funding this work.

References

[1.] Dobrin, P.B., Littooy, F.N., Endean, E.D., Mechanical factors predisposing to intimal hyperplasia and medial thickening in autogenous vein grafts, Surgery, 105, 393–400, 1989.
[2.] Fung, Y.C., Biomechanics: Mechanical Properties of Living Tissue, Springer, New York, 1993.
[3.] Zhaoa,_J., Andreasenb, J.J., Yanga, J., Rasmussenc, B.S., Liaoa, D., Gregersena, H., Manual pressure distension of the human saphenous vein changes its biomechanical properties—implication for coronary artery bypass grafting, Journal of Biomechanics, 40, 2268–2276, 2007.
[4.] Donovan, D.L., Schmidt, S.P., Townshend, S.P., Njus, G.O., Sharp, W.V., Material and structural characterization of human saphenous vein, Journal of Vascular Surgery, 12, 531-537, 1990.
[5.] Stooker, W., Gok, M., Sipkema, P., Niessen, H.W.M., Baidoshvili, A., Westerhof, N., Jansen, E.K., Wildervuur, C.R.H., Eijsman, L., Pressure-diameter relationship in the human greater saphenous vein, Annals of Thoracic Surgery, 76, 1533-1538, 2003.
[6.] Molnar, G.F., Nemes, A., Kekesi, V., Monos, E., Nadasy, G.L., Maintained geometry, elasticity and contractility of human saphenous vein segments stored in a complex tissue culture medium, European Journal of Vascular and Endovascular Surgery, 40, 88-93, 2010.
[7.] Psaila, J.V., Melhuish, J., Viscoelastic properties and collagen content of the long saphenous vein in normal and varicose veins, British Journal of Surgery, 76, 37-40, 1988.
[8.] Saravanan, U., On large elastic deformation of prestressed right circular annular cylinders, International Journal of Non-Linear Mechanics, 46(1), 96-113, 2011.

The Scleral Inflation Response of Mouse Eyes to Increases in Pressure.

Kristin M. Myers
Assistant Professor, Columbia University
500 W. 120th St., 234 S.W. Mudd, M.C. 4703, New York, NY 10027

Frances Cone
Research Specialist, Glaucoma Research Lab
Wilmer Eye Institute, Johns Hopkins University
400 N. Broadway St. , Baltimore, MD 21287

Harry Quigley
Director, Glaucoma Service
Director, Dana Center for Preventive Ophthalmology
Wilmer Eye Institute, Johns Hopkins University
600 North Wolfe St., Wilmer 122, Baltimore, MD 21287-9019

Thao D. Nguyen
Assistant Professor, Johns Hopkins University
125 Latrobe, 3400 N. Charles Street, Baltimore, MD 21218

ABSTRACT

Glaucoma is one of the leading causes of blindness in the world. Evidence suggests that the stress generated in the eye wall by an elevated intraocular pressure plays a role in damaging the vision-transmitting retinal ganglions cells. However, the relationship between the connective tissue's mechanical properties and how it affects the cellular function is not understood. The purpose of this study was to measure the inflation response of intact C57/BL6 (control) mouse sclera to increases in intraocular pressure, comparing old (11 month) and young (2 month) animals. Mouse eyes were enucleated, mounted by the cornea to a custom fixture, cannulated and immersed in PBS. An active feedback, pressure-controlled syringe pump inflated the cannulated eyes in a series of load-unload and ramp-hold creep tests. A CCD video camera attached to a microscope imaged the expanding scleral surface at 0.5Hz. Scleral displacement was measured with a digital image correlation program. After testing, fresh tissue thickness measurements were taken on scleral slices at multiple locations. An optimized inverse finite element analysis was performed to fit a non-linear anisotropic material model to the experimental data, and material parameters are compared between groups.

INTRODUCTION

The connective tissue of the eye wall serves to protect delicate internal ocular structures and to resist deformation caused by fluctuations in intraocular pressure (IOP). The load-bearing eye wall consists of the anterior cornea and the posterior sclera. Both cornea and sclera are collagen rich structures with preferentially aligned collagen lamellae dictating its mechanical response. Experimental and theoretical studies that explored the mechanical characteristics of these connective tissues give evidence to suggest

T. Proulx (ed.), *Mechanics of Biological Systems and Materials, Volume 2*, Conference Proceedings of the Society for Experimental Mechanics Series 9999, DOI 10.1007/978-1-4614-0219-0_11,

that scleral material properties may influence the onset of glaucoma [1-4]. Open-angle glaucoma is the progressive loss of vision due to the death of retinal ganglion cells (RGCs). These neurons bundle through the back of the sclera via the optic nerve head (ONH) to transmit visual information to the brain. It is known that intraocular pressure (IOP) plays an important role in the damage to RGCs [1-3], however the relationship between the mechanical properties of the scleral connective tissue and how it affects RGC death is not well understood. Despite recent therapeutic advances to reduce vision loss in glaucoma patients, the rate of blindness remains high [5, 6].

The focus of this paper is to discuss an experimental and modeling study that quantifies scleral material properties measured for a bulge inflation response of an excised and intact mouse sclera. The mouse model was pursued because of the ability to mechanically test tissue that has been altered through age, disease, and genetic manipulation, and to determine how each of these factors influences the deformation response of scleral tissue to increases in IOP. The experimental methodology for the mouse sclera bulge tests has been described in detail in Myers et al. [7], and the nonlinear, anisotropic, distributed fiber material model that was used to fit the data has been described in Nguyen and Boyce [8]. This paper details the inverse finite element method and optimization scheme that was employed to fit the material model to measured scleral edge displacements for different aged wild-type C57BL/6 mouse sclera. The resulting best-fit material parameters were compared between different aged animals.

METHODS

Inflation tests were conducted on intact ex-vivo mouse eyes using a method previously described elsewhere [7]. Three 2 month old and three 11 month old C57BL/6 mouse scleral specimens were tested. Briefly, eyes were enucleated, cleaned of orbital fat and muscle, mounted by the cornea to a custom fixture, cannulated and immersed in PBS. A programmable pressure-controlled syringe pump inflated the cannulated eyes in a series of load-unload tests to 2kPa (15mmHg) and ramp-hold creep tests to 2, 4 and 6kPa. A CCD video camera (Pt. Grey Research GRAS-20S4M/C) attached to a microscope (Zeiss Stemi) imaged the expanding scleral surface at 0.5Hz (5μm/pixel). Scleral displacement for locations adjacent to the ONH to the holder was measured with a DIC program (Vic 2D, Correlated Solutions Inc.). After testing, fresh tissue thickness measurements were taken on scleral slices at multiple locations under a light microscope. Because mouse scleral tissue has a varying thickness profile and contains a stress concentration at the optic nerve head, a finite element (FE) model of the scleral specimens was developed to determine the complex stress state of the posterior portion of the eye. To quantify material properties, parameters of a nonlinear, anisotropic, distributed fiber material model [8] were optimized such that the FE model predictions matched the deformation response of the expanding scleral edge during inflation.

The finite element (FE) model was constructed for each specimen using the scleral geometry measured from the DIC camera at a reference pressure of 0.8kPa (6mmHg). After extracting the coordinates of the reference position of the scleral edge, a geometric representation was constructed by first fitting a conic section to one half of the measured scleral edge. A second conic section was fit to the inner scleral surface based on the measured thickness profile. Lastly, the inner and outer surface geometries were revolved 90° to form a quarter model of the scleral shell, with the optic nerve head (ONH) located at 100μm from the apex. The sclera was modeled as a composite material with a nearly incompressible Neo-Hookean matrix embedded with regions of non-linear transversely isotropic and anisotropic fibers. To summarize the details of the material model [8], the matrix was characterized by a shear μ_{matrix} and bulk κ_{matrix} modulus, with κ_{matrix} orders of magnitude larger than μ_{matrix}. The non-linear response of the collagen fibers was modeled with a phenomenological exponential function with model parameters [β, $\alpha_o = 4\alpha\beta$], with β charactering the strain-stiffening and α_o representing the initial stiffness of the fibers.

Two regions of preferentially aligned collagen were defined for the mouse sclera to be within 100μm of the ONH and 500μm of the limbus. For these regions, the anisotropic contribution of the free energy density was calculated from the orientation average of the free-energy density of the fibers weighted by a von Mises probability distribution function. The model parameters [n_{ONH}, n_{Limbus}] from this relationship represented dispersion parameters for these anisotropic regions, where n=0 represents a transversely isotropic case and larger values indicate an increase in circumferentially oriented collagen fibers. The ONH was modeled as a Neo-Hookean material, with a shear modulus $0.1\mu_{matrix}$.

Symmetric boundary conditions were applied on the x=0 and z=0 planes of the shell for the FE calculation, and clamped boundary conditions were placed on the y=0 plane to represent the glued limbus. The scleral model was meshed using trilinear hexahedral elements, with 3 elements through the thickness. Fig. 1 shows the scleral geometry, FE mesh, and boundary conditions for each of the specimens C56BL/6 mouse scleral specimens tested and modeled. FE calculations were performed using Tahoe and the coupled optimization scheme was performed using Dakota. Both programs are open-source codes developed at Sandia National Laboratories. The optimization scheme used a global pattern search algorithm to find the best-fit material model parameters [μ_{matrix}, β, α_o, n_{ONH}, n_{Limbus}], which minimized the difference between the experimental measurements and the model predictions at 10 evenly spaced pressure states between 0.8 and 6kPa.

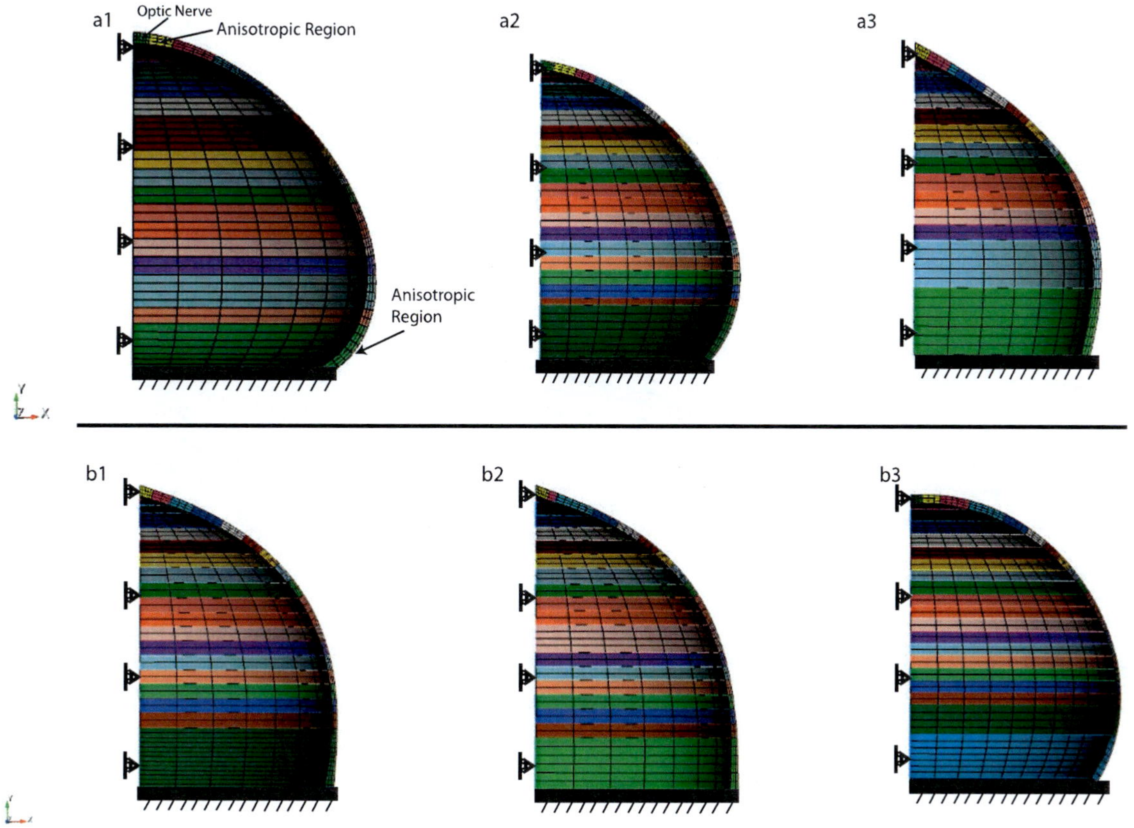

Fig. 1 Finite element meshes for a1-a3) three 2 month old C57BL/6 scleral specimens b1-b3) three 11 month old C57BL/6 scleral specimens

RESULTS

Fig. 2 reports the best-fit model predictions for each of the specimens presented in Fig. 1. The model fits are shown with the experimental data for scleral edge displacements measured at evenly spaced intraocular pressures from 0.8 to 6kPa. The corresponding averaged material parameters for the younger and older scleral tissue are reported in Fig. 3. The model successfully captured the deformation profile of the inflating scleral surface as well as the stiffening behavior of the loading response to increases in pressure. The fiber modeling parameters [β, α_o] were able to characterize this strain stiffening, with model parameters indicating that the older tissue may contain stiffer collagen or elastin fibers when compared to the younger tissue. However, the difference in [β, α_o] between different aged tissues was not statistically significant. Therefore, more trials are needed to explore this hypothesis. During inflation testing, the scleral tissue from the limbus to approximately 0.05mm posterior had the largest displacement gradient when compared to other scleral regions. The anisotropic dispersion model parameter n_{Limbus} was able to capture this behavior by including fibers preferentially aligned in the circumferential direction. Lastly, the model parameter n_{ONH} was nearly zero for both younger and older tissue, indicating that a transversely isotropic model was sufficient to capture the displacement behavior in this scleral region. Histology studies are needed to confirm these material parameters findings.

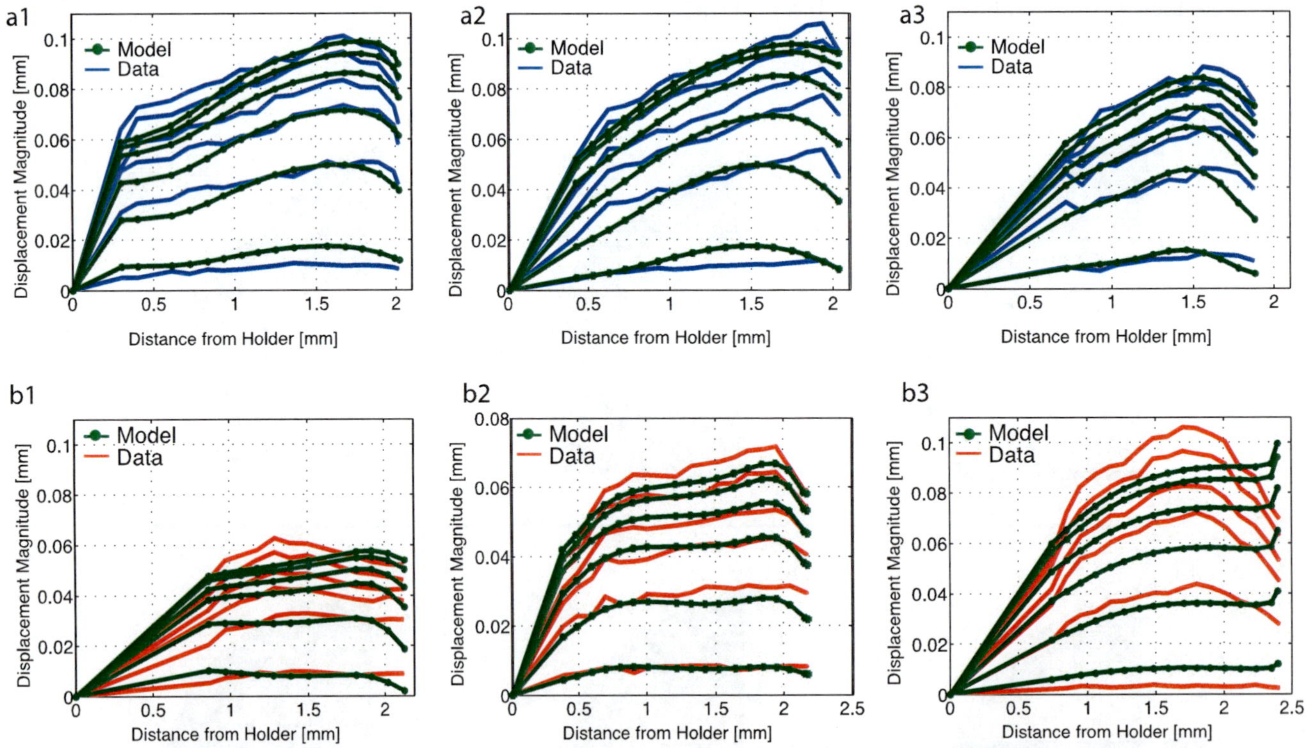

Fig. 2 Scleral edge displacement and the corresponding best-fit model prediction at evenly spaced intraocular pressures from 0.8 and 6kPa for a1-a3) three 2 month old C57BL/6 scleral specimens b1-b3) three 11 month old C57BL/6 scleral specimens

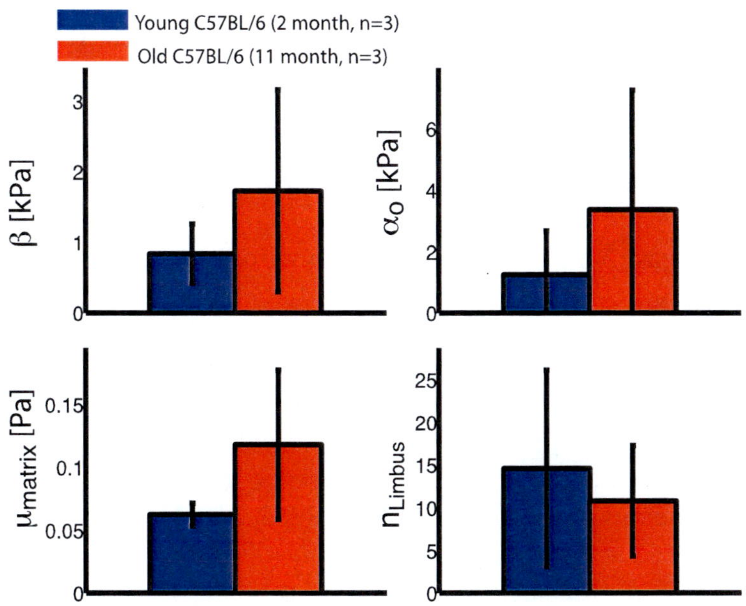

Fig. 3 Averaged material parameters for the 2 and 11 month old C57BL/6 scleral specimens

DISCUSSION

The nonlinear fiber composite model captured the inflation behavior of the mouse sclera reasonably well, with a region of circumferentially-aligned fibers in the limbal region. The anisotropic material parameter n_{ONH}, which represents the degree of circumferential fibers for the scleral region immediately adjacent to the ONH, was zero for both the younger and older mouse tissue. This result is inconsistent with evidence that shows that collagen and elastin fibers are preferentially aligned around the ONH [9]. Further histological and experimental evidence is needed to determine the roll of these fibers in the mechanical response of the mouse scleral tissue. In addition, it was hypothesized that parameters associated with the collagen fibers would be stiffer for the older tissue when compared to the younger tissue due to the decrease in collagen turn-over rate with age. The trend for the averaged model parameters associated with collagen behavior did indicate stiffening with age. However, additional trials will be needed to confirm this trend. Current efforts include improving the inflation testing set-up and investigating scleral tissue from glaucoma models.

REFERENCES

1. Burgoyne CF, Downs JC, Bellezza AJ, Suh JKF, Hart RT, The optic nerve head as a biomechanical structure: a new paradigm for understanding the role of iop-related stress and strain in the pathophysiology of glaucomatous optic nerve head damage, Prog Ret Eye Res, 24, 39-73 2005.

2. Ethier CR, Predicted outcome from hypotensive therapy for glaucomatous optic neuropathy, Can J Ophthalmol, 41, 9–11, 2006.

3. Nickells RW, From ocular hypertension to ganglion cell death: a theoretical sequence of events leading to glaucoma, Can J Ophthalmol, 421, 278–287, 2007.

4. Downs JC, Thomas KA, Hart RT, Burgoyne CF, Elastic and viscoelastic material properties of peripapillary sclera in normal and early glaucomatous monkey eyes, ARVO Abstract Invest Ophthalmol Vis Sci 42, S831 2001.

5. Quigley HA, Broman AT, The number of people with glaucoma worldwide in 2010 and 2020. British J. Ophthalmol, 90, 262–267, 2006.

6. Hattenhauer MG, Johnson DH, Ing H, Herman DC, Hodge DO, Yawn BP, Butterfield LC, Gray DT, The probability of blindness from open angle glaucoma, Ophthalmol, 85, 1277–1282, 2001.

7. Myers KM, Cone FE, Quigley HA, Gelman S, Pease ME, Nguyen TD, The in vitro response of mouse sclera, Exp Eye Res, 91, 6, 866-875, 2010.

8. Nguyen TD, Boyce BL, An Inverse Finite Element for Determining the Anisotropic Properties of the Cornea, Biomech Model Mechanbiol, DOI: 10.1007/s10237-010-0237-3ng, 2010.

9. Gelman S, Cone FE, Pease ME, Nguyen TD, Myers K, Quigley HA, The presence and distribution of elastin in the posterior and retrobulbar regions of the mouse eye, Exp Eye Res, 90, 2, 210-215, 2010.

Microstructure and Mechanical Properties of Dungeness Crab Exoskeletons

Jie Lian and Junlan Wang
Department of Mechanical Engineering, University of Washington, Seattle, Washington, USA

ABSTRACT

In this paper, we report our recent studies of the microstructure and mechanical properties of Dungeness crab exoskeleton. Nanoindentation and scanning probe microscopy (SEM) were carried out on different parts of the crab (claw and body shell), and different parts of the exoskeleton (exocuticle and endocuticle). Clearly defined four-layer structure was observed in the exoskeleton. In the cross sections of the claw and body shell, the mechanical properties show a decreasing trend from the exocuticle to endocuticle. In both cross sections and in-plane sections, a higher property was measured in the claw than the body shell. Finally, super low mechanical properties were obtained for nanoindentation on the in-plane sections of the claw's endocuticle than the cross-sections. The chitin-protein volume fraction was found to be the major factors affecting the mechanical properties in the different body parts as well as different layers of the exoskeleton.

1. INTRODUCTION

In biological systems, some light-weight structures possess attractive mechanical properties, such as high strength and durability. For example, in many crustaceans (lobsters, crabs and shrimps) and insects, their external covering – exoskeleton exhibits good mechanical rigidity and strength. In mollusk shells, their interior shinny mother of pearl layer – nacre, displays superior mechanical properties, such as toughness, impact resistance, and strength. In building future energy-saving transportation systems, the design of such lightweight materials that can adopt the deformation and fracture modes of biological systems will be highly desired. For designing these artificial, bio-mimetic materials, it is crucial to understand the structure-function relationship in the biological systems. Different from inorganic engineering materials, the exoskeletons and nacres are hierarchically organized from the molecular to macro level. In addition, these hard biological composites consist of organic matrix mineralized by inorganic components with various shapes, microstructures and compositions.

In literature, there have been some studies on the strengthening and toughening mechanisms of nacre using the three and four-point bending tests [1-3], tensile and compression tests[1,4], and nanoindentation experiments[3,5]. The outstanding mechanical properties in nacre is usually attributed to its hierarchically organized structure and 95 vol.% mineral component (Aragonitic $CaCO_3$) with only a small percent of organic biopolymer. Different from nacre, the exoskeletons of the crustaceans mainly comprise the organic component (primarily chitin) with various degree of mineralization (calcium carbonate) in different layers. The mechanical properties of crustacean exoskeletons have been studied by tensile and

T. Proulx (ed.), *Mechanics of Biological Systems and Materials, Volume 2*, Conference Proceedings of the Society for Experimental Mechanics Series 9999, DOI 10.1007/978-1-4614-0219-0_12, © The Society for Experimental Mechanics, Inc. 2011

compression tests [6-9], bending tests[9,10], microindentation tests [8,11], as well as nanoindentation tests[12]. The exoskeleton displays quite different mechanical properties in each discrete layer, various parts and directions. This requires understanding of the microstructure in different locations of the exoskeleton, and anisotropic properties in different loading directions.

In this study, Dungeness crab exoskeletons were selected for the experimental investigations. The exoskeleton of Dungeness crab share similar microstructure with other crustaceans. The exoskeleton can be divided into four discrete layers: epicuticle, exocuticle, endocuticle and membranous layers, in the order from outermost to innermost [13]. Because of the anisotropic structure of biological materials at multiple levels, the mechanical properties are usually also anisotropic. For the crustacean's exoskeleton, there have been some studies on microstructure, organic/inorganic compositions and mechanical properties [8,9,11,14,15]. However, to our best knowledge, the anisotropic mechanical properties of crustacean's exoskeleton have not been thoroughly studied. Nanoindentation is an effective technique to characterize the anisotropic mechanical properties in different directions. In this study, the hierarchical structure and anisotropic mechanical properties of Dungeness crab exoskeletons were investigated. It is aimed to gain insights of the anisotropy property – structure relation of this hierarchically organized architecture, and inspire the biomimetic design in improving the mechanical properties of structural materials.

2. EXPERIMENTAL METHOD

All test specimens were taken from live Dungeness crab, obtained from the local seafood market. The specimens were cut from the flat areas of the body shell and the tip area of the claw, with muscles removed. Specimens were tested in fresh condition and in the air dry state. The mechanical characterizations were carried out within 48 hours after the exoskeleton of the crab was cut. Samples were prepared in cross sections and in-plane sections. Fig. 1(a) shows the whole crab Cancer magister. Fig. 1 (b) and (c) show the claw and body shell for the study respectively. Naturally fractured sections of the exoskeleton were used for SEM to study the microstructure. A combined energy dispersed X-ray spectrometer (EDS) system with SEM was used to analyze the mineral composition on both cross-sections and in-plane sections (sections parallel to the exoskeleton surface). For nanoindentation studies, samples were polished using sand paper and alumina powder with particle size down to 0.05 μm. Epoxy mounting was applied to hold the thin layer of cross sections during polishing. The nanoindentation tests were carried out on a Ubi1 nanomechanical test instrument (Hysitron, Inc., MN). A three-sided diamond pyramidal Berkovich tip was used as the indenter tip. The nominal radius of curvature of the Berkovich tip is around 150 nm. During indentation, the load and displacement were continuously monitored by a three-plate capacitive force/displacement transducer. Standard Oliver and Pharr method is used to calculate the reduced modulus and hardness from the load-displacement information [16].

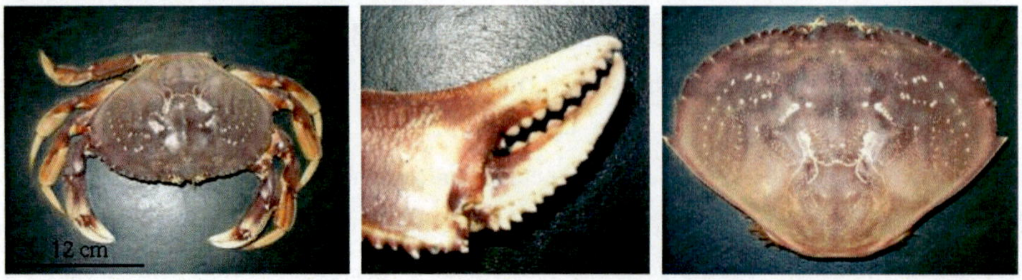

(a) (b) (c)

Fig. 1 (a) Image of Dungeness crab, with dimension about 9.5 inch in length; (b) claw and (c) body shell of the crab

3. EXPERIMENTAL RESULTS

3.1 Microstructure

The microstructure of the exoskeleton displays an anisotropic nature, which can be distinguished from the SEM images. From the cross sectional views in Fig. 2, the four discrete layers in the claw and body shell can be clearly observed. Among the four layers, the epicuticle and membrane are the thinnest, while the exocuticle is thicker, and the endocuticle is the thickest. Fig. 3 show the SEM images of the cross sections of the endocuticle in the claw and body shell, respectively. Layered structures are observed to be stacked periodically. Such periodically stacked layers were previously named "Bouligand" structure, and have been observed by researchers in both exocuticle and endocuticle of different types of crustaceans [11,12]. In the micro and nanometer scale, the "Bouligand" structure is featured by helicoidal stacking sequence of fibrous chitin-protein planes, deposited with calcium carbonate minerals. When the stacking of the fibrous planes completes a 180 ° rotation, a single layer is formed. The length of the single layer is an indication of the stacking density. Smaller length corresponds to a higher packing density. Clearly, the stacking density in the body shell is much higher than that in the claw. While the cross sections of the exoskeleton were manifested by the "Bouligand" structure, the microstructure of the in-plane sections is quite different. Fig. 4 show the SEM images of the in-plane sections of the endocuticle in the claw and body shell. There are pore canals and pore canal tubules within the layer plane. Pore canals are those oval-shaped holes formed by intersections of the bundles of parallel chitin-protein fiber branches. The pore canal tubules lie inside the pore canals, which go through the thickness of the exoskeleton. These tubules play an important role in the transport of ions and nutrition during the formation of the new exoskeleton after the crabs molt [8]. The claw and body shell share similar porous structure but with a quite different porosity (higher in body shell).

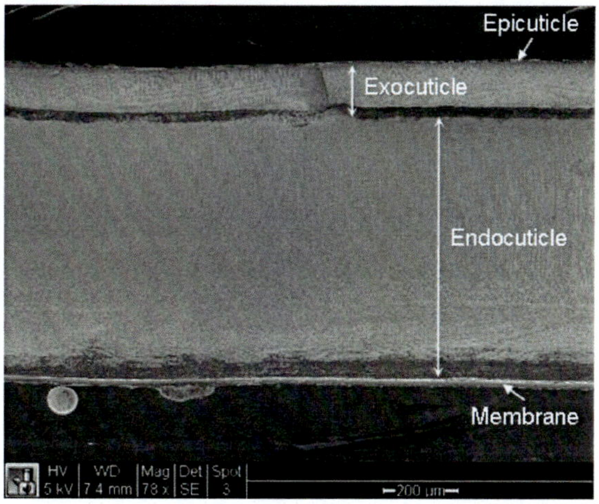

Fig. 2 SEM images of cross sections of the polished exoskeleton in the claw

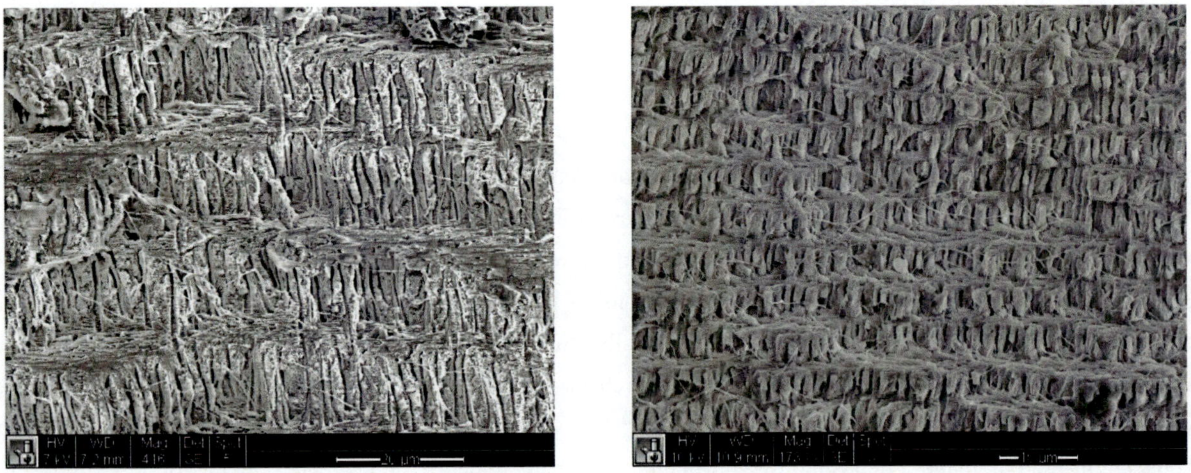

(a) (b)

Fig. 3 SEM images of cross sections of endocuticle in: (a) claw with layer spacing around 10 μm, (b) body shell with layer spacing around 4 μm

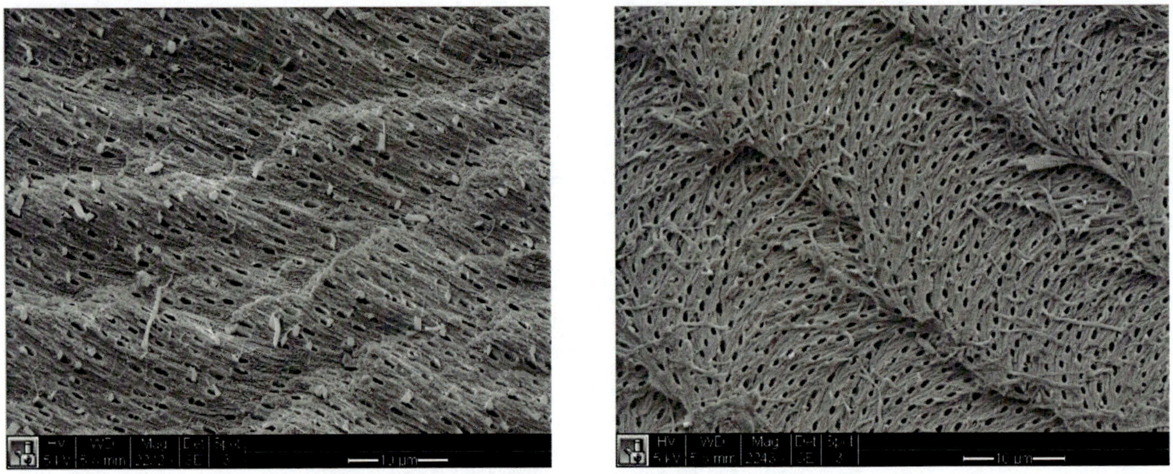

Fig. 4 SEM images of in-plane sections of endocuticle in: (a) claw, (b) body shell

3.2 Mechanical properties

To study the mechanical properties in the exoskeleton, two groups of nanoindentation experiments were carried out on two orthogonal planes, i.e. the cross sections and the in-plane sections. The applied loads range from 300 μN to 2000 μN. A few comparisons of the mechanical properties were made on different parts of the crab (claw and body shell), different layers of the exoskeleton (exocuticle and endocuticle), and different sections (cross-sections and in-plane).

Different from microindentation experiment, in nanoindentation, the typical indent size is less than the layer spacing (a few micrometers), thus nanoindentation on the cross sections represents the local mechanical properties of one single layer rather than the overall mechanical properties of the multi-layered structure. The hardness measured throughout the four discrete layers is compared in Fig. 5 (a) and (b) for the claw, and body shell, respectively. In both the claw and body shell, the epicuticle was observed to be the hardest part in the exoskeleton. The hardness shows a decreasing trend from the external to the internal surfaces. Overall, a higher hardness was measured in the claw than the body shell.

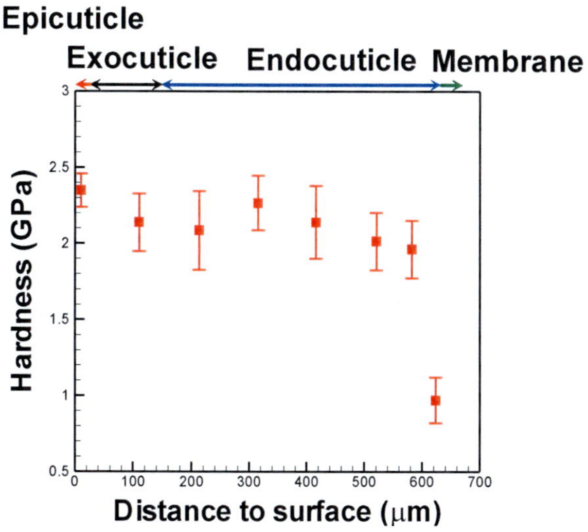

(a)

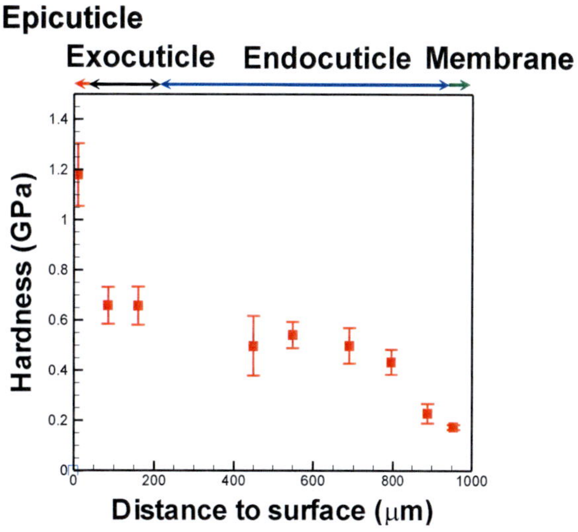

(b)

Fig. 5 Hardness measured on the cross sections of: (a) claw, and (b) body shell

Distinct load-displacement response and mechanical properties were observed in the in-plane sections compared to the cross sections. This indicates the anisotropy of the properties in this biological composite. The hardness is compared in Fig. 6 (a) and (b). In the exocuticle of both claw and body shell, the hardness was measured to be slightly higher in the in-plane section than the cross section. In the endocuticle of the body shell, the in-plane section also yields higher hardness than the cross sections. However, opposite results were obtained for the endocuticle of the claw. The hardness was measured to be much lower in the in-plane section.

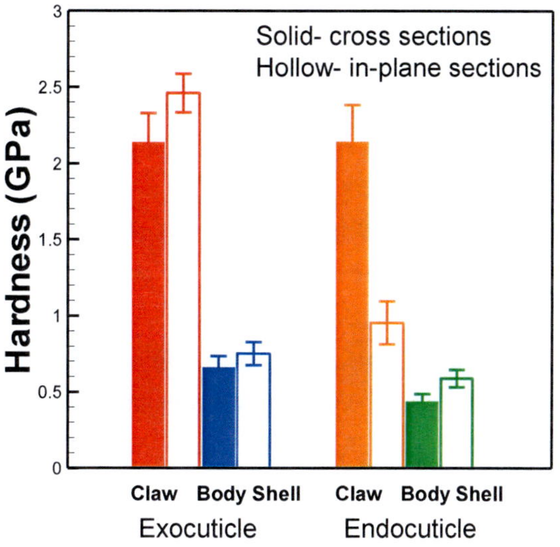

Fig. 6 Hardness comparison between in-plane and cross sections of the claw and body shell

3.3 Organic and inorganic components composition

In order to understand the microstructure and mechanical property relationship, it is crucial to analyze the organic and inorganic components composition of the exoskeleton. In this study, the mineral components were analyzed from EDS tests in the exocuticle and endocuticle layers of the claw and body shell. The characterization areas in all specimens were 5 μm × 5 μm. Different types of minerals were characterized, with calcium found to be the major component, while sodium and magnesium found to be the minor components. In the claw, the calcium takes 17.78 at.% of the exocuticle and 17.32 at.% of the endocuticle. In the body shell, the calcium takes 26.47 at.% of the exocuticle and 28.12 at.% of the endocuticle. The results indicate that the extent of calcification is very close in the exocuticle and endocuticle. However, the amount of calcium is lower in the claw than the body shell, thus calcination may not be the reason for the measured stronger mechanical property in the claw.

The organic component chitin could be the important strengthening factor of the cuticle layers as it has relatively high strength compared to the protein matrix [17]. For example, Wegst *et al.* reported that the stiffness of chitin was in the range of 10 –130 GPa [18,19]. The stiffness of the protein matrix is only about 0.1 GPa to a few GPa, which is much lower than the stiffness of chitin. In the exocuticle and endocuticle layers, chitin usually exists in the form of chitin-protein fibers oriented within the in-plane sections, and vertically aligned as the pore canal fibers throughout the thickness. Higher volume of the chitin-protein fibers or pore canal fibers could cause higher strength in the biological composites. From the cross sections of the exocuticle and endocuticle, it was observed that the exocuticle has a higher chitin-protein fibers volume fraction than the endocuticle. Moreover, it was observed that the claw has a higher chitin-protein fiber volume fraction in the endocuticle than the body shell does. Therefore, the higher volume fraction of the chitin-protein fibers is believed to be the major reason for the higher hardness in the claw.

4. CONCLUSION

In this study, the microstructure and mechanical properties of Dungeness crab exoskeleton were investigated. The microstructure was studied by SEM, and the mineral composition was studied by EDS and compared among the exocuticle and endocuticle of the claw and body shell. Nanoindentation experiments were carried out on the cross sections and the in-plane sections of the exoskeleton surfaces. From the microstructural studies, four main discrete layers were observed, i.e, epicuticle, exocuticle, endocuticle and memebrane layers. In the cross sections, the packing density of the exocuticle is much higher than that of the endocuticle. And the packing density of the body shell is much higher than that of the claw. In the in-plane sections, porous structure exists in both claw and body shell. The pore canal fibers are located inside the oval-shaped pore canals, and are aligned vertically through the thickness of the exoskeleton. The porosity in the claw's endocuticle is observed to be the highest. From the nanomechanical characterization, the exocuticle has higher hardness than the endocuticle. This is attributed to the higher chitin fiber volume fraction in the exocuticle. For nanoindentation on the in-plane sections, the claw's exocuticle and body shell's exotucile and endocuticle have higher mechanical properties than that in the cross-sections. However, in the endocuticle of the claw, the mechanical properties are much lower on the in-plane than the cross sections.

References

1. Currey, J.D., *Mechanical-Properties of Mother of Pearl in Tension.* Proceedings of the Royal Society of London Series B-Biological Sciences, 1977. 196(1125): p. 443-&.
2. Jackson, A.P., J.F.V. Vincent, and R.M. Turner, *The Mechanical Design of Nacre.* Proceedings of the Royal Society of London Series B-Biological Sciences, 1988. 234(1277): p. 415-&.
3. Wang, R.Z., Suo, Z., Evans, A.G., Yao, N., and Aksay, I.A., *Deformation mechanisms in nacre.* Journal of Materials Research, 2001. 16(9): p. 2485-2493.
4. Meyers, M.A., Lin, A.Y.M., Chen, P.Y., and Muyco, J., *Mechanical strength of abalone nacre: Role of the soft organic layer.* Journal of the Mechanical Behavior of Biomedical Materials, 2008. 1(1): p. 76-85.
5. Katti, K.S., B. Mohanty, and D.R. Katti, *Nanomechanical properties of nacre.* Journal of Materials Research, 2006. 21(5): p. 1237-1242.
6. Hepburn, H.R., Joffe, I., Green, N., and Nelson, K.J., *Mechanical-Properties of a Crab Shell.* Comparative Biochemistry and Physiology, 1975. 50(Na3): p. 551-&.
7. Joffe, I., Hepburn, H.R., Nelson, K.J., and Green, N., *Mechanical-Properties of a Crustacean Exoskeleton.* Comparative Biochemistry and Physiology, 1975. 50(Na3): p. 545-&.
8. Chen, P.Y., Lin, A.Y.M., Lin, Y.S., Seki, Y., Stokes, A.G., Peyras, J., Olevsky, E.A., Meyers, M.A., and
9. Taylor, J.R.A., J. Hebrank, and W.M. Kier, *Mechanical properties of the rigid and hydrostatic skeletons of molting blue crabs, Callinectes sapidus Rathbun.* Journal of Experimental Biology, 2007. 210(24): p. 4272-4278.
10. Melnick, C.A., Z. Chen, and J.J. Mecholsky, *Hardness and toughness of exoskeleton material in the stone crab, Menippe mercenaria.* Journal of Materials Research, 1996. 11(11): p. 2903-2907.
11. Raabe, D., C. Sachs, and P. Romano, *The crustacean exoskeleton as an example of a structurally and mechanically graded biological nanocomposite material.* Acta Materialia, 2005. 53(15): p. 4281-4292.
12. Cribb, B.W., Rathmell, A., Charters, R., Rasch, R., Huang, H., and Tibbetts, I.R., *Structure, composition and properties of naturally occurring non-calcified crustacean cuticle.* Arthropod Structure & Development, 2009. 38(3): p. 173-178.
13. Roer, R. and R. Dillaman, *The Structure and Calcification of the Crustacean Cuticle.* Amer. Zool., 1984. 24(4): p. 893-909.
14. Al-Sawalmih, A., Li, C., Siegel, S., Fabritius, H., Yi, S., Raabe, D., Fratzl, P., and Paris, O., *Microtexture and Chitin/Calcite Orientation Relationship in the Mineralized Exoskeleton of the American Lobster.* Advanced Functional Materials, 2008. 18: p. 3307-3314.
15. Sachs, C., H. Fabritius, and D. Raabe, *Hardness and elastic properties of dehydrated cuticle from the lobster Homarus americanus obtained by nanoindentation.* Journal of Materials Research, 2006. 21(8): p. 1987-1995.
16. Oliver, W.C. and G.M. Pharr, *An improved technique for determining hardness and elastic modulus using load and displacement sensing indentation experiments.* Journal of Materials Research, 1992. 7(6): p. 1564-1583.
17. Cheng, L., L. Wang, and A.M. Karlsson, *Image analyses of two crustacean exoskeletons and implications of the exoskeletal microstructure on the mechanical behavior.* Journal of Materials Research, 2008. 23(11): p. 2854-2872.
18. Wegst, U.G.K. and M.F. Ashby, *The mechanical efficiency of natural materials.* Philosophical Magazine, 2004. 84(21): p. 2167-2186.
19. Vincent, J.F.V., *Arthropod cuticle: a natural composite shell system.* Composites Part A: Applied Science and Manufacturing 2002. 33(10): p. 1311-1315.

The Mechanical Properties of Tendril of Climbing Plant

Liou*, N.-S. and Ruan, G.-W.
Department of Mechanical Engineering, Southern Taiwan University
No. 1, Nan-Tai Street, Yongkang Dist., Tainan City 710, Taiwan R.O.C.
email: nliou@mail.stut.edu.tw

Abstract
Tendril is a threadlike specialized stem, leaf or petiole used by climbing plants for support and attachment. The spiral contraction which occurs after a tendril has caught an object is of great use to the climbing plant. The spiral contraction can drags up the shoot thus there is no waste of growth. Besides, a far more important use rendered by the spiral contraction of the tendrils is that they are thus made highly elastic so that the force is distributed between several adjacent attached tendrils; and this renders the whole far stronger than it otherwise would be, as the tendrils cannot break separately. It is this elasticity that protects the plants from being torn away from their supports during stormy weather or pulled by object. In this study, the mechanical properties of Luffa tendrils were investigated. By using tensile experiments and finite element analyses, the time dependent mechanical properties of attached mature tendrils of Luffa under tensile loads were investigated. The improvement of supporting capability of Luffa tendrils due to spiral contraction was quantified. The results of the finite element analyses showed the maximum tensile force on a tendril can be reduced more than 58% due to distributing force to 4 adjacent attached tendrils when applied concentrated force on the stem.

Introduction
Climbing plants may be divided into four classes [1]. Plants of the first class ascend merely by the aid of hooks; and those of the second by rootlets. The plants of these two classes do not exhibit any special movements. Plants of the third class twine, without aid of any other movement, spirally around a support. Plants of the fourth class endowed with irritable organs such as tendrils, which when they touch any object clasp it; such organs consist of modified leaves, branches, or flower-peduncles. Tendrils are formed by the modification of leaves with their petioles, flower-peduncles or branches. They are sensitive to contact and used exclusively for climbing. If the extremity of a tendril has come into contact with an object, it bends around and attaches to the object and the tendril contracts spirally; on the other hand, a tendril does not contract, but soon withers away and drops off if it seizes nothing. The spiral contraction of tendrils, usually starting in half a day after their extremities have caught some object, shortens the tendrils and renders them elastic. In the first few days after the attachment, a tendril remains weak and brittle, but it rapidly becomes highly elastic and acquires great strength so that when the main stem is pulled the force is distributed between several attached tendrils.

The mechanical properties of mature Luffa tendrils were investigated in this study. The Luffa tendril, borne at the junction of stem and leaf petiole, has 2~6 branches. All the branches are sensitive on all sides. The tendril will bend in about 10 minutes after being lightly rubbed, or after coming into contact with an object. By using tensile experiments and finite element analyses, the time dependent mechanical properties of attached mature tendrils of Luffa under tensile loads were investigated. The improvement of supporting capability of Luffa tendrils due to spiral contraction was quantified.

Material and method
The tendrils of Luffa are considered linear viscoelastic materials and Wiechert model with 3 Maxwell spring-dashpot elements was used to approximate the time dependent mechanical behavior of Luffa tendrils. The relaxation moduli of 7 parameter Wiechert models can be expressed as follows:

T. Proulx (ed.), *Mechanics of Biological Systems and Materials, Volume 2*, Conference Proceedings of the Society for Experimental Mechanics Series 9999, DOI 10.1007/978-1-4614-0219-0_13, © The Society for Experimental Mechanics, Inc. 2011

$$E_{\infty}(t) = E_0\left(1 - g_1 e^{-t/\tau_1} - g_2 e^{-t/\tau_2} - g_3 e^{-t/\tau_3}\right). \tag{1}$$

In equation (1), E_{∞}, E_0 are elastic moduli, g_i (i=1; 2; 3)and τ_i (i=1; 2; 3) are relaxation factors and times respectively. The linear viscoelastic material parameters of tendrils were determined by using the optimization-based inverse iterative finite element analysis procedure [2]. The generalized pattern search method [3-5] was used for the optimization process. The finite element analyses were performed by using ABAQUS^{TM} and the iteration loop was performed in MATLAB^{TM}. The generalized pattern search method is a direct search derivative-free algorithm which can minimize an objective function with respect to the unknown parameters. The linear viscoelastic material parameters of tendrils were determined in two steps. Firstly, the instantaneous elastic modulus and Poisson's ratio were determined by minimizing the squared error between the simulated and experimental load measured immediately after the sudden tensile displacement was applied. Secondly, the time dependent parameters were determined by minimizing the squared error between the simulated and experimental load vs. time curves when the ensile displacement was fixed.

In order to get the instantaneous force-tensile displacement data and force-time relaxation data of tendrils under tensile load for the inverse iterative finite element analysis procedure, the ramp-hold tensile test was performed on the tendril specimen cut from fresh mature tendrils of Luffa (Fig. 1a). The tensile test was carried out by using the Instron tensile test machine (Fig. 1b). The tensile displacement is 1 cm with displacement range(?) of 1mm/sec. The tensile force and displacement were recorded for a total time of 610 sec (Fig. 1c).

(a) (b) (c)

Figure 1: (a) Typical tendril of Luffa (b) Specimen under tensile ramp-hold test (c) Experimental force vs. time data

The 3D Finite element model generated from the image of the tendril specimen was used for the inverse iterative finite element analysis procedure (Fig. 2a). In order to investigate how the force was distributed between attached tendrils when the main stem was pulled, finite element models of a Luffa stem with 3 and 5 tendrils as shown in Fig. 2b,c were made. An upward enforced displacement at a constant displacement rate of 7 cm/sec was applied at the middle section of the stem and all tendrils were anchored at the end points. The corresponding force at the location where enforced displacement was applied and reaction forces at the end point of tendrils vs. displacement were used to evaluate how the force was distributed between attached tendrils when the main stem was pulled.

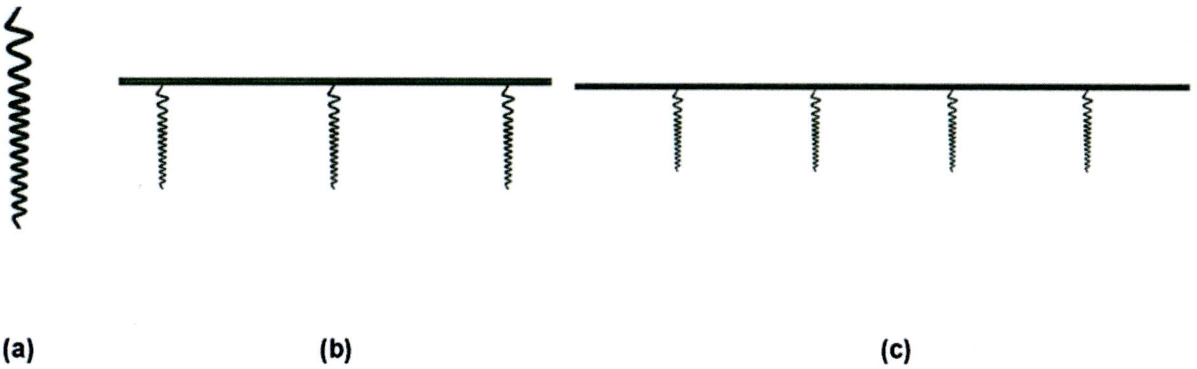

Figure 2: Finite element models of (a) tendril specimen (b) Luffa stem with three tendrils and (c) Luffa stem with four tendrils

Results

The values of instantaneous elastic modulus and Poission's ratio of Luffa tendril specimen estimated by the inverse iterative finite element analysis procedure are 213 MPa and 0.33 respectively. The values of time dependent viscoelastic parameters g_i (i=1; 2)and τ_i (i=1; 2) are 0.153, 0.265, 10 (sec) and 430 (sec) respectively. The force-time curve computed from finite element simulation with fitted linear viscoelastic and the force-time curve of ramp-hold experiment are shown in Fig. 3.

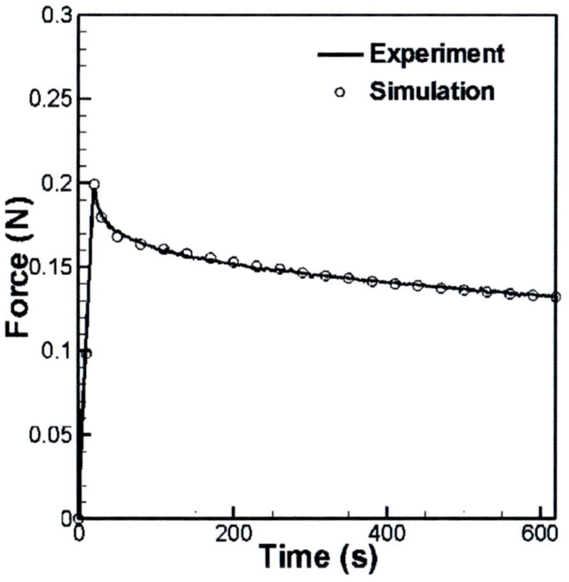

Figure 3: The force-time curve computed from finite element simulation with fitted linear viscoelastic and force-time curve of ramp-hold experiment

Fig. 4a shows the force-displacement curve of single tendril specimen under enforced displacement at a constant displacement rate of 7 cm/sec. The slightly upward curvy trend of the force-displacement curve shows the tendril can be considered a nonlinear spring and the spring constant increases as the tensile displacement increases. Fig. 4b shows the reaction forces at fixed point of tendrils of the Luffa stem model with 3 tendrils under enforced displacement (displacement rate = 7 cm/sec). Due to the symmetry of the model, only reaction forces of the center and one side of a tendril are shown. It can be seen, from the reaction forces of tendrils at 7 cm enforced displacement, that the center tendril carries 78% of pull load applied on the stem section. The reaction forces at fixed point of tendrils of the Luffa stem model with 4 tendrils under enforced displacement (displacement rate = 7 cm/sec) is shown in Fig. 4c. Due to the symmetry of the model, only two reaction forces of tendrils in one side are shown. It can be seen, from the reaction forces of tendrils at 14 cm enforced displacement, that the tendril close to the middle point of the stem section carries only 42% of pull load applied on the stem section.

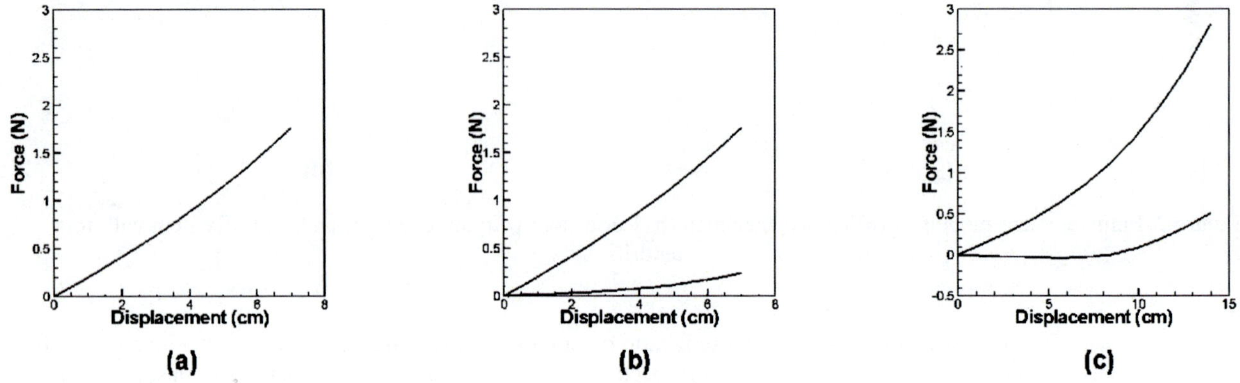

Figure 4: the force is distributed between attached tendrils (a), (b) and (c)

Discussion and conclusion

The mechanical properties of mature Luffa tendrils were identified by using the optimization-based inverse iterative finite element analysis procedure. The simulation of Luffa stem model under enforced displacement pull revealed the spiral contraction of tendrils can effectively distribute the pull force on the stem to several tendrils attached to the stem. The tendril specimen used for this study was cut from the portion of tendril with same spiral direction. In the real world, since neither the support nor the stem can rotate, the coils of the spiral tendril go from a right-handed helix to a left-handed helix or vice versa. The phenomenon of helix hand reversal was called perversion [1, 6] How perversion of tendril affects the spring constant will be investigated in the future.

References

1. Darwin, C., *The Movements and Habits of Climbing Plants.* 1875, John Murray: London.
2. Liu, K.F., M.R. VanLandingham, and T.C. Ovaert, *Mechanical characterization of soft viscoelastic gels via indentation and optimization-based inverse finite element analysis.* Journal of the Mechanical Behavior of Biomedical Materials, 2009. 2(4): p. 355-363.
3. Torczon, V., *ON the convergence of pattern search algorithms.* Siam Journal on Optimization, 1997. 7(1): p. 1-25.
4. Audet, C. and J.E. Dennis, *Analysis of generalized pattern searches.* Siam Journal on Optimization, 2003. 13(3): p. 889-903.
5. Lewis, R.M. and V. Torczon, *Pattern search methods or linearly constrained minimization.* Siam Journal on Optimization, 2000. 10(3): p. 917-941.
6. Goriely, A. and M. Tabor, *Spontaneous Helix Hand Reversal and Tendril Perversion in Climbing Plants.* Physical Review Letters, 1998. 80(7): p. 1564-1567.

A Multiscale Triphasic Biomechanical Model for Tumors' Classification

K. Barber and C.S. Drapaca, PhD
Department of Engineering Science and Mechanics, The Pennsylvania State University,
University Park, PA 16802, USA

ABSTRACT
The aim of this paper is to formulate a novel mathematical model that will be able to differentiate not only between normal and abnormal tissues, but, more importantly, between benign and malignant tumors. We present some very promising preliminary results of a multiscale triphasic model for biological tissues that couple the electro-chemical processes that take place in tissue's microstructure and tissue's biomechanics. The multiscaling is based on a recently developed homogenization technique for materials with evolving microstructure.

1. INTRODUCTION

The ever-growing field of non-invasive diagnostic technologies is continually providing new insights into *in vivo* biological processes, requiring joint efforts among researchers in medicine, science, and engineering. One of these emerging technologies, Magnetic Resonance Elastography (MRE), uses an imaging technique to measure the elasticity of biological tissues subject to mechanical stresses [4, 5]. The resulting strains are measured using magnetic resonance imaging and the related elastic modulus is computed from models of tissues mechanics. The elastic modulus contains important information about the pathology of the imaged tissues. Thus, MRE can help in tumor detection, determination of characteristics of disease, and in assessment of rehabilitation.

It was noticed experimentally that most biological tissues have incompressible viscoelastic features: they have a certain amount of rigidity that is characteristic of solid bodies, but, at the same time, they flow and dissipate energy by frictional losses as viscous fluids do [1, 2]. The incompressibility assumption for soft tissues is based on the fact that most tissues are made primarily of water. In addition, since the displacements in MRE are very small (on the order of microns), a linear constitutive law is usually assumed. However, despite the richness of the data set, the variety of processing techniques and the simplifications made in the biomechanical model, it remains a challenge to extract accurate results at high resolution in complex, heterogeneous tissues from the intrinsically noisy data. Therefore, any improvement in the MRE data processing with the help of biomechanics and computational methods will be of significant importance to modern medicine.

Fig.1: (a) Benign tumor: the fibrous connective tissue capsule (orange) separates the inside benign cells (black boundaries) from the outside normal cells (yellow). (b) Malignant tumor: the irregularly-shaped cancer cells (red boundaries) are anisotropic and diffuse. (inspired from [10])

The aim of this paper is to formulate a new mathematical model that will be able to differentiate not only between normal and abnormal tissues, but, more importantly, between benign (not cancerous) and malignant (cancerous) tumors. As it can be seen in Figure 1, benign tumors are localized, self-contained (encapsulated), with smooth boundaries, and tend to be more isotropic. On the other hand, malignant tumors are not localized, diffuse, have irregular boundaries, and are anisotropic. The biomechanical models used so far in MRE are classic macroscopic models that do not incorporate any relevant information about the biochemical and mechanical processes that take place in tissue's microstructure and thus fail to properly classify

T. Proulx (ed.), *Mechanics of Biological Systems and Materials, Volume 2*, Conference Proceedings of the Society for Experimental Mechanics Series 9999, DOI 10.1007/978-1-4614-0219-0_14, © The Society for Experimental Mechanics, Inc. 2011

tissues according to their pathology. In this paper we present a novel multiscale model that has the potential to differentiate not only between healthy and diseased tissues but also between benign and malignant tumors. We model the tissue as a triphasic material of porous, viscoelastic solid filled with interstitial fluid and dissolved ions [3, 6]. We will use a recently developed homogenization technique for materials with evolving microstructure [7–9] and present some preliminary results.

2. MULTISCALE TRIPHASIC MODEL

Inspired by [6], we assume that the microstructure of a biological tissue is a mixture of an intrinsically incompressible, isotropic, porous-permeable linear elastic solid phase containing the extracellular matrix and the (linear viscoelastic) solid cells, an intrinsically incompressible, isotropic, Newtonian viscous fluid phase containing the interstitial fluid, and another fluid phase, the ion phase, with, for now, two monovalent species. The tissue as a whole is electrically neutral and for now we assume electric and thermal effects are negligible with respect to chemical and mechanical effects. For simplicity of presentation, we consider the problem to be one-dimensional and neglect reaction processes. With these simplifications, the momentum equation and the continuity equations for the mixture and ions (see for example [6] for a detailed presentation of these equations in the general case) are:

$$
\begin{aligned}
&\frac{\partial p}{\partial x}=\frac{\partial}{\partial x}\left(E\frac{\partial u}{\partial x}\right)\\
&\frac{\partial u}{\partial t}-\frac{\partial}{\partial x}\left(\frac{E}{\alpha}\left(1-\varphi_0^s\right)\frac{\partial u}{\partial x}+\frac{RT}{\alpha}(1-\phi)\left(1-\varphi_0^s\right)\left(c_p+c_m\right)\right)=0\\
&\frac{\partial c_p}{\partial t}-\frac{\partial}{\partial x}\left(D_p\frac{\partial c_p}{\partial x}-c_p\frac{\partial u}{\partial t}\right)=0\\
&\frac{\partial c_m}{\partial t}-\frac{\partial}{\partial x}\left(D_m\frac{\partial c_m}{\partial x}-c_m\frac{\partial u}{\partial t}\right)=0
\end{aligned}
\qquad (2.1)
$$

where p,u,c_p,c_m are the unknown fluid pressure, solid displacement, and, respectively, the molar concentrations of the two ionic species, and $E,D_p,D_m,R,T,\alpha,\phi,\varphi_0^s$ are (in order) the following physical parameters: the Young's modulus of the solid, the diffusion coefficients of the two ionic species, the universal gas constant, the absolute temperature, a constant related to the hydraulic permeability, an osmotic coefficient, and the initial solidity.

As in [7-9], we introduce a characteristic macroscopic length L, a characteristic microscopic length l such that $\varepsilon=\frac{l}{L}\ll 1$, and define characteristic lengths associated to the diffusion of the species in the pore $l_p=\left(H_p\right)^{1/2}l,\ l_m=\left(H_m\right)^{1/2}l$.

Thus the characteristic diffusion times are: $\tau_p=\frac{\left(l_p\right)^k L^{2-k}}{D_{p0}},\tau_m=\frac{\left(l_m\right)^j L^{2-j}}{D_{m0}},0\le k,j\le 2$.

We introduce the following non-dimensional functions:

$$
\begin{aligned}
&\tilde{u}=\frac{u}{L},\tilde{c}_p=\frac{c_p}{c_0},\tilde{c}_m=\frac{c_m}{c_0},\tilde{p}=\frac{p}{p_0}\\
&\tilde{x}=\frac{x}{L},\tilde{t}=\frac{t}{\tau},\tilde{t}_p=\frac{t}{\tau_p},\tilde{t}_m=\frac{t}{\tau_m},\tau=\frac{l^2}{E_0}\\
&\tilde{E}=\frac{E}{E_0},\tilde{D}_p=\frac{\left(H_p\right)^{k/2}}{D_{p0}}D_p,\tilde{D}_m=\frac{\left(H_m\right)^{j/2}}{D_{m0}}D_m
\end{aligned}
\qquad (2.2)
$$

where $c_0, p_0, E_0, D_{p0}, D_{m0}$ are reference quantities. Thus the non-dimensional equations corresponding to system (2.1) are:

$$\begin{aligned}
&\frac{\partial \tilde{p}}{\partial \tilde{x}} = \frac{E_0}{p_0}\frac{\partial}{\partial \tilde{x}}\left(\tilde{E}\frac{\partial \tilde{u}}{\partial \tilde{x}}\right)\\
&\frac{\partial \tilde{u}}{\partial \tilde{t}} - \frac{\partial}{\partial \tilde{x}}\left(\frac{\varepsilon^2}{\alpha}\left(1-\varphi_0^s\right)\frac{\partial \tilde{u}}{\partial \tilde{x}} + \varepsilon^2\frac{RTc_0}{\alpha E_0 \tilde{E}}(1-\phi)\left(1-\varphi_0^s\right)\left(\tilde{c}_p + \tilde{c}_m\right)\right) = 0\\
&\frac{\partial \tilde{c}_p}{\partial \tilde{t}_p} - \frac{\partial}{\partial \tilde{x}}\left(\varepsilon^k \tilde{D}_p \frac{\partial \tilde{c}_p}{\partial \tilde{x}} - \tilde{c}_p \frac{\partial \tilde{u}}{\partial \tilde{t}}\right) = 0\\
&\frac{\partial \tilde{c}_m}{\partial \tilde{t}_m} - \frac{\partial}{\partial \tilde{x}}\left(\varepsilon^j \tilde{D}_m \frac{\partial \tilde{c}_m}{\partial \tilde{x}} - \tilde{c}_m \frac{\partial \tilde{u}}{\partial \tilde{t}}\right) = 0
\end{aligned} \tag{2.3}$$

We assume that all processes happen on the same time scale so $\tau = \tau_p = \tau_m$. From (2.2) we note that this establishes a relationship among reference diffusion coefficients, reference Young's modulus, and length scales. System (2.3) is completed with zero jump conditions at the boundaries of the domains occupied by the three phases. We employ now the method of homogenization in domains with evolving microstructure proposed in [7-9]. The method is based on the assumption that there exists a reference configuration that is periodic. By mapping the current configuration with evolving and deforming microstructure into the reference configuration with periodic geometry, the classic homogenization technique can be used to get the corresponding macroscopic equations, and then map them to the current configuration. In our simplified case of linear elasticity and small strains, the equations corresponding to system (2.3) in the periodic reference configuration look like (2.3) except that the convection terms in the last two equations are missing. The macroscopic equations are therefore similar to the ones given in [7-9].

3. RESULTS

We consider a very simple case where we are interested only in the solid displacement due to either microscopic or macroscopic diffusion. The periodic domain is the interval $(0,1)$. Let $L = 1, l = 10^{-4}, H_p = H_m = 0.1, \tau = 1$. For now we are not interested in the effective physical parameters and therefore we take them equal to unity. Thus $\tilde{c}_p = \tilde{c}_m \equiv \tilde{c}$. We distinct two cases:

1. $\tilde{D}_p = \tilde{D}_m = 1 \Rightarrow k = j = 0$ (macroscopic diffusion)
2. $\tilde{D}_p = \tilde{D}_m = 10^{-9} \Rightarrow k = j = 2$ (microscopic diffusion)

In both cases we assume the following initial and boundary conditions:

$$\begin{aligned}
&\tilde{u}(\tilde{x},0) = \tilde{u}(0,\tilde{t}) = \tilde{u}(1,\tilde{t}) = 0\\
&\tilde{c}(\tilde{x},0) = \mathrm{e}^{-10(\tilde{x}-0.5)^2}, \tilde{c}(0,\tilde{t}) = \tilde{c}(1,\tilde{t}) = \mathrm{e}^{-5/2}
\end{aligned} \tag{3.1}$$

The solution in case 1 to the macroscopic problem is shown in Figures 2 and 3. Note that the solid phase shrinks in the regions where the concentrations are increased. In case 2 there is no macroscopic diffusion since the initial condition for the concentrations depends only on the macroscopic variable $\tilde{x}$ and not on the microscopic variable $\tilde{x}/\varepsilon$ which is the variable used in the diffusion equation in this case. Thus the solution in case 2 is $\tilde{c}(\tilde{x},\tilde{t}) = \tilde{c}(\tilde{x},0), \tilde{u}(\tilde{x},\tilde{t}) = \tilde{u}(\tilde{x},0) = 0$.

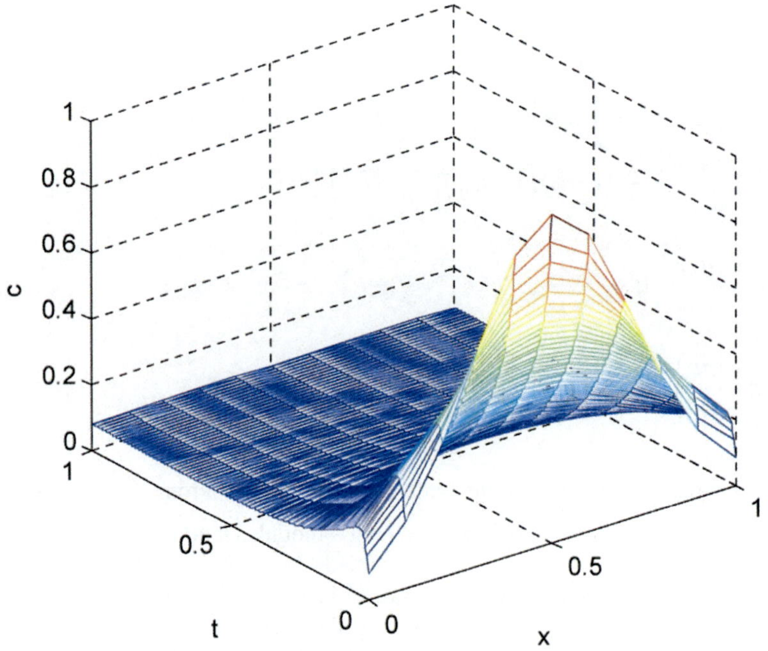

Fig.2: Concentration field in case 1.

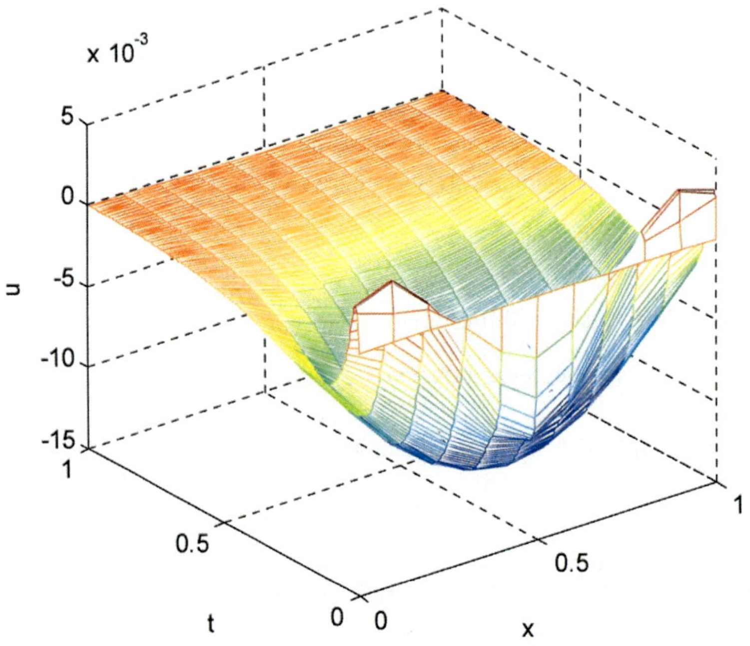

Fig.3: Displacement field in case 1.

One possible, very simplistic, scenario for tumor modeling and classification could therefore be that benign tumors experience only microscopic diffusion without macroscopic deformation of the solid phase (cells), while malignant tumors are characterized by macroscopic diffusion and deformation of the cells. In particular, if the concentrations increase, the cells will shrink and possibly become stiffer which will agree with recent results (see for example [11]) that claim that cancer tumors are indeed much harder than benign tumors. Finally, a proper analysis of the diffusion processes and scales may capture important information about the moment when a benign tumor transforms into cancer (we plan to investigate this in our future work).

3. CONCLUSION

In this paper we proposed a novel multiscale triphasic model for biological tissues that is based on the method of homogenization of domains with evolving microstructure. Our very preliminary results show promise in obtaining better models for tumor classification for robust and reliable imaging elastography.

REFERENCES:

[1] F. A. Duck, Physical Properties of Tissues- A Comprehensive Reference Book, 6th edition, Sheffield, UK: Academic, 1990.

[2] Y.C. Fung, Biomechanics - Mechanical Properties of Living Tissues, 2nd edition, Springer, New York, 1993.

[3] W.Y. Gu, W.M. Lai and V.C. Mow, A mixture theory for charged-hydrated soft tissues containing multi-electrolytes: passive transport and swelling behaviors, J. Biomech. Engrg. 120, 169-180, 1998.

[4] R. Muthupillai, D.J. Lomas, P.J. Rossman, J.F. Greenleaf, A. Manduca and R.L. Ehman, Magnetic resonance elastography by direct visualization of propagating acoustic strain waves, Science 269, 1854-1857, 1995.

[5] R. Muthupillai, P.J Rossman, D.J. Lomas, J.F. Greenleaf, S.J. Riederer and R.L. Ehman, Magnetic resonance imaging of transverse acoustic strain waves, Magn. Reson. Med. 36, 266-274, 1996.

[6] D.N. Sun, W.Y. Gu, X.E. Guo, W.M. Lai and V.C. Mow, A mixed _nite element formulation of triphasic mechano-electrochemical theory for charged, hydrated biological soft tissues, Int. J. Numer. Meth. Engng. 45, 1375-1402, 1999.

[7] M.A. Peter, Homogenization in domains with evolving microstructure, C.R. Mechanique 335, 357-362, 2007.

[8] M.A. Peter, Homogenization of a chemical degradation mechanism inducing an evolving microstructure, C.R. Mecanique, 335, 679-684, 2007.

[9] M.A. Peter, M. Bohm, Multiscale modeling of chemical degradation mechanisms in porous media with evolving microstructure, Multiscale Model.Simul. 7(4), 1643-1668, 2009.

[10] B. Alberts, A. Johnston, J. Lewis, M. Raff, K. Roberts, P. Walter, Molecular Biology of the Cell, 5th Ed, 2007.

[11] C.S. Drapaca, A.J. Palocaren, Biomechanical modeling of tumor classification and growth, Rev.Roum.Sci.Tech.-Mec.Appl., 55(2), 115-124, 2010.

Developing Hyper-Viscoelastic Constitutive Models of Porcine Meniscus from Unconfined Compression Test Data

[1]Liou*, N.-S., [2]Jeng, Y.-R., [1]Chen, S.-F., [1]Ruan, G.-W. and [2]Wu, K.-T.
[1]Department of Mechanical Engineering, Southern Taiwan University
No. 1, Nan-Tai Street, Yongkang Dist., Tainan City 710, Taiwan R.O.C.
[2]Department of Mechanical Engineering, National Chung Cheng University
168 University Road, Minhsiung Township, Chiayi County 621, Taiwan, ROC
email: nliou@mail.stut.edu.tw

Abstract
This study examined the time dependent mechanical properties of porcine meniscus under unconfined compression condition. The ramp-hold compression tests with different ramping displacement rates and long-term compression tests with very low displacement rate were performed on porcine meniscus specimens. The time dependent mechanical response of porcine meniscus was described by hyper-viscoelastic constitutive model due to its simplicity and can be easily implemented to finite element models. Inverse iterative finite element analysis procedure was used to obtain the material parameters of hyper-viscoelastic models of meniscus in two stages. In the first stage, the long-term hyperelastic material parameters of meniscus were obtained from the compression test data with very low displacement rate. The long-term hyperelastic material parameters were used as known parameters in the second stage, and the time dependent material parameters were determined from the ramp-hold compression tests with different ramping displacement rates. It can be seen from the force-time curves of ramp-hold experimental data that the mechanical properties of meniscus is highly time-dependent. The instantaneous modulus is more than 50 times higher than the long-term elastic modulus. The third-order reduce polynomial hyper-viscoelastic model established in this study can well describe the mechanical behavior of porcine meniscus under compression load. This model can be used in the finite element study of meniscus biomechanics.

Introduction
The menisci play important roles in the biomechanics of knee. Beside improving the passive stability of the knee and helping joint lubrication, the meniscus also acts as shock absorber. The functionality of the menisci and their role in biomechanics of the knee such as lubrication and load transmission has been discussed by many researchers. Many studies have investigated the biomechanical behavior of knee meniscus to better recognize traumatic and pathologic processes [1-3]. The meniscus can be considered as a viscoelastic material which may experience different stresses such as tensile, compressive and shear stresses. The studies about the effect of these stresses on meniscal tissue reveal that the biomechanical properties of meniscal tissue are anisotropic and inhomogeneous. The mechanical responses of meniscus depend on testing direction (axial, circumferential and radial). Besides, the mechanical properties of meniscus vary with location (anterior, central and posterior). The study of human meniscus under confined compression showed that the posterior portion of the meniscus had the highest aggregate modulus and the anterior portion had the lowest aggregate modulus [4], and the study of bovine media meniscus showed that the surface region of meniscus has a significantly higher compressive modulus than the interior of the tissue [5]. The results of these studies show that meniscus is a functional gradient material thus it is not possible to use one set of material constant to describe the mechanical response of meniscal tissue at all locations. While the equilibrium or long-term mechanical behavior of meniscus has been well describe in literature, the time-dependent behavior under large compressive deformation has received less attention. This study examined the time dependent mechanical properties of porcine meniscus under unconfined compression condition. Inverse iterative finite element analysis procedure was used to obtain the material parameters which are suitable for finite element study of meniscus biomechanics. The results of this study can be used as baseline for further development of functional gradient constitutive models for meniscus.

Material and method
Cubic meniscus specimens (2x2x2 mm) were cut from the central part of interior portion of the medial meniscus of 4 porcine stifles (n=3 per meniscus; Fig. 1a,b). The harvested meniscus specimens were stored at 4^0C in phosphate buffered saline prior

T. Proulx (ed.), *Mechanics of Biological Systems and Materials, Volume 2*, Conference Proceedings of the Society for Experimental Mechanics Series 9999, DOI 10.1007/978-1-4614-0219-0_15, © The Society for Experimental Mechanics, Inc. 2011

to mechanical testing. Two types of unconfined compression tests (Fig. 1c) were performed to obtain the time dependent and long-term compressive modulus of the specimens. Firstly, ramp-hold compression tests with different compression rates were applied to specimens. Specimens were compressed at displacement rates of 0.02 and 0.2 mm/sec up to 10% engineering strain and hold for 600 seconds. The force-time curves were recorded at 10 Hz sampling rate. Secondly, Specimens were compressed at displacement rates of 0.0002 mm/sec up to 0.4 engineering strain to obtain long-term equilibrium force-displacement curves. Experiments were performed at room temperature with specimens in PBS bath. Mechanical tests were done within 24 hours after specimens were prepared by using Instron material testing machine.

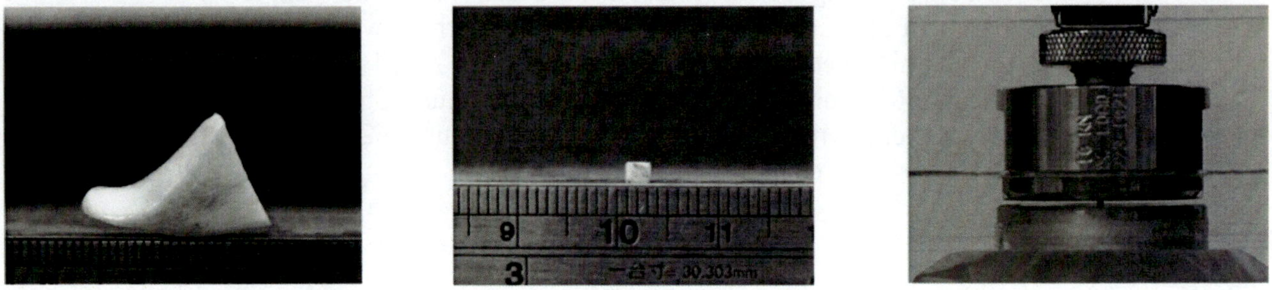

(a) (b) (c)

Figure 1: (a) Central part of the medial meniscus, (b) meniscus specimen and (c) specimen under unconfined compression test

The hyperviscoelastic model was used to describe the time dependent mechanical properties of interior part of meniscus. The following polynomial strain energy function of the hyperviscoelastic model was used in this study.

$$W=\int_0^t \left\{ \sum_{i+j=1}^{N} \left[C_{ij0}\left(1-\sum_{k=1}^{n} g_k\left(1-e^{-(t-\tau)/\tau_k}\right)\right)\right] \times \frac{d}{d\tau}\left[(J_1-3)^i (J_2-3)^j\right]\right\} d\tau. \quad (1)$$

In equation (1), g_k and τ_k are relaxation coefficients and characteristic times respectively, N is the order of polynominal in strain invariants, J_1, J_2 are the first and second strain invariants, C_{ij0}'s are the instantaneous hyperelastic parameters. In this study, 3rd order non-compressible reduce polynomial hyperviscoelastic model with 2 linear viscoelastic time constants were used. The characteristic times ($\tau_1 = 2.5$ sec, $\tau_1 = 25$ sec) of time constants were chosen. These assumptions left C_{10}, C_{20}, C_{30}, g_1 and g_2 to be determined. The hyperviscoelastic material parameters of meniscus were determined by using optimization-based inverse iterative finite element analysis procedure [6]. The generalized pattern search method [7-9] was used for the optimization process. The finite element analysis were performed by using ABAQUS™ and the iteration loop were performed in MATLAB™. The generalized pattern search method is a direct search derivative-free algorithm which can minimize an objective function with respect to the unknown parameters. The hyperviscoelastic material parameters of meniscus were determined in two steps. Firstly, the long-term hyperelastic parameters were determined by minimizing the squared error between the simulated and experimental forces of long-term equilibrium force-displacement curve. Secondly, the time dependent parameters were determined by minimizing the squared error between the simulated and experimental forces of ramp-hold compression with different compression rates.

Results

The stress-strain curves of meniscus tissue samples (n=12) under very low displacement rate was shown in Fig. 2. This stress-strain relation was considered as the long-term unconfined compressive behavior of interior part of meniscus.

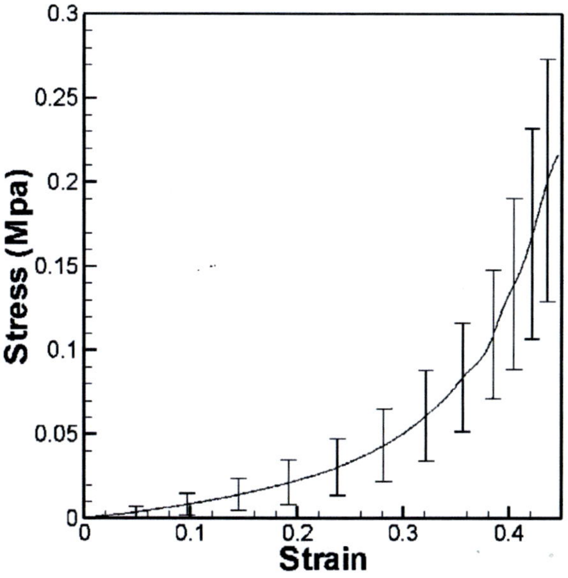

Figure 2: Stress-strain curve of meniscus under 0.0002 mm/sec displacement rate

The experimental time dependent stress-time curves of meniscus under ramp-hold compression tests are shown in Fig. 3. It can be seen that the mechanical behavior of meniscus is highly time dependent.

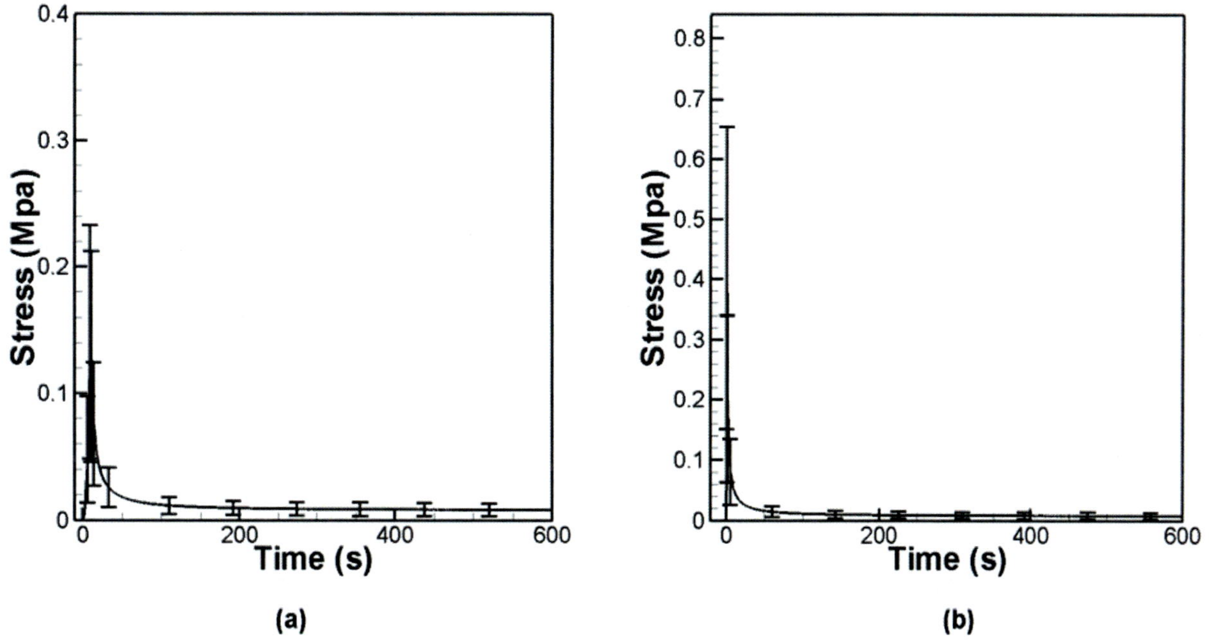

Figure 3: The experimental time dependent stress-time curves of meniscus under ramp-hold compression tests (a) 0.02 and (b) 0.2 mm/sec displacement rates

The fitted long term hyperelastic parameters $C_{10\infty}$, $C_{20\infty}$ and $C_{30\infty}$ of the 3rd order reduce polynomial hyperelastic model were 1.19e-2, 7.62e-3 and 5.25e-3 (MPa) respectively. The stress-strain curve plot by using fitted hyperelastic parameters and experimental data are shown in Fig. 4a. It can be seen that 3rd order reduce polynomial hyperelastic model can well describe the long-term compressive stress-strain relation of meniscus tissue up to compressive strain of 0.4. By taking the fitted long-term hyperelastic parameters as known factors, the viscoelastic parameters g_1 and g_2 fitted from the experimental time dependent stress-time curves of meniscus are 0.8 and 0.19 respectively with the characteristic times ($\tau_1 = 2.5\ \text{sec}$, $\tau_1 = 25\ \text{sec}$) were chosen. The stress-time curves plot by using fitted viscoelastic parameters and experimental data are shown in Fig. 4b,c.

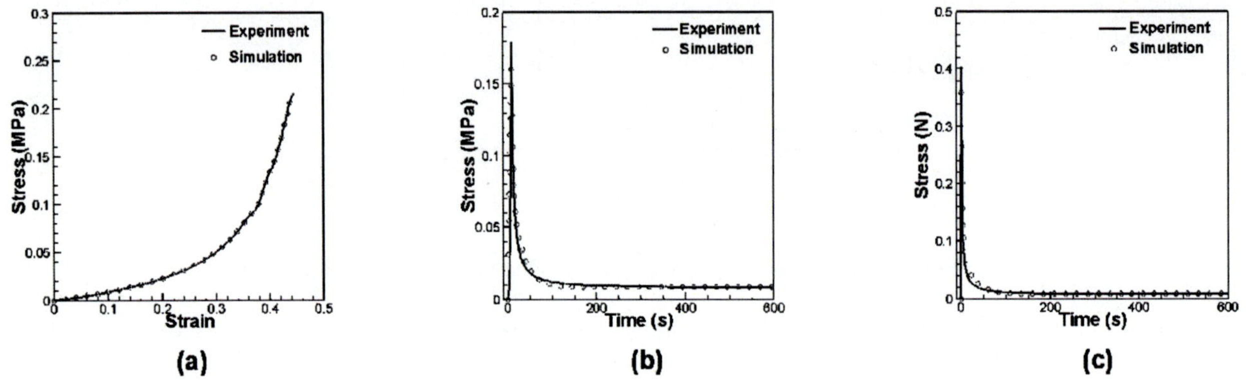

Figure 4: (a) The stress-strain curve plot by using fitted hyperelastic parameters of 3rd order reduce polynomial hyperelastic model and experimental data. The stress-time curves plot by using fitted viscoelastic parameters and ramp-hold experimental data under (b) 0.02 and (c) 0.2 mm/sec displacement rates.

Discussion and conclusion

This study showed that the mechanical behavior of interior part of meniscus is very time-dependent under compression. The result of ramp-hold experiment with displacement rate of 0.2 mm/sec (strain rate = 0.1 s^{-1}) shows the time dependent stress for meniscus tissue samples were 0.1 MPa and 0.002 MPa respectively at the begin and end (600 sec holding time) of ramp-hold experiment. The values of these moduli imply the instantaneous modulus is more than 50 times higher than the long-term modulus of interior of meniscus. Although the time-dependent mechanical response of specimens taken from the interior of center part of meniscus can not accurately reflect the mechanical behavior of meniscus tissue at other locations during physiological loading conditions since meniscus is heterogeneous material. The findings of this study still provide significant baseline information for development of more complicate constitutive models such as functional gradient constitutive models for meniscus. Furthermore, the material parameters obtained by inverse iterative finite element analysis procedure are suitable and ready to be used for finite element study of meniscus biomechanics such as load transmission and shock absorption or other knee biomechanics.

Acknowledgement

This work was supported by teaching and research improvement project of Southern Taiwan University.

References

1. Sweigart, M.A., C.F. Zhu, D.M. Burt, P.D. deHoll, C.M. Agrawal, T.O. Clanton, and K.A. Athanasiou, *Intraspecies and interspecies comparison of the compressive properties of the medial meniscus.* Annals of Biomedical Engineering, 2004. 32(11): p. 1569-1579.
2. Pena, E., B. Calvo, M.A. Martinez, D. Palanca, and M. Doblare, *Finite element analysis of the effect of meniscal tears and meniscectomies on human knee biomechanics.* Clinical Biomechanics, 2005. 20(5): p. 498-507.
3. Barber, F.A., M.A. Herbert, F.A. Schroeder, J. Aziz-Jacobo, and M.J. Sutker, *Biomechanical Testing of New Meniscal Repair Techniques Containing Ultra High-Molecular Weight Polyethylene Suture.* Arthroscopy-the Journal of Arthroscopic and Related Surgery, 2009. 25(9): p. 959-967.
4. Hacker, S.A., S.L.-Y. Woo, J.S. Wayne, and K.M. K., *Compressive properties of the human meniscus.* Trans. Orthop. Res.

Soc., 1992. 17: p. 672.
5. Lai, J.H. and M.E. Levenston, *Meniscus and cartilage exhibit distinct intra-tissue strain distributions under unconfined compression.* Osteoarthritis and Cartilage, 2010. 18(10): p. 1291-1299.
6. Liu, K.F., M.R. VanLandingham, and T.C. Ovaert, *Mechanical characterization of soft viscoelastic gels via indentation and optimization-based inverse finite element analysis.* Journal of the Mechanical Behavior of Biomedical Materials, 2009. 2(4): p. 355-363.
7. Torczon, V., *ON the convergence of pattern search algorithms.* Siam Journal on Optimization, 1997. 7(1): p. 1-25.
8. Audet, C. and J.E. Dennis, *Analysis of generalized pattern searches.* Siam Journal on Optimization, 2003. 13(3): p. 889-903.
9. Lewis, R.M. and V. Torczon, *Pattern search methods or linearly constrained minimization.* Siam Journal on Optimization, 2000. 10(3): p. 917-941.

The Mechanical Performance of Teleost Fish Scales

Deju Zhu[1], Franck Vernerey[2], and Francois Barthelat*[3]

[1] Postdoctoral research fellow, Department of Mechanical Engineering, McGill University, Montreal, QC, Canada, E-mail: deju.zhu@mail.mcgill.ca
[2] Assistant Professor, Ph.D., Department of Civil, Environmental, Architectural Engineering, University of Colorado, Boulder, CO, USA, E-mail: franck.vernerey@colorado.edu
[3] Assistant Professor, Ph.D., Department of Mechanical Engineering, McGill University, Montreal, QC, Canada, corresponding author, E-mail: francois.barthelat@mcgill.ca

ABSTRACT

High-performance natural materials and system are now serving as models for new engineering designs. In this work we have investigated the structure and mechanics of a single teleost fish scale from striped bass *Morone saxatilis* as part of larger project on novel flexible protective systems inspired from fish skin. These scales are about 300 microns thick and consist of a hard outer bony layer supported by a cross ply of collagen fibrils. Basic properties were obtained from tensile tests on single scales. While the bony layer is brittle, the collagen layer can undergo large deformations, eventually failing by extensive fiber pullout. Perforation tests with a sharp needle on a single scale resting on a soft substrate were also used to assess the performance and mechanics of a scale under a predator's bite. We found that multiple small circumferential cracks developed on the top surface of the bony layer near the penetration site, while four major radial cracks formed through the thickness of bony layer.

Keywords: fish scale; collagen; penetration; radial cracks

1. Introduction

The presence of scales in fish skin can be related to the protective functions of the vertebrate integument [1]. The extensive integumental skeleton of the "primitive" bony fishes obviously had efficient protective properties but such a heavy kind of armour hampered the locomotion ability of these fishes. During fish evolution, the increase of swimming performance is linked to a general trend in the reduction of the integumental skeleton [2]. The thin flexible scales of the living teleosts represent an outcome of this morphogenetic shift [3]. Fish scales exhibit a great variability in shape, size and arrangement. The general classification includes cosmoid, ganoid, placoid, and leptoid (including cycloid and ctenoid scales) [4]. While the full range of fish-scale function is not currently known, preliminary observations showed that it performs extremely well in a variety of tasks. In addition to having excellent hydrodynamics properties [5], fish scales provide a protective layer resisting penetration and providing a physical barrier against attack from predators. At larger lengths, the arrangement of the scales provides a flexible skin that allows for changes in shape. In fact, the scaled skin has been showed to play a critical structural role in fish locomotion by regulating wave propagation [6] and by acting as an external tendon [7], storing mechanical energy in order to make swimming more efficient. The structure of scaled skin at the macro level has probably inspired the scaled armor used by ancient Roman military to provide resistance to penetration while retaining relative freedom of movement. More recently, scaled skin probably also inspired more modern armor systems. While these personal armors share some mechanisms with natural fish scale, no systematic biomimetic "transfer of technology" was attempted so far, partly because a fundamental understanding of the mechanics of fish skin is still lacking.

There is very limited literature available in terms of mechanical properties of scales. Ikoma et al. [8] characterized the structure of elasmoid scales from sea bream (*Pagrus Major*), a type of scale composed of mineralized collagen fibers. Tensile test on individual scales showed nonlinearity and a progressive failure, with a relatively

T. Proulx (ed.), *Mechanics of Biological Systems and Materials, Volume 2*, Conference Proceedings of the Society for Experimental Mechanics Series 9999, DOI 10.1007/978-1-4614-0219-0_16,

high modulus (2.2 GPa) and tensile strength (90 MPa). Bruet et al. [9] presented a multiscale experimental and computational approach that reveals the materials design principles present within individual ganoid scales from the 'living fossil' *Polypterus senegalus,* and investigated the mechanistic origins of penetration resistance (approximating a biting attack). The results showed that one overarching mechanical design strategy is the juxtaposition of multiple distinct reinforcing layers, each of which has its own unique deformation and energy dissipation mechanisms. Later on Wang et al. [10] investigated the anisotropic design principles of the same type of fish scale, which possesses a quad-layered structure at the micrometer scale with nanostructured material constituting each layer, their findings indicated the potential to use anisotropy of the individual layers as a means for design optimization of hierarchically structured material systems for dissipative armor. In both research the penetration resistance was evaluated by nanoindentation experiment, the penetration depth is limited to 60 nm, however the biting attack from predators is much more severe and might fully penetrate the entire scale and the underneath flesh.

The present study investigated the structure and mechanics of a single teleost (ctenoid) fish scale from striped bass (*Morone saxatilis*) as a part of larger project on novel flexible protective systems inspired from fish skin [11]. The paper is organized as follows. In the next section, we describe the hierarchical structure of teleost fish as well as individual scales. Section 3 evaluates the mechanical properties of the entire scale and its collagen layer. Section 4 then concentrates on the evaluation of the penetration resistance of a single scale by emulating a predator's attack and the investigation of the failure mechanisms. Finally, section 5 provides concluding remarks.

2. The Structure of Teleost Fish Scale

Many biological materials draw their remarkable propertied from their complex, hierarchical structures [12, 13] and fish skin is no exception to this general rule. The three main hierarchical levels of teleost fish skin are depicted on Fig.1. They include macroscopic level (fish skin), millimeter level (individual scale) and micrometer level (cross ply structure and individual mineralized collagen fibrils).

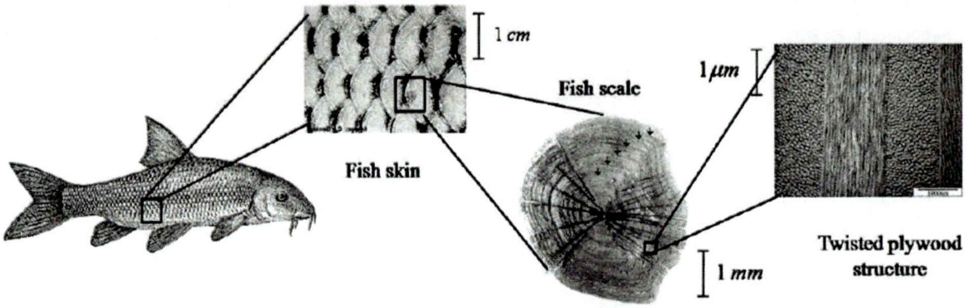

Fig. 1 Overview of the hierarchical structure of teleost fish scales

Besides its organized hierarchical nature, the structure of fish-scales is distinguished by the presence of both an organic and mineral phase. In nature, this composite often leads to unique combinations of stiffness, hardness and toughness (as in bone, teeth, and seashells) [12, 14]. Due to the variety of building blocks across different length-scales, the mechanical properties of fish-scale may be modified by slight changes at any level of its structure. As a result, by retaining the same overall structure, this material is able to adapt and remain optimal in many different environments.

Four areas were recognized on the surface of ctenoid scale from striped bass studied: anterior, two laterals, and posterior (Fig. 2a). In the anterior area, it features uniformly distributed granules and the comb-like ctenii on the margin of the scale. The posterior area consists of ridges (radii) and grooves which form circular rings around a center called the 'focus' (F) [15]. The focus is the part of the scale that developed first during ontogenesis. The position of the focus on the scale remains the same through the life history of the scale [16].

For fishes in the teleostei family, in which the striped bass is part of, it was found that there are two main structures that could be found; either fully orthogonal arrangement of successive collagen layers or intermediate orthogonal progression of the layers [17]. Further study of the basal plate (collagen layer) shows that the scale from striped bass is made up of multiple layers of collagen fibrils arranged in a orthogonal "plywood" manner, but in a curved structure around the scale's focus (Fig. 2b, c). The cross-section of the scale is made up of two major layers: the basal plate (collagen layer) and the external plate (bony layer), each of which is about 100 micron thick. Previous research has shown that the basal plate is mainly composed of collagen fibrils (Fig. 2c, d), more specifically collagen type-I. The external layer on the other hand is a calcified layer in which hydroxyapatite crystals are found [18]. This material is commonly found in materials such as teeth and bone in the human body. Therefore we can say the external plate (bony layer) acts as a shield that protects the fish from external attacks. The basal plate (collagen layer) on the other hand is mainly responsible to support and strengthen the scale [19].

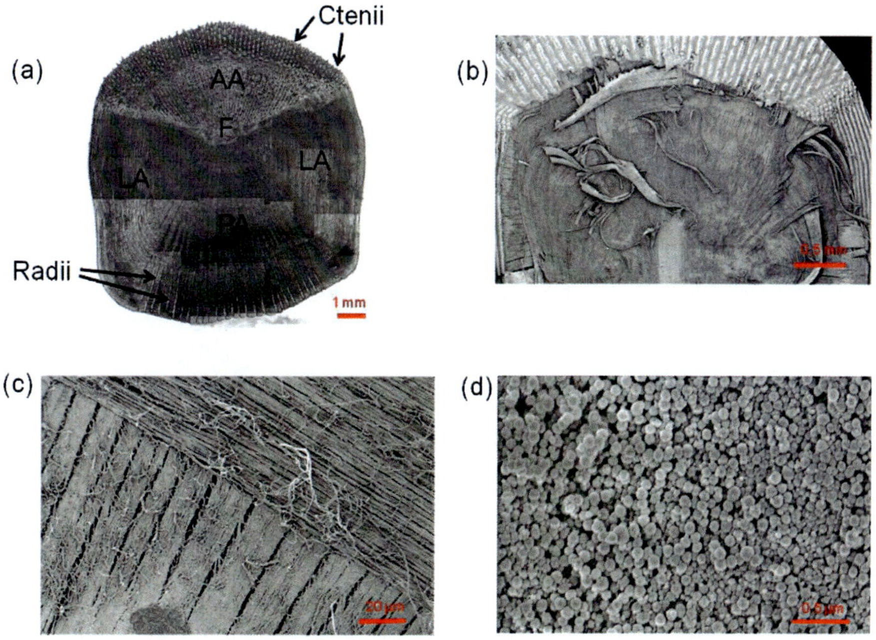

Fig. 2 (a) top view; (b) back view; (c) plywood-like structure in collagen layer; (d) collagen fibrils. AA = anterior area; LA = lateral area; PA = posterior area; F= focus.

3. Tensile Behavior

3.1 Specimen preparation

The scales were first removed by hand from a fresh striped bass. No tools were used in order to avoid damaging the scales. A simple "lift and pull" approach was enough to detach the scales from the fish's body. The scales were stored in a freezer at -20 °C until tested. Before the test, the scales were removed from the freezer and put in water bath for about 5 minutes to unfreeze and hydrate, and then cut into dog-bone-shaped specimens with a 1.5 mm cross-section width by using a multi-tube rotary hole-punch and a dissecting scissor. One more step was further taken to obtain the collagen specimen by carefully peeling the collagen layer from the scale. The thickness of individual collagen specimen might not be exactly identical to each other after the peeling process. The thickness of individual specimen was measured at three locations, i.e. the both ends and the middle of the gage section; and the averaged value was used for the stress calculation. In order to comprehend the laminate-orthogonal structure of the collagen layer of the scale, tensile tests have been done on scale and collagen specimens in three directions regarding the scale's longitudinal direction: 0°, 45°, and 90°. We should point out that the thickness of the scale specimen may vary up to 30% in the longitudinal direction (0°) along the gage section of 4 mm; while it varies much less (~ 10%) in the directions of 45° and 90°, however the thickness of collagen specimen is almost constant along the gage section in all directions. The average thickness of the scale and collagen specimens is approximately 0.256 ± 0.032 mm, 0.054±0.017 mm, respectively.

3.2 Experimental procedure

The scale sample was mounted on a miniature loading stage (Ernest F. Fullam Inc., Latham, NY), which consists of two crossheads mounted on two threaded rods such that the crossheads move at the same rate, but in opposite directions [20]. The sample was fixed on the crossheads by using two plate grips with screws. Additionally, very fine sandpaper was glued on the plate grips to provide better load transfer from the crossheads to the sample. A little amount of superglue was applied on the clamp regions of the sample to prevent any slippage between the sandpaper/grips and the test sample. The loading stage and the in-placed sample were placed under an optical microscope (BX-51M, Olympus, Markham, Canada) equipped with a CCD camera (RETIGA 2000R, Qimaging, Surrey, Canada); and then the specimens were loaded in tension to complete failure at loading speed of 0.005 m/s. The load was measured by using a 100N load cell. Images were taken throughout the entire test every 10 seconds using the CCD camera.

3.3 Tensile tests results

The stress-strain curves of entire scale and collagen samples in three directions (0°, 45°, 90°) resulting from the tensile tests are depicted in Fig. 3, showing quite different tensile behavior of these two types of samples. The tensile stress-strain curves of entire scale samples were initially linear with a corresponding Young's modulus (*E*) of 859 ± 81 MPa, 638 ± 103 MPa, and 801 ± 240 MPa for the 0°, 45°, and 90° direction (Fig. 4a), respectively, followed by a nonlinear region with lower stiffness. The first stress drop corresponds to the failure of bony layer; afterward the collagen layer takes the entire load with considerable ductile deformation before the complete failure with a maximum strain up to 0.5. Corresponding optical images of the fracture surface indicates that sliding of collagen lamellae and pulling out of individual collagen fibrils are responsible for the large deformation. The tensile strength of entire scale is 32.1 ± 3.7 MPa, 47.0 ± 9.1 MPa, and 51.7 ± 8.5 MPa for the direction of 0°, 45°, and 90°, respectively (Fig. 4b).

By contrast, the collagen samples showed a linear elastic response (*E* ~462 ± 72 MPa, 429 ± 75 MPa and 452 ± 28 MPa for 0°, 45°, and 90° (Fig. 4a)), followed by a region of large inelastic yielding before starting to fail at the maximum stress (tensile strength) of 63.5 ± 6.5 MPa, 66.7 ± 5.5 MPa and 64.5 ± 7.1 MPa for the direction of 0°, 45°, and 90°, respectively (Fig. 4b). The maximum strain at complete failure is at the range of 0.3~0.5, which provide the fish scale with excellent plasticity. The similarity in the Young's modulus and tensile strength of 0°, 45°, and 90° directions further proved the hypothesis of orthogonal plywood-like structure in the collagen layer, but in a curved structure around the scale's focus. One should note that the tensile strength of collagen sample is higher than that of entire scale, indicating although the bony layer is stiffer, but weaker than the collagen layer in tension.

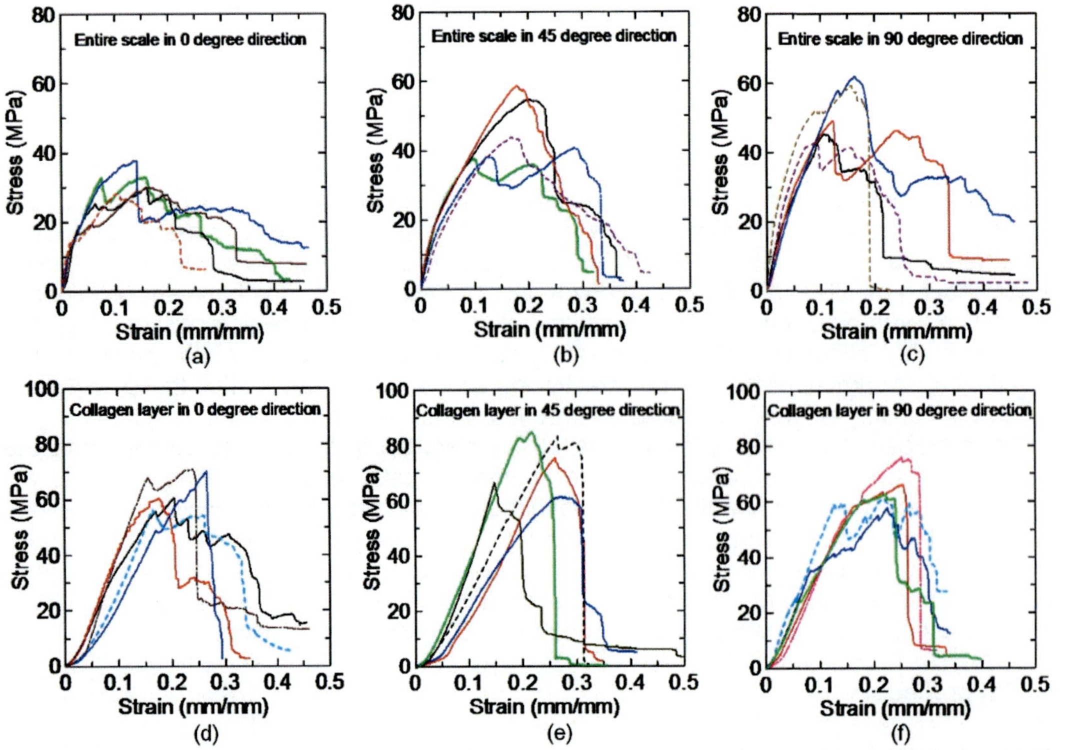

Fig. 3 Tensile stress-strain curves: (a, b, c) entire scale; (d, e, f) collagen layer in 0°, 45°, and 90° directions

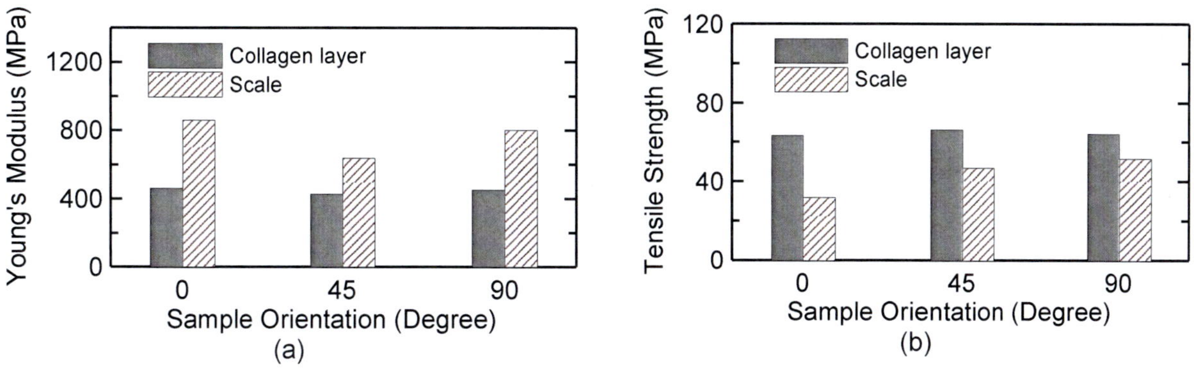

Fig. 4 Mechanical properties of entire scale and collagen layer: (a) Young's modulus; (b) tensile strength

4. Penetration Performance

4.1 Experimental procedure

In order to evaluate the penetration resistance of a single scale, by emulating a predator's attack: a sharp tooth penetrating the scale resting on a soft substrate, penetration tests have been conducted on the same miniature loading stage (Ernest F. Fullam Inc., Latham, NY). A load cell with a capacity of 22.5 N was fitted on the frame for load measurement. A steel needle, which simulates a sharp tooth of a predator, was clamped on one crosshead with a cover plate and two bolts. A block of soft silicone, which simulates the flesh of fish to support the scale, was supported by a "U" shaped fixture. The steel needle will travel towards the central area of the scale at a speed of 0.005 mm/s until the scale has been penetrated (Fig. 5).

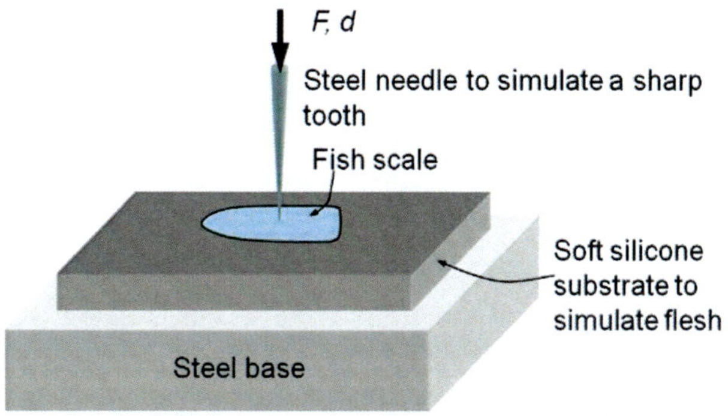

Fig. 5 Schematic diagram of single scale penetration test setup

4.2 Penetration test results and failure mechanisms

In the case of pull penetration the load-displacement curve consists of three distinct regions: i) linear region followed by first load drop, ii) secondary load increase with peak load, iii) post peak region (Fig. 6). When the needle tip (indenter) starts to make contact with the bony layer, the scale and soft substrate deform to resist the penetration. Although the load increases linearly up to the first peak, the actual behavior of the system (scale and substrate) is far more complex than linear elastic deformation. When the load increases to 1.0 N, only the top surface of the bony layer is damaged. However when load increase to 1.5 N, the indenter has partially penetrated into the bony layer and generated a crater with a diameter of 45 micron. The linear response up to approximately 2.5 N is resulted from the combination of the weakening due to damage/cracking development in bony layer and the strengthening of collagen layer and substrate. When the bony layer completely fails, the load decreases a little amount (5~10%) and increases again because the underneath collagen layer and substrate provide additional resistance. A small hole appears on the top surface at the center of the penetration size (not complete through the bony layer), further contact of the indenter with the scale enlarges the hole due to the expanding diameter of the indenter, multiple circumferential cracks develops around the penetration site, and four radial cracks (Fig. 7a) due to flexural stress emerge from the rim of the hole and propagate towards the edges of the scale while the indenter penetrates through the scale. As long as the indenter fully penetrates into the substrate, the load decreases again and causes extensive delamination and damage in the collagen layers (Fig. 7b). In the partial penetration case, the tests were stopped right after the load started to decrease at about 2.5 N. The SEM images of partially penetrated scale show a small hole (shear plug) at the center of the penetration site as well as the four flexural cracks through the thickness of the bony layer, but no damage in the collagen layer (Fig. 7c, d), indicating the first load drop in the load-displacement curves is due to the failure of bony layer. The multiple circumferential cracks are still visible and less extensive compared to the fully penetrated sample.

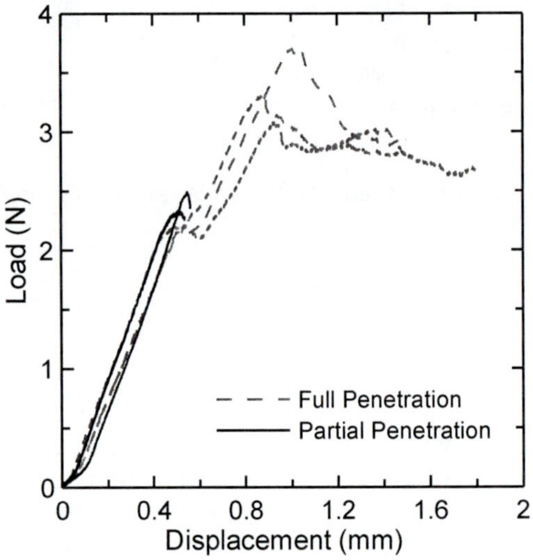

Fig. 6 Experimental curves of penetration tests

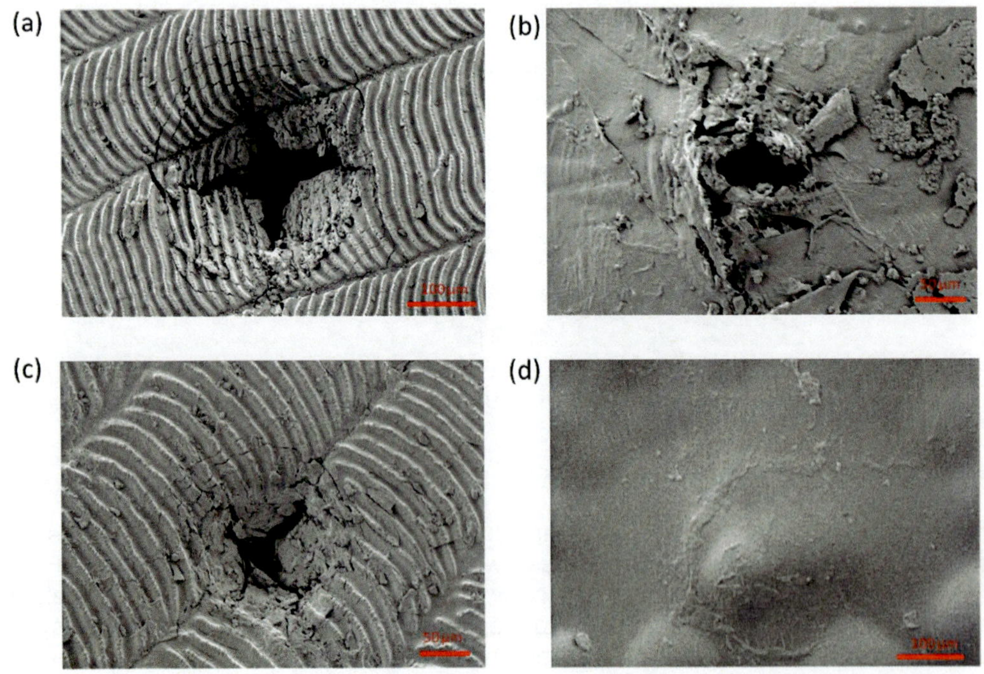

Fig. 7 SEM images of tested scales: (a, b) full penetration; (c, d) partial penetration

5. Conclusions

This article presented the structure and mechanical properties of a single teleost (ctenoid) fish scale from striped bass. The penetration resistance and failure mechanism of a scale under concentrated loading (similar to a predator's bite) was investigated by penetration tests with the assistance of SEM images.

In particular, we have the following observations:

a) The collagen layer is made up of multiple layers of collagen fibrils arranged in an orthogonal "plywood" manner, but in a curved structure around the scale's focus.
b) The bony layer failed with the formation of four radial cracks through the thickness and multiple circumferential cracks on the top surface of the scale around the penetration site. The collagen layer and substrate deferred the formation and development of the radial cracks.
c) The mechanical properties of scale and collagen layer do not strongly depend on the sample direction of 0°, 45°, and 90°. The bony layer is stiffer, but weaker than the collagen layer in tension.

Reference

1. Bereiter-Hahn J, Matoltsy A, and Sylvia Richards K, Biology of the Integument, Volume 2 Vertebrates. 1986: Berlin: Springer-Verlag.
2. Zylberberg L, Geraudie J, Meunier F, and Sire J. Biomineralization in the integumental skeleton of the living lower vertebrates. Bone, 4, 171–224, 1992.
3. Bereiter-Hahn J, and Zylberberg L. Regeneration of teleost fish scale. Comparative Biochemistry and Physiology a-Physiology, 105, 625-641, 1993.
4. Kardong KV, Vertebrates: comparative anatomy, function, evolution. 2008: McGraw-Hill New York.
5. Sudo S, Tsuyuki K, Ito Y, and Ikohagi T. A study on the surface shape of fish scales. Jsme International Journal Series C-Mechanical Systems Machine Elements and Manufacturing, 45, 1100-1105, 2002.
6. Long JH, Koob TJ, Irving K, Combie K, Engel V, Livingston N, Lammert A, and Schumacher J. Biomimetic evolutionary analysis: testing the adaptive value of vertebrate tail stiffness in autonomous swimming robots. Journal of Experimental Biology, 209, 4732-4746, 2006.
7. Hebrank MR, and Hebrank JH. The mechanics of fish skin - lack of an external tendon role in 2 teleosts. Biological Bulletin, 171, 236-247, 1986.
8. Ikoma T, Kobayashi H, Tanaka J, Walsh D, and Mann S. Physical properties of type I collagen extracted from fish scales of Pagrus major and Oreochromis niloticas. International Journal of Biological Macromolecules, 32, 199-204, 2003.
9. Bruet BJF, Song JH, Boyce MC, and Ortiz C. Materials design principles of ancient fish armour. Nature Materials, 7, 748-756, 2008.
10. Wang LF, Song JH, Ortiz C, and Boyce MC. Anisotropic design of a multilayered biological exoskeleton. Journal of Materials Research, 24, 3477-3494, 2009.
11. Vernerey FJ, and Barthelat F. On the mechanics of fishscale structures. International Journal of Solids and Structures, 47, 2268-2275, 2010.
12. Meyers MA, Lin AYM, Seki Y, Chen PY, Kad BK, and Bodde S. Structural biological composites: An overview. Jom, 58, 35-41, 2006.
13. Fratzl P, and Weinkamer R. Nature's hierarchical materials. Progress in Materials Science, 52, 1263-1334, 2007.
14. Barthelat F. Biomimetics for next generation materials. Philosophical Transactions of the Royal Society a-Mathematical Physical and Engineering Sciences, 365, 2907-2919, 2007.
15. Jawad LA. Comparative scale morphology and squamation patterns in triplefins (Pisces: Teleostei: Perciformes: Tripterygiidae). Tuhinga, 16, 137–168, 2005.
16. Liu C, and Shen S. Lepidology of the mugilid fishes. Journal of Taiwan Museum, 44, 321–357, 1991.
17. Bigi A, Burghammer M, Falconi R, Koch MHJ, Panzavolta S, and Riekel C. Twisted plywood pattern of collagen fibrils in teleost scales: an X-ray diffraction investigation. Journal of structural biology, 136, 137-143, 2001.
18. Schonborner AA, Boivin G, and Baud CA. Mineralization processes in teleost fish scales. Cell and Tissue Research, 202, 203-212, 1979.
19. Meunier FJ, and Castanet J. Organisation spatiale des fibres de collagène de la plaque basale des écailles des Téléostéens. Zoologica Scripta, 11, 141-153, 1982.
20. Barthelat F, Tang H, Zavattieri PD, Li CM, and Espinosa HD. On the mechanics of mother-of-pearl: a key feature in the material hierarchical structure. Journal of the Mechanics and Physics of Solids, 55, 306-337, 2007.

Effects of Processing Conditions on Chitosan-Hydroxyapatite Biocomposite Mechanical Properties

Chad S. Korach[1], Gary Halada[2], and Hani Mubarez[1]
[1]Department of Mechanical Engineering, Stony Brook University
Stony Brook, NY, 11794-2300 USA
[2]Department of Materials Science and Engineering, Stony Brook University
Stony Brook, NY, 11794-2275 USA

ABSTRACT
Hydroxyapatite (HAp) based composites are used in numerous biomedical applications from bone pastes and cements to tissue scaffolds. Chitosan (Ch) is a ubiquitous biocompatible polymer derived from chitin which is soluble in acidic solutions and can be found in natural organisms such as shellfish. Past work has demonstrated that composites of HAP-Ch have high mechanical strength when processing conditions are optimized. Here, we present the effects of changes in the HAP/Ch mass fractions, the use of malic and citric acids on chitosan processing, and of nanoscale HAp particles on the composite strength and modulus. SEM is used to analyze the fracture surfaces and identify matrix-particle bonding morphology. Chemical variations are measured within the material structure by the use of FTIR and EXAFS and related to the composite processing changes.

1. Introduction
Hydroxyapatite-chitosan composites are unique biopolymer composites that have been explored recently for various biomedical applications. The composites are biodegradability and biocompatibility lending to the biomedical interests. Hydroxyapatite (HAp), a calcium phosphate compound $Ca_{10}(PO_4)_6(OH)_2$, is a hard and brittle ceramic that has been used extensively in many biomedical applications due to its osteoconductivity, biocompatibility, non-toxicity, non-inflammatory, and non-immunogenic nature [1]. HAp has been used for treating bone defects. Additionally, HAp is one of the main substances forming mineralized tissue in nature (the others being calcite and aragonite); therefore it is highly biocompatible and has been used as bone substitutes [2]. However, due to HAp's hardness and brittleness, it is usually used as composite filler that incorporates a compliant polymer to compensate for the brittle ceramic HAp phase [3].

An organic polymer that has been explored for this purpose is chitosan [4]. Chitosan (Ch), $C_6H_{11}NO_4$, is a partially deacetylated derivative of the naturally occurring chitin. Ch and chitin can be found ubiquitously in shellfish [5] and other naturally occurring organic structures. Additionally, it is biocompatible, biodegradable, nontoxic, and has anti-bacterial properties [4]. This naturally occurring polymer readily dissolves in many acidic solutions and is soluble in dilute acids with pH < 6 where the free amino groups in Ch are protonated in the acid medium [6]. In this paper we characterize HAp/Ch composites by analyzing the chemical structure to determine if chemical bonding is occurring at the particle-matrix interface, which is a critical location for the strength of composites. Mechanical flexural testing is used to assess the composite strength which have been shown previously [7] to have strong interfacial bonding.

2. Materials and Methods

2.1. Materials
Chitosan powder of medium molecular weight (molecular weight ranges from 190,000 - 310,000) with 75-85% degree of deacetylation was used for all manufactured materials. Reagent grade synthetic HAp powder with molecular weight of 502.31 was also used. DL-Malic acid powder with molecular weight of 134.09 was used for dissolution of chitosan. All reagents were obtained from Sigma Aldrich (USA). Deionized water was used throughout the experiment.

2.2. Sample Preparation
Composite samples were made by creating a chitosan sol using 0.9g of Ch and 1.14g of malic acid in deionized water. The chitosan-malic acid sol formed a gel after constant stirring. HAp was added to the gel in HAp:Ch ratios ranging from 12 to 33, and mixed vigorously to combine and distribute the particles. The HAp/Ch/Malic acid compound is spread to a thickness of ~2 mm on aluminum plates to dry for 24 hours in laboratory air. Samples easily release from the metal plates and are then cut into rectangular samples for three-point bending tests as per ASTM D790. Top and bottom surfaces are polished with

T. Proulx (ed.), *Mechanics of Biological Systems and Materials, Volume 2*, Conference Proceedings of the Society for Experimental Mechanics Series 9999, DOI 10.1007/978-1-4614-0219-0_17,

standard metallographic polishing pastes to planarize the surfaces. Sample thickness is typically 1.5 mm after preparation.

Samples for performing FTIR spectroscopy were made separately so as to isolate the chemical contributions between components. First, 0.1g of malic acid was dissolved in 12.5g of water. Once dissolved, 0.134g of Ch was added into the malic acid solution. This solution was vigorously stirred until the Ch was homogenously dissolved. A droplet of the solution was placed on a glass slide with a pipette. The slide was left at room temperature to dry for 12 hours. After drying, a thin film of Ch dissolved in malic acid solution remains. A second type of sample was made by adding 0.015g of HAp powder to the Ch and malic acid solution. After vigorous stirring of the HAp until homogenous distribution is reached, a droplet of the resulting HAp/Ch/malic acid solution was placed on a glass slide. The slide was left at room temperature to dry for 12 hours. The result after drying is a thin film of HAp/Ch/malic acid, where a low particle loading exists. These samples allow for detailed chemical analysis of the HAp/Ch interface.

2.3. FTIR Characterization

The chemical interactions between the components of the composite are investigated by Fourier transform infrared spectroscopy (FTIR). FTIR spectroscopy was performed using a Thermo Nicolet Magna 860 and Continuum IR Microscope at the National Synchrotron Light Source (NSLS) at Brookhaven National Laboratory (Upton, NY) using beam line U2B. Spectra were taken with 128 scans, a spectral resolution of 4 cm^{-1}, an aperture size of 15 μm x 15 μm, and a MCT detector. The FTIR spectroscopy was performed for the Ch in malic acid sample with no HAp, and also for the Ch in malic acid with HAp sample. For the latter, spatial mapping of the spectroscopy was done starting from a location away from a HAp particle and performing 7.5 μm steps to move closer to the HAp particle (see Figure 1).

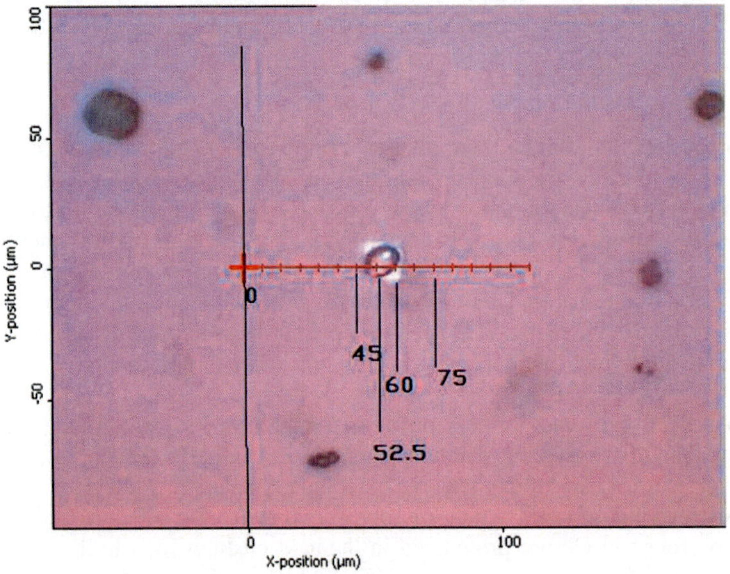

Figure 1: Optical image of the HAp/Ch/malic acid film sample used to for spatially resolved FTIR. The FTIR was started at the location on the left shown by the + mark. Equidistant steps of 7.5 μm moved in the direction toward the right. Between 45 and 60 μm is the HAp particle.

2.4 EXAFS Characterization

Extended range x-ray absorption fine structure (EXAFS) spectroscopy was performed on Ch/HAp composites (same type fabricated for mechanical testing) using BNL-NSLS beam line X15B. Samples are placed in a He-filled hutch where a maximum of 15 keV energy beam is incident on the sample. A fluorescence detector absorbs ejected photoelectrons which are recorded as a function of energy.

2.5 Mechanical Testing

Three point bending tests are performed on the composite samples following ASTM D790 standard. The tests determine flexural strength and modulus of the composites. Specimen sizes are ~38 x 13.5 x 1.5 mm (L x W x H). Support geometry followed ASTM D790. The span was set for 25.4 mm and a crosshead rate of 0.66 mm/min. was used.

3. Results and Discussion

3.1 Ch and Malic Acid Chemical Interaction

The FTIR spectra of the chitosan dissolved in the malic acid solution is shown in Figure 2. Peaks which are characteristic of Ch, such as the amide I absorbance band at 1660 associated with the C=O in the acetylated unit of the Ch [8], are no longer present. This is a possible indication that the malic acid reacts with the amino group of the N-acetyl-D-glucosamine units. Another characteristic peak of Ch is the $-NH_2$ peak at 1590 [9]. This peak has shifted to 1575 indicating the malic acid reacted with the amino group in the deacetylated units. Other characteristic peaks are observed to have no change, such as the bridge C-O-C stretch at 1155 and the skeletal C-O stretch at 1081 and 1035 [10]. This possibly indicates that the malic acid did not interact with the bridge in between the chitosan units. The peak at 3396 relates to the -OH stretch vibration in the hydroxyl.

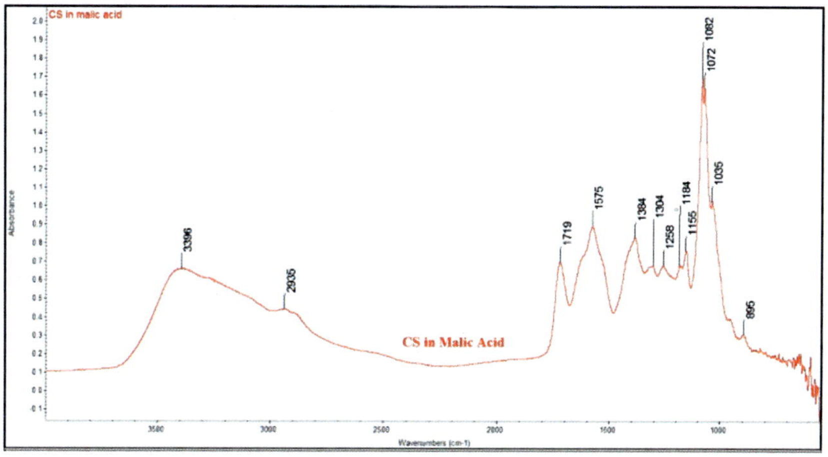

Figure 2: FTIR spectra from a film of Chitosan dissolved in malic acid.

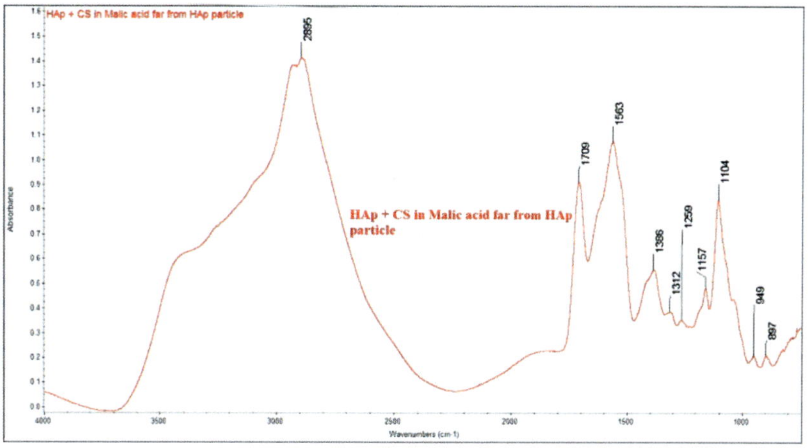

Figure 3: FTIR spectra of HAp/Ch/malic acid film at a position within the chitosan matrix far from a HAp particle (position 0 μm in Figure 1).

3.2 HAp/Ch/Malic Acid Chemical Interactions

The chemical interactions between HAp, Ch, and malic acid were investigated by examining the FTIR spectroscopy of the low-dispersion HAp in the chitosan and malic acid matrix. The FTIR spectra at locations farthest from the HAp particle exhibited similar characteristics regardless of the particle distance. Figure 3 shows a typical FTIR spectrum away from the HAp particle. To note, the FTIR spectra of the HAp/Ch/malic acid far away from the HAp particle is different from that of Ch in malic acid alone. The 1718 C=O peak for the Ch in malic acid has shifted to 1709. Similarly, the 1575 $-NH_2$ peak has shifted to 1563, and the 1082 skeletal C-O stretch shifted to 1104. The fact that all these changes occurred far from the HAp particle suggests that material from the HAp particles and is interacting with the chitosan and malic acid throughout the matrix. The change in the amino group of the chitosan indicates that HAp is reacting with the chitosan in that functional group. However, since the malic acid also reacted with the amino groups in the Ch/malic acid film, the C=O bond in the malic acid was affected too. Currently, it is unclear why the 1082 C-O skeletal stretch shifted to 1104.

When imaging directly a HAp particle in the Ch/malic acid matrix, the resulting FTIR spectra, when compared with that of the matrix far from the particle (cf. Fig. 3 with 4), shows that there are additional small peaks between 1500 and 1600. Also, there is a new peak at 969. The small peaks from 1500 to 1600 superimposed on the larger 1562 peak suggests the presence of carboxyl. The fact that these peaks appeared only when imaging the HAp particle directly suggests that HAp is interacting with the carboxyl in the malic acid. The carboxyl may be the chemical bond that attaches the HAp particle to the Ch matrix. The new peak found at 969 may be related to the *ν1* P-O symmetric stretch [11], identifying chemical changes in the HAp at the surface. The chemical analyses have shown that there are chemical bonds between the HAp and Ch matrix are used to manufacture the composites. The strength of the bonds will provide important indicators of the composite mechanical properties.

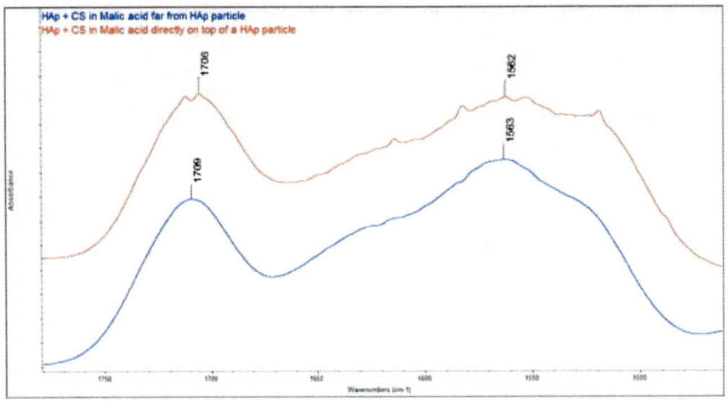

Figure 4: A close up view (1470-1780 cm^{-1}) of the FTIR spectra for the matrix region of the HAp/Ch/malic acid film compared with the spectra taken directly on a HAp particle (position 52.5μm in Figure 1) in the film.

3.3 EXAFS Characterization

Absorption data were collected for an energy range of ~2.1 to 2.75 keV. This range includes the pre-edge and near-edge data as well as the EXAFS data for phosphorous and apatite. Data are plotted as the log of the ratio of the incident to transmitted adsorption versus the beam energy. Figure 5 shows the results for the HAp powder overlaid with the HAp/Ch composite; both data sets have been normalized. Each result represents the average of 10 independent adsorption scans. Disruptions in the extended range indicate changes in the apatite structure due to the addition of chitosan. Utilizing the ATHENA software [12], the background subtracted adsorption data plotted in k-space shows similar scattering for both the HAp and HAp/Ch composite (Fig. 6), though shifts and changes in the peak data are present, indicating a structural change in the apatite crystal.

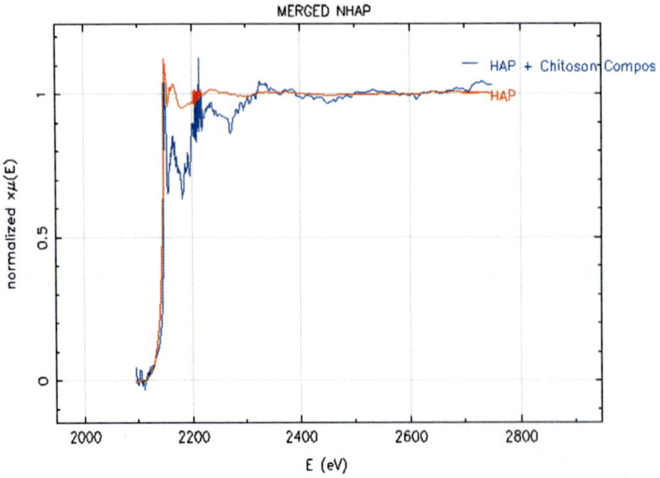

Figure 5: Extended x-ray adsorption fine structure (EXAFS) spectra (normalized adsorption coefficient v. energy) of HAp powder and HAp/Ch composite. Near edge at ~2140 eV is similar for both materials, though differences in the extended range are evident indicating changes in the apatite structure with the addition of chitosan.

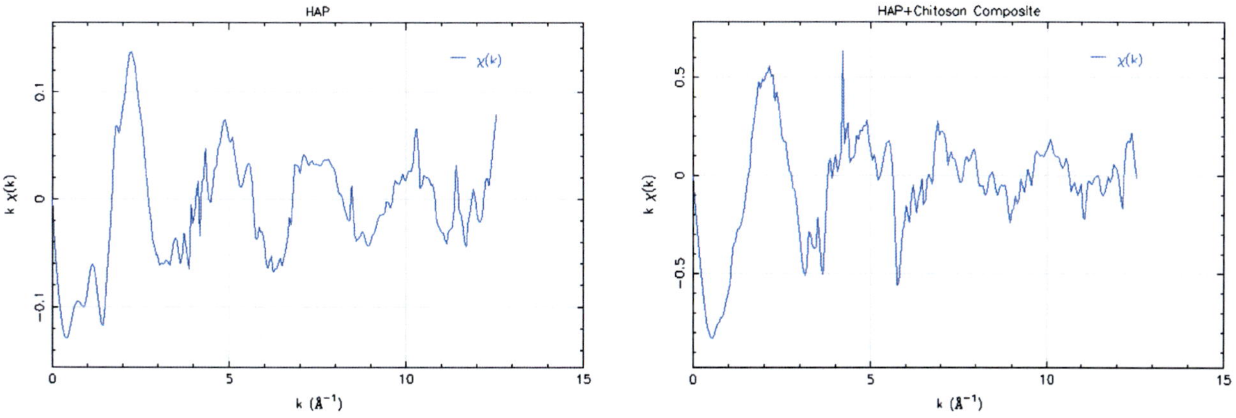

Figure 6: Extended x-ray adsorption fine structure (EXAFS) spectra (background subtracted adsorption v. wave number) of HAp powder (left) and HAp/Ch composite (right). Scattering is similar for both materials, though differences in the peak shapes and shifts exist indicating changes in the structure.

3.4 Mechanical Characterization

The three point bending tests yielded load versus cross-head displacement results that are plotted as flexural stress versus strain (Fig. 7). From ASTM D790, the toe region at the beginning of the test was subtracted out of the displacement data before conversion to strain. The flexural strength is calculated using the maximum load at failure, which was fracture of the sample, and is 14.7 MPa. The result is similar to those of Sun et al. [3], whom found strength values of calcium phosphate-chitosan cements ranging from 4 to 14 MPa. The slope of the loading curve was calculated from the data as 46 N/mm, which yields from bending theory at flexural modulus of 3.1 GPa. Previous studies [7] have shown by SEM that fracture of Ch/HAp composites occurs within the HAp particles and the matrix and not at the particle interface, indicating that a strong matrix-particle bond exists.

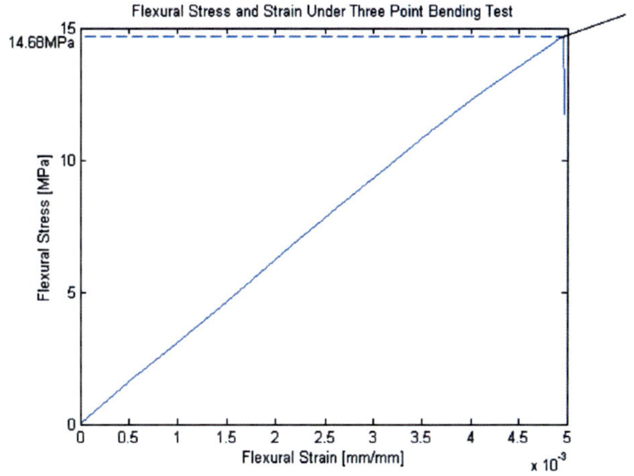

Figure 7: Flexural stress versus strain for three point bending test of the HAp/Ch composite. Flexural strength of 14.7 MPa and modulus or 3.1 GPa were measured from the data. Failure occurred due to fracture of the sample at center span.

4. Conclusions and Future Work

Composites of hydroxyapatite powders and chitosan were manufactured and characterized for their chemical structure and mechanical properties. FTIR spectroscopy shows that the amino group both in the acetylated and deacetylated units in chitosan and the carboxyl in malic acid are the chemical glue that holds the Ch/HAp composite together. EXAFS data supports the results that there are strong bonds between the chitosan and HAp indicated by changes in the apatite structure for the composite. Three point bend tests of the composite found a flexural modulus of 3.1 GPa and strength of 14.7 MPa which are similar to values reported in the literature for Ch/HAp composites. The study to date indicates that the strength of the composites is due to chemical interactions between the HAp filler and chitosan matrix. Future work will entail extraction of changes in bond distances from the EXAFS data compared with mechanical testing of composites made with different mixing ratios. Since the composites are intended for implantable applications, stabilization of the chitosan matrix will be performed and samples tested in aqueous environments to determine the effect on mechanical properties.

Acknowledgements
CSK and HM respectfully acknowledge the support from NSF through grant CMMI-0626025 for HM as an REU student in summer 2010.

References

1. L. Kong, Y. Gao, W. Cao, Y. Gong, N. Zhao, X. Zhang, "Preparation and characterization of nano-hydroxyapatite/chitosan composite scaffolds", *J. Biomed. Matls. Res. Part A*, **75**(2), 275 (2005)
2. H. Zhao, L. Ma, C. Gao, J. Shen, "Fabrication and properties of mineralized collagen-chitosan/hydroxyapatite scaffolds", *Polymers for Advanced Technologies*, **19**(11), 1590 (2008)
3. L. Sun, H.H.K. Xu, S. Takagi, L.C. Chow, "Fast Setting Calcium Phosophate Cement-Chitosan Composite: Mechanical Properties and Dissolution Rates", *J. of Biomatls. Appl.*, **21**, 299 (2007)
4. I. Yamaguchi, S. Iizuka, A. Osaka, H. Monma, J. Tanaka, "The effect of citric acid addition on chitosan/hydroxyapatite composites", *Colloids and Surfaces A*, **214**(1-3), 111 (2003)
5. E. Khor, L.Y. Lim, "Implantable Applications of Chitin and Chitosan", *Biomaterials*, **24**(13), 2339 (2003)
6. S.V. Madihally and H.W. Matthew, "Porous chitosan scaffolds for tissue engineering", *Biomaterials*, **20**(12), 1133 (1999)
7. K. Nelson, "Processing and Characterization of Chitosan and Hydroxyapatite filled Polymer Composites", Master's Thesis, Stony Brook University (2009)
8. A. Pawlak and M. Mucha, "Thermogravimetric and FTIR studies of chitosan blends", *Thermochimica acta*, **396**(2), 153 (2002)
9. E. Piron and A. Domard, "Interaction between chitosan and uranyl ions. Part 2. Mechanism of interaction", *International Journal of Biological Macromolecules*, **22**,33 (1998)
10. B. Geng, Z. Jin, T. Li, X. Qi, "Preparation of chitosan-stabilized Fe nanoparticles for removal of hexavalent chromium in water", *Science of the Total Environment*, **407**(18), 4994 (2009)
11. A. Stoch, A. Brozek, S. Blazewicz, W. Jastrzebski, J. Stoch, A. Adamczyk, I. Roj, "FTIR study of electrochemically deposited hydroxyapatite coatings on carbon materials", *Journal of Molecular Structure*, **651-653**, 389 (2002)
12. B. Ravel and M. Newville, "ATHENA, ARTEMIS, HEPHAESTUS: data analysis for X-ray absorption spectroscopy using IFEFFIT", *J. Synchrotron Rad.*, **12**(4), 537 (2005)

Mechanical Properties of a Nanostructured Poly (KAMPS)/aragonite Composite

Chad S. Korach
Department of Mechanical Engineering,
Stony Brook University, Stony Brook, NY, 11794, USA
Ranjith Krishna Pai
Center for Functional Nanomaterials, Brookhaven National Laboratory, Upton, NY 11973, USA
International Iberian Nanotechnology Laboratory, 4715-310 Braga, Portugal

ABSTRACT
The mechanical properties of a new poly(KAMPS)/aragonite composite are investigated. The composite is fabricated using biomimetic pathways and formed entirely from dilute aqueous solutions. Nanorods of aragonite are formed within a matrix of poly(KAMPS)-based polyelectrolyte and a nanostructred material with rod widths of 120nm and polymer-filled spacings of 10-20nm is created. Nanoindentation is used to measure mechanical properties of the composite material. The elastic modulus of 44GPa and hardness of 2.8GPa are found similar to that of nacre and exhibit a hardening mechanism at the nanoscale. The new biomimetic composite has application in the biomedical and dental fields.

1. Introduction

Well-controlled nanostructures obtained via polymer-mediated synthesis have generated considerable scientific and technological interest.[1] The ability to control the nucleation and growth of hierarchical structures often leads to fascinating shapes combined with remarkable mechanical properties but requires a high level of control over structure, size, morphology, and orientation by assembly at organic surfaces.[2] Calcium carbonate ($CaCO_3$) is one of the most studied systems for its pivotal role in understanding the natural mechanism of biomineralization and for designing new biomimetic composite materials[3,4]. A classic and widely studied [5-10] example of a biocomposite is the nacre of abalone shell, composed of hexagonal platelets of aragonite (a polymorph of $CaCO_3$), 10-20 μm wide and, 0.2-0.9 μm thick, arranged in a continuous, parallel lamina. These layers are separated by sheets of organic matrix (10-50 nm thick) composed of elastic biopolymers such as chitin, lustrin and silk-like proteins [11]. This mixture of brittle platelets and thin layers of elastic biopolymers make the material strong [12] and resilient (shock absorbent). Strength and resilience are also likely to be due to adhesion by the "brickwork" arrangement of the platelets, which inhibits transverse crack propagation. This design at multiple length scales increases its hardness enormously, making the biocomposite similar to that of silicon [13].

The intimate association of organic/inorganic materials of nacre and its structure-function relationship has inspired a large class of biomimetic-advanced materials.[14] The addition of organic polymers to inorganic components markedly improves the ability to absorb energy during deformation of composites.[15] Traditional ceramics are brittle, and have been improved upon with new ideas such as reversible toughening mechanisms, and large fracture tolerance,[16,17] leading to the development of ceramics with an order larger toughness. The toughness of these materials though remains lower than that of steel. Studying the materiomics of natural biocomposites may create the ability to, by mimicking their nanostructuring and mechanisms, fabricate ceramics that are 50 times tougher.[18,19] Biomimetic strategies have been proposed to develop materials with mechanical characteristics similar to nacre [20-26], and there have been many materials fabricated to either mimic nacre or to mimic the mechanisms of nacre to create toughening. A comprehensive review can be found in [27]. However, none of them can truly recreate the similar aragonite 'bricks' with the remarkable mechanism of platelet sliding, because most conventional processing techniques simply do not offer the nanoscale level of control needed to create a highly regular bricks-and-mortar-type arrangements. Biomimicry of the toughening mechanism of nacre has been performed at larger scales where control of the processing proves simpler, and has demonstrated successfully the toughening. Here, a simple, economic, and one step strategy to engineer polyelectrolyte-based composite materials that mimics both nanoscale structural and mechanical properties of nacre is introduced.

Previously, it was found that on the molecular level, calcium-mediated sacrificial bonds increase stiffness and enhance energy dissipation in bone. [28] The ability of a polymeric component to infer a large fracture or adhesion energy is related to the bonding to the more rigid components and the ability to sustain a significant elongation without complete breakage. Among the various kinds of polymeric components employed as adhesion based-assembly in bioinspired morphosynthesis, 2-acrylamido-2-methyl-1-propane sulfonic acid (AMPS) based polyelectrolyte is considered to be extraordinarily effective.

T. Proulx (ed.), *Mechanics of Biological Systems and Materials, Volume 2*, Conference Proceedings of the Society for Experimental Mechanics Series 9999, DOI 10.1007/978-1-4614-0219-0_18,

These types of polyelectrolyte are used to control the colloidal stability and rheological responses of a wide range of industrially important products and processes, including in the food [29] and paper industries [30] and in enhanced oil [31] and water purification. [32] They respond rapidly to changes in pH, ionic strength, and temperature; and form strong gels are of particular interest for medical and pharmaceutical applications.[33,34] In addition to their technical applications and particular solution properties, polyelectrolytes modify the crystallization behaviour of minerals formed from solutions. A particular interest of ours is the way in which such polyelectrolyte controls $CaCO_3$ crystallization and the ability to readily form sacrificial bonds at the interface between the inorganic components.[35-37]

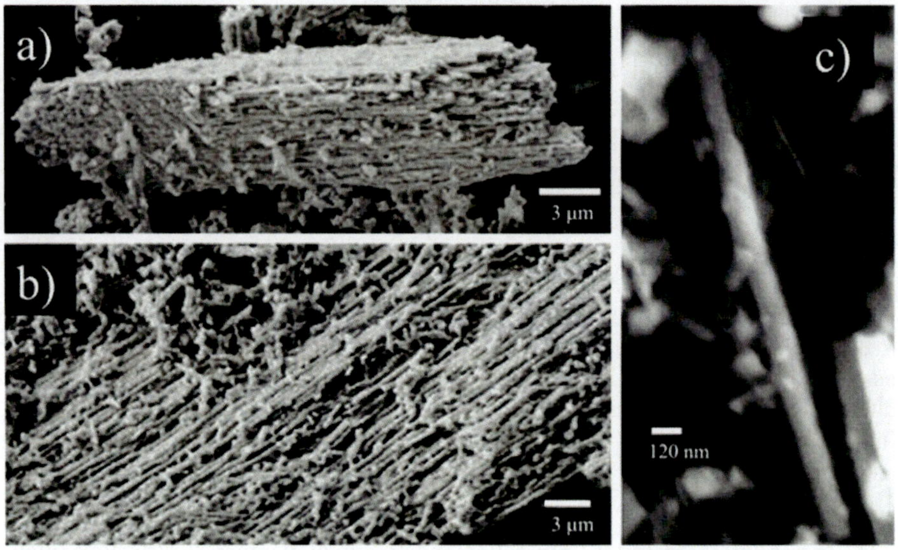

Fig. 1. Scanning electron microscopy images of nacre-like carbonate composite material. a) side-view of the as-prepared sample obtained in the presence of poly (KAMPS), b) higher magnification SEM image show that the "brickwork" arrangement of the aragonite nanorods and c) a single aragonite nanorod.

2. Synthesis and Characterization

The scanning electron microscopy (SEM) images shown in Fig.1 illustrates the nacre-like carbonate composite (polymer adhesion by the "brickwork" arrangement of the nanorods) attainable by double decomposition reaction. Details of the synthesis can be found in [38]. The morphology of the composite was further examined by transmission electron microscopy (TEM) under high magnification. The composite microstructure observed by TEM looks like intertwining sheaves (Fig. 2a). Sheaves are formed of nanorods lying parallel to one another, separated by thin layers of polyelectrolyte (10-27nm), composed to form nacre-like carbonate composite (Fig. 2c). From the analysis of TEM images from Fig. 2a and c, it was found that nanorods have average widths of 120 ± 2 nm and lengths of up to 10 μm, which is in good agreement with the dimensions of the seashell nacre. All nanorods have similar width and appear to grow in the [010] direction as demonstrated by its selected area electron diffraction (SAED) pattern (Fig. 2b). From SAED, and XRD diffraction experiments [38], an orthorhombic unit cell was proposed for the nacre-like carbonate composite with space group Pmcn (62) with lattice constants of a=0.496 nm, b=0.796 nm, and c=0.574 nm (JCPDS card no. 76-0606), indicating that the prepared nacre-like carbonate composite is single crystals of aragonite. All together, the characterization data confirm the co-existence of uniform intertwining sheaves of aragonite nanorods and polymer layers, analogues of a nacreous biocomposite material.

Salient points regarding the synthesis and characterization are summarized as follows:

- The growth process from amorphous calcium carbonate (ACC) nanoparticles to the final single crystal was investigated by TEM by monitoring the early stages of the crystal growth and found to fuse to form aragonite mesocrystals (aligned single crystals).
- The growth mechanism of the nacre-like carbonate composite is analogous to the construction of natural nacre aragonite platelets. [39, 40]
- Faces of $CaCO_3$ are highly charged, because they are composed of only calcium or carbonate ions, which are organized in alternating layers.[41] This was confirmed by FTIR, and the adsorption of positively charged NH_3-quantum dots confirmed the presence of polyelectrolyte layers, at the interface between the aragonite nanorods.[38]

- The poly (KAMPS) forms strong gels, which results in bonding at the interface between the aragonite nanorods. and leads to an adhesion of the "brickwork" arrangement of the aragonite nanorods that mimics the natural structure of nacre.
- The interfaces between the two rigid (inorganic) components determine the mechanical properties of the composite, which will be enhanced if the inorganic are modified by polymers.[42, 43]
- After washing by milliQ water and anhydrous ethanol, respectively, ~5 wt% polyelectrolyte is still present in the nacre-like carbonate composite.

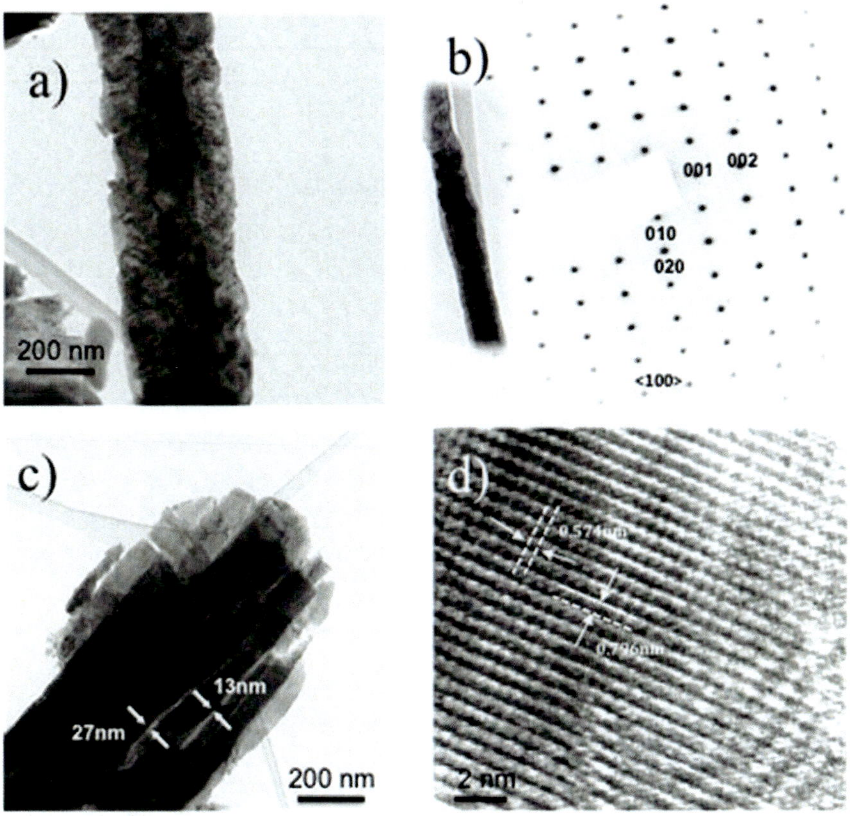

Fig.2. Transmission electron microscopy (TEM) images of nacre-like carbonate composite material shows a) intertwining sheaves of aragonite nanorods, b) selected-area electron diffraction pattern from a single aragonite nanorod, c) nanorods lying parallel to one another, separated by thin layer of polyelectrolyte, d) High Resolution TEM image of a selected surface area of the single crystalline aragonite nanorods.

3. Nanoindentation

3.1. Experimental

All mechanical property measurements were performed by nanoindentation with a Berkovich diamond using a NanoTest nanomechanical system (Micro Materials, UK). A fused silica substrate was used to calibrate the diamond area function and machine parameters. The composite material was embedded in an epoxy resin (Buehler Epoxicure) and polished using a 3 step process after curing. Two individual domains of the composite were located with sizes of 100-200 μm and indentations were performed across the entire domain length, including the epoxy resin region for comparative purposes. The loading and unloading rates were 0.1 mN/sec, with a maximum load of 2 mN. A 10 sec. hold at the maximum load is used to measure material creep effects. The slope of the initial unloading curve represents the elastic response of the material and was analyzed by the Oliver and Pharr method [44] to determine hardness and elastic modulus.

3.2. Results and Discussion

The tightly packed aragonite nanorods and polyelectrolyte as highly regular layer-by-layer microstructure form desirable interfaces (Fig. 1a, b and Fig. 2a,c), which might be critical to maintaining mechanical integrity for bone substitution.[27] Nanoindentation was used to assess, at the microstructural level, material properties of the nacre-like carbonate composite. Load-displacement curves from the embedded composite are shown in Fig. 3 left where typical maximum depths were 150-

200 nm. The Oliver and Pharr method was used to analyze the load-displacement data, with the resulting reduced elastic modulus (*E*) and hardness (*H*) shown in Table, Fig. 3 right. The average value for the reduced elastic modulus was 43.4 ± 6.1 GPa, and the hardness was 2.8 ± 0.5 GPa. The modulus values are found to be larger than those of other nacre-like biocomposite materials which ranged from 1 GPa for a "stack like" $CaCO_3$ [26] measured by indentation to 11 GPa for layered polyelectrolyte clay materials[20] measured in tension. The results were similar to the 45 GPa modulus measured in tension for nano-clay composite sheets, [45] but lower than that of aragonite (value of 80 GPa). It is interesting that our nacre-like carbonate composite modulus results proved similar to that of natural nacre [54] (30-45 GPa).

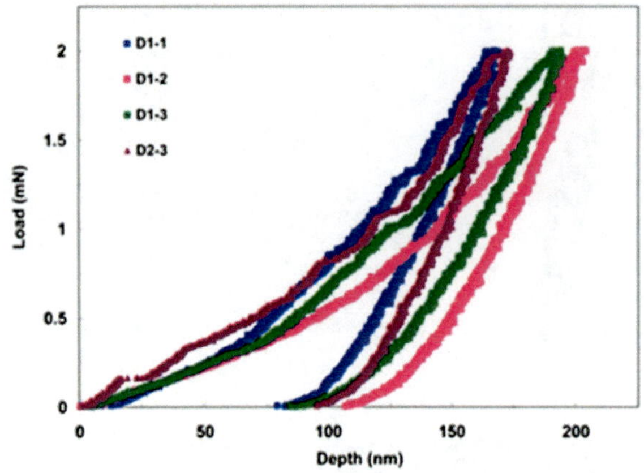

	Hardness (GPa)	Elastic Modulus (GPa)
Domain 1		
1	3.33	43.3
2	2.14	35.3
3	2.49	33.7
Domain 2		
1	1.85	46.9
2	3.29	45.3
3	2.83	48.3
4	3.57	49.9
5	2.81	51.4
6	2.73	39.8
7	3.09	40.0
Average	2.81	43.4
s.d.	0.54	6.0

Fig.3. Left: Nanoindentation load-displacement curves for the carbonate composite material, Table right: Material characterization results for the Hardness and Reduced Elastic Modulus performed by nanoindentation.

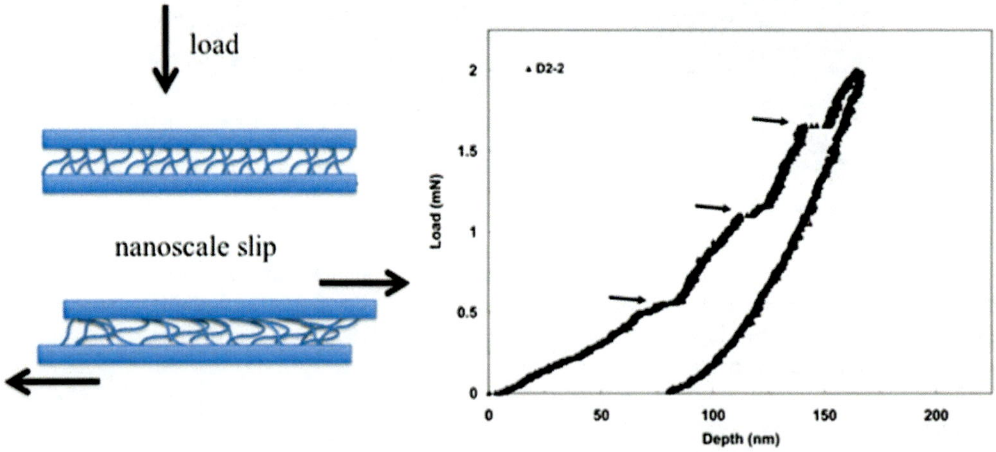

Fig.4. Left, schematic demonstrate: The poly (KAMPS) at the interface has strong adhesion to the aragonite nanorods. Under load, large extension occur and cohesion is maintained between aragonite nanorods over a displacement of 10 nm. The shear strains at the interface are locally much higher and the poly (KAMPS) may contribute to the hardening observed at the microscale, where nanoscale slip or compaction events (arrow on the right) occur while loading as observed in the load-displacement curve for the composite material, Right: Load-displacement curve for composite material showing nanoscale slip or compaction events (arrows) occurring while loading.

During the loading portion of the indentation, evidence of pop-in is observed and is shown in Fig. 4 right. Arrows indicate locations where a jump in displacement was observed at a given load. This behaviour has also been seen in the indentation of nacre [46] where it is attributed to the movement of the interfaces between nacre's tablets. This is an indicator that our nacre-like carbonate composite responds in a similar manner, where the material interfaces may compact or slip and then stiffen (Fig. 4 left), creating a hardening in the material. The displacement distance is estimated from Fig. 4 to be ~10 nm for each event. The behaviour was observed to become more rapid with increasing load, an indication that the mechanism may be dependent on a critical load or stress, and likely that the indenter was 'sensing' interfaces deeper into the material. The observed slip results were highly repeatable and found to occur in most of the *P-h* curves. The exact nature of this mechanism may be related to the polymer chain length or the polymer-mineral bonds holding the composite together, and is worth further investigation, for example it is unknown at this point if the displacement is due to purely compaction or a combined compaction-sliding event. From TEM, the interface surfaces do not exhibit nanoscale asperities which have been associated with hardening in nacre from sliding [9] nor dove-tailed interlocking [10,19,47], though the polymer adhesion at the interface [48] is believed to play a major role. Due to the fact that the poly-KAMPS based composite has different structure than nacre (rods versus plates), the structural significance of platelet rotation is lost. To note, the composite did not exhibit creep behaviour at the maximum load hold (see Fig. 5).

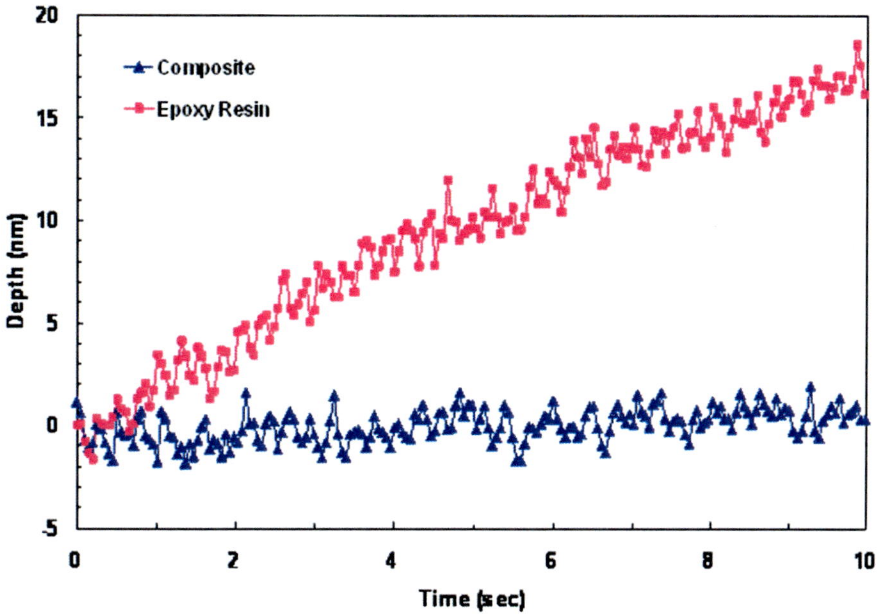

Fig. 5. Creep hold data from nanoindentation of nacre-like carbonate composite material and the epoxy used for sample molding.

4. Conclusions

In summary, we have obtained an adhesion by the "brickwork" arrangement of the aragonite nanorods that mimics both structural and mechanical properties of natural nacre by a rapid, economical and scalable one-step methodology in the presence of polyelectrolyte. Structural and spectroscopic analysis reveals that the uniform intertwining sheaves of aragonite nanorods with a unique orthorhombic shape were generated via an oriented attachment mechanism, where ACC nanoparticles in the early stage combine and fuse to form single crystals. The new nacre-like carbonate composite has a modulus and hardness on a similar order to nacre and other bio-composites, exhibits limited creep, and demonstrates a nanoscale deformation that may associate to hardening. Our nacre-like carbonate composite with a small amount of polyelectrolyte on the surface can be potentially used as fillers in biocomposites with excellent bioactivity and high mechanical properties. Finally, we note that our nacre-like carbonate composite material mimicking some of the mechanisms of deformation and structure of nacre could lead to new high-performance materials, thereby greatly intensifying their range of application.

Acknowledgements

Research was carried out at the Center for Functional Nanomaterials, Brookhaven National Laboratory, which is supported by the U.S. Department of Energy, Office of Basic Energy Sciences (Contract No. DE-AC02-98CH10886). Use of the National Synchrotron Light Source, Brookhaven National Laboratory, was supported by the U.S. Department of Energy, Office of Science, Office of Basic Energy Sciences, Under Contract No. DE-AC02-98CH10886. RKP acknowledges financial support from the International Iberian Nanotechnology Laboratory (INL) in Braga, Portugal.

References

[1] E. M. Pouget, P. H. H. Bomans, J. Goos, P. M. Frederik, G. de With, N. Sommerdijk, *Science* **2009**, *323*, 1455.
[2] W. Fan, M. A. Snyder, S. Kumar, P. S. Lee, W. C. Yoo, A. V. McCormick, R. L. Penn, A. Stein, M. Tsapatsis, *Nature Materials* **2008**, *7*, 984.

[3] S. Mann, *Nature* **1993**, *365*, 499.
[4] F. C. Meldrum, H. Colfen, *Chemical Reviews* **2008**, *108*, 4332.
[5] J. D. Currey, J. D. Taylor, *Journal of Zoology* **1974**, *173*, 395.
[6] J. D. Currey, *Proceedings of the Royal Society of London Series B-Biological Sciences* **1977**, *196*, 443.
[7] A. P. Jackson, J. F. V. Vincent, R. M. Turner, *Proceedings of the Royal Society of London Series B-Biological Sciences* **1988**, *234*, 415.
[8] S. Kamat, X. Su, R. Ballarini, A. H. Heuer, *Nature* **2000**, *405*, 1036.
[9] R. Z. Wang, Z. Suo, A. G. Evans, N. Yao, I. A. Aksay, *Journal of Materials Research* **2001**, *16*, 2485.
[10] F. Barthelat, H. Tang, P. D. Zavattieri, C. M. Li, H. D. Espinosa, *Journal of the Mechanics and Physics of Solids* **2007**, *55*, 306.
[11] F. Song, A. K. Soh, Y. L. Bai, *Biomaterials* **2003**, *24*, 3623.
[12] E. Oroudjev, J. Soares, S. Arcdiacono, J. B. Thompson, S. A. Fossey, H. G. Hansma, *Proceedings of the National Academy of Sciences of the United States of America* **2002**, *99*, 6460.
[13] M. Suzuki, K. Saruwatari, T. Kogure, Y. Yamamoto, T. Nishimura, T. Kato, H. Nagasawa, *Science* **2009**, *325*, 1388.
[14] I. A. Aksay, M. Trau, S. Manne, I. Honma, N. Yao, L. Zhou, P. Fenter, P. M. Eisenberger, S. M. Gruner, *Science* **1996**, *273*, 892.
[15] L. J. Bonderer, A. R. Studart, L. J. Gauckler, *Science* **2008**, *319*, 1069.
[16] J. B. Thompson, J. H. Kindt, B. Drake, H. G. Hansma, D. E. Morse, P. K. Hansma, *Nature* **2001**, *414*, 773.
[17] B. L. Smith, T. E. Schaffer, M. Viani, J. B. Thompson, N. A. Frederick, J. Kindt, A. Belcher, G. D. Stucky, D. E. Morse, P. K. Hansma, *Nature* **1999**, *399*, 761.
[18] H. D. Espinosa, J. E. Rim, F. Barthelat, M. J. Buehler, *Progress in Materials Science* **2009**, *54*, 1059.
[19] F. Barthelat, *Bioinspiration & Biomimetics*, *5*, 035001.
[20] Z. Y. Tang, N. A. Kotov, S. Magonov, B. Ozturk, *Nature Materials* **2003**, *2*, 413.
[21] E. Munch, M. E. Launey, D. H. Alsem, E. Saiz, A. P. Tomsia, R. O. Ritchie, *Science* **2008**, *322*, 1516.
[22] P. Podsiadlo, A. K. Kaushik, E. M. Arruda, A. M. Waas, B. S. Shim, J. D. Xu, H. Nandivada, B. G. Pumplin, J. Lahann, A. Ramamoorthy, N. A. Kotov, *Science* **2007**, *318*, 80.
[23] S. Deville, E. Saiz, R. K. Nalla, A. P. Tomsia, *Science* **2006**, *311*, 515.
[24] B. R. Heywood, S. Mann, *Advanced Materials* **1994**, *6*, 9.
[25] A. Sellinger, P. M. Weiss, A. Nguyen, Y. F. Lu, R. A. Assink, W. L. Gong, C. J. Brinker, *Nature* **1998**, *394*, 256.
[26] Y. L. Hu, Y. J. Ma, Y. Zhou, F. D. Nie, X. H. Duan, C. H. Pei, *Journal of Crystal Growth*, *312*, 831.
[27] G.M. Luz, J.F. Mano, *Philosophical Transactions of the Royal Society a-Mathematical Physical and Engineering Sciences* **2009**, *367* (1893), 1587.
[28] G. E. Fantner, T. Hassenkam, J. H. Kindt, J. C. Weaver, H. Birkedal, L. Pechenik, J. A. Cutroni, G. A. G. Cidade, G. D. Stucky, D. E. Morse, P. K. Hansma, *Nature Materials* **2005**, *4*, 612.
[29] S. B. Rossmurphy, *Journal of Rheology* **1995**, *39*, 1451.
[30] M. Sjoberg, L. Bergstrom, A. Larsson, E. Sjostrom, *Colloids and Surfaces a-Physicochemical and Engineering Aspects* **1999**, *159*, 197.
[31] M. H. Yang, *Polymer Testing* **2001**, *20*, 635.
[32] R. S. Armentrout, C. L. McCormick, *Macromolecules* **2000**, *33*, 2944.
[33] S. H. Gehrke, *Advances in Polymer Science* **1993**, *110*, 81.
[34] T. Tanaka, I. Nishio, S. T. Sun, S. Uenonishio, *Science* **1982**, *218*, 467.
[35] R. K. Pai, S. Hild, A. Ziegler, O. Marti, *Langmuir* **2004**, *20*, 3123.
[36] R. K. Pai, S. Pillai, *Crystal Growth & Design* **2007**, *7*, 215.
[37] R. K. Pai, S. Pillai, *Journal of the American Chemical Society* **2008**, *130*, 13074.
[38] R. K. Pai, L. Zhang, D. Nykpanchuk, C.S. Korach, M. Cotlet, *Advanced Biomaterials* **2011**, Accepted for publication.
[39] K. Oaki, H. Imai, *Angewandte Chemie-International Edition* **2005**, *44*, 6571.
[40] M. Rousseau, E. Lopez, P. Stempfle, M. Brendle, L. Franke, A. Guette, R. Naslain, X. Bourrat, *Biomaterials* **2005**, *26*, 6254.
[41] A. W. Xu, M. Antonietti, H. Colfen, Y. P. Fang, *Advanced Functional Materials* **2006**, *16*, 903.
[42] M. Wang, *Biomaterials* **2003**, *24*, 2133.
[43] M. Wang, W. Bonfield, *Biomaterials* **2001**, *22*, 1311.
[44] W. C. Oliver, G. M. Pharr, *Journal of Materials Research* **1992**, *7*, 1564.
[45] A. Walther, I. Bjurhager, J. M. Malho, J. Pere, J. Ruokolainen, L. A. Berglund, O. Ikkala, *Nano Letters*, *10*, 2742.
[46] F. Barthelat, C. M. Li, C. Comi, H. D. Espinosa, *Journal of Materials Research* **2006**, *21*, 1977.
[47] K.S. Katti, D.R. Katti, S.M. Pradhan, A. Bhosle, *Journal of Materials Research* **2005**, *20* (5), 1097.
[48] B. Mohanty, K.S. Katti, D.R. Katti, *Mechanics Research Communications* *35* (1-2), 17.

Nanoscale Fracture Resistance Measurement of a Composite Bone Cement

M. Khandaker and S. Tarantini
Department of Engineering and Physics, University of Central Oklahoma, Edmond, OK 73034

ABSTRACT

R-curve characterizes the resistance to fracture of a propagating crack on a brittle material. Ceramic filler material can be added to the PMMA bone cement to improve its resistance to fracture of a propagating crack. The motivation of this study was to improve the fracture resistance of the composite bone cement by incorporating bioactive ceramic nanoparticles with the bone cement. The hypothesis of this study was that extrinsic toughening characteristics of composite cement affect bone-composite fracture toughness. Two specific objectives were achieved during the study: (1) the fracture properties of composite cements were evaluated using in-situ field emission scanning electron microscope (FESEM), (2) R-curve affects on the fracture toughness of composite cements were examined using in-situ FESEM based R-curve measurement techniques. ASTM standard bow tie single edge notch tension specimen was fabricated using commercial PolyMethylMethAcrylate (PMMA) bone cement, bone cement with MgO micro and nanoparticles. The in-situ FESEM SENT fracture tests were conducted on cement specimens at a speed of 1.1 µm/sec. using Evex SEM tensile test stage. This study found that the initiation stress intensity factor of the CBC specimen was higher compared to MgO additives incorporated CBC specimen. Higher energy is required for CBC specimen compared to the composite CBC specimens to propagate crack up to failure. MgO particles contribute to the origin of voids in the cement structure which affect the structural integrity MgO incorporated PMMA bone cement.

BACKGROUND

The debonding of the PMMA cement from bone or implant is frequently reported in literature [1-8]. In the case of total hip replacements, implant loosening occurs due to debonding of either the bone-cement or prosthetic-cement interfaces [2, 4]. A current trend of biomaterial research is focused on the use of nanomaterials to solve the debonding problem by improving the fracture properties of composite cements [9-13]. The suitability of a nanocomposite cement to solve the debonding problem requires complete understanding of the fracture behavior of bone-nanocomposite cement or prosthetic-cement interfaces. Though the incorporation of nanoparticles to conventional cements showed improved mechanical, biological and chemical properties [9-16], the influences of nanoparticles on fracture toughness of bone-nanocomposite interfaces is not understand yet. The problem of debonding between bone and conventional cement cannot be solved just by enhancing the mechanical properties of the cement by incorporating nanoparticles to cement. The debonding problem can be solved by proper selection of the nanomaterials and accurate controlling the surface area of interface in such a way that these nanomaterials can provide extra resistance to the propagation of the crack at bone-nanocomposite or prosthetic-cement interfaces.

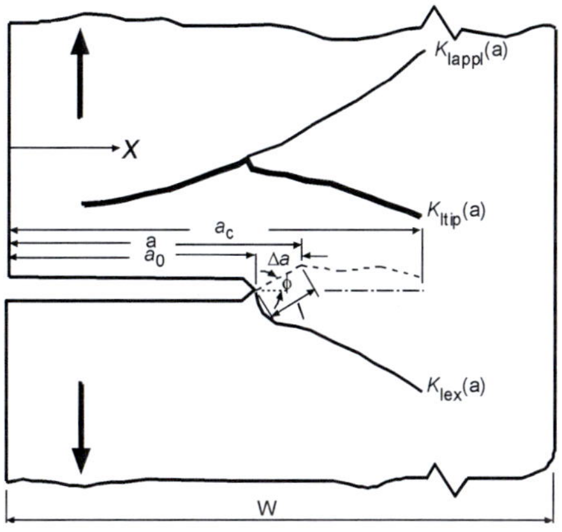

Figure 1 Representation of a typical R-curve for crack extension length, Δa. In the figure, K_{Iappl}, K_{Iex} and K_{Itip} are the applied, extrinsic and total crack tip SIF, respectively. a_0 is pre-crack length at which $K_{Itip} = K_{I0}$ and ac critical crack length at breakage where $K_{Itip} = K_{IC}$. K_{I0} and K_{IC} is known as initiation and fracture toughness, respectively.

In brittle material, the behavior of increased fracture toughness with crack propagation is known as fracture toughening or R-curve behavior [17]. Figure 1 shows a typical R-curve for a brittle material. Recently, several research groups reported improvement of the fracture toughness of conventional bone cements by incorporating nanoparticles to cement [18, 19]. Chan research group [16] reported improved fracture toughness of bioactive composite cement (bis-GMA based monomer blend) by addition of Schott glass and SiO2 nanoparticles.

T. Proulx (ed.), *Mechanics of Biological Systems and Materials, Volume 2*, Conference Proceedings of the Society for Experimental Mechanics Series 9999, DOI 10.1007/978-1-4614-0219-0_19,

Fracture toughening behavior (e.g. particle bridging, debonding at the poles of particle/matrix interface, crack trapping and crack deflection around the particles) was identified as the cause for this improved fracture toughness [16]. But the improvement of interface fracture toughness of composite bone cements due to incorporation of bioactive nanoparticles to cement is not known yet.

Research found that size, quantity and dispersion of additive particles in cement controls the fracture properties of composite cements [18]. Effects of specimen geometry (e.g. width, thickness, initial crack length) on fracture toughness of various bones were reported in literature [24-27]. The type of loading (tensile, bending) used in mechanical testing also affects fracture toughness of bone-cement interfaces [28]. The early part of the R-curve (Figure 1), typically less than 500 μm in length, determines the reliability and strength of bone and cement materials. It is difficult to measure this length accurately using conventional microscopic (e.g. micro Raman spectroscopy, Confocal Microscope, Epifluorescence Stereo- microscope). Accurate measurement of R-curve is reported in literature using microtensile or microbend tester by in-situ environmental field emission scanning electron microscope (FESEM) [29, 30]. Also, digital image correlation (DIC) method was utilized by researchers to evaluate crack tip displacement and stress field in biological tissues [31-33]. In this study in-situ FESEM experimental technique will be used to measure the mixed mode fracture resistance curve of various composite bone cements.

The most challenging issue associated with commercially available PMMA bone cements such as Cobalt (Biomet, Inc.), Simplex (Stryker, Inc.), Palacos (Heraeus company) for the application of total joint replacement is their poor osseointegration (incorporation of the cement with surrounding bone tissues) [5]. Problems about infection and loosening of the bone cements at the bone-cement interface have been reported in literature [6]. One way to reduce infection and loosening would be to promote osteoblast cell growth around the cemented surfaces. Such cells can eliminate contact between the bone and the environment and restricting contamination at the cemented prosthetic joint. Another way to reduce loosening would be to increase mechanical interlock between bone and cement [7-8]. This can be done by enhancing the surface roughness of the PMMA cement. Several research groups found improvement of PMMA bone cements surface roughness properties by incorporating different kinds of bioactive filler materials with the PMMA cement [6, 9-12]. Recently researchers found MgO, Silver, hydroxyapatite nanoparticle filler materials improve bone healing properties of conventional PMMA bone cements [6, 9-10, 12-14]. The influences of the inclusion of these additives with bone cement on the fracture strength were not investigated yet. Such study is required for the suitability of using the additives with the bone cement. Ricker et al.[12] research on PMMA cement showed increased surface roughness, and enhanced cell adhesion of mouse osteoblast cell on PMMA due to the inclusion of MgO additives with PMMA. The suitability of incorporating MgO additive with bone cement requires complete understanding of the failure mechanism of bone/MgO additives incorporated bone cement interfaces. No study has been conducted to evaluate the effect of MgO additives on the mechanical integrity between bone and PMMA cement with MgO additives. In this present study, three kinds of bond cement were investigated. Cobalt™ HV bone cement (referred in this literature as CBC), a commercial orthopedic bone cement, was used as control PMMA bone cement. CBC with 36 μm and 22 nm sizes MgO powders were used as the experimental bone cement specimen.

MATERIALS AND METHODS

SAMPLE PREPARATION

According the Biomet, Inc recommendation, 10 grams of poly Methyl MethAcrylate (PMMA) beads was as added to 5 ml of benzoyl peroxide monomer to prepare the Cobalt" HV bone cement (CBC) specimen. Both 36 μm and 22 nm sizes MgO powders were purchased from Sigma-Aldrich. In order to create samples of μMgO-CBC and nMgO-CBC specimens, 10 percent (w/w) of the 36 μm and 22 nm size MgO additives powders (0.5 gm) were mixed with the PMMA (4.5 gm) and added to the monomer (5 ml). A glass mold as shown in Figure 2(a) was used to prepare different kinds of dog bone bone-cement blocks. The cement was added to the top half of the glass molding chamber while the cement was in its doughy phase. A set of weights equivalent to 80 kPa pressure (clinically applied range [17]) were applied to the samples during the curing process. The pressure was initiated at exactly three minutes after the onset of mixing and was sustained throughout the curing period. The prepared specimens are shown in Figure 2(b). The center notch was created using a low-speed diamond saw cutter (Buehler isomet 11-1180-100). All specimens were prepared according to ASTM 399 standard.

(a)

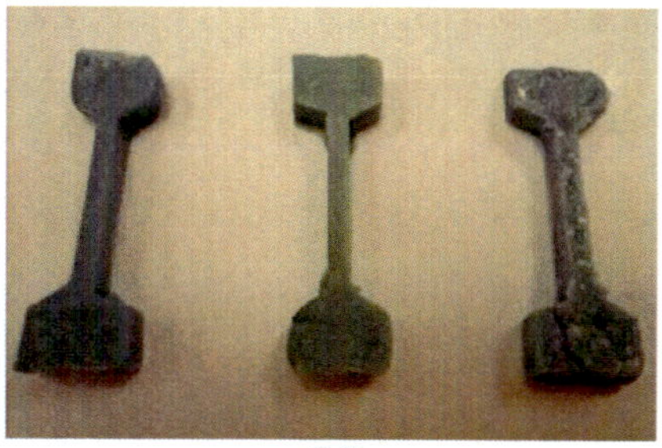

(b)

Figure 2 (a) Fabricated dog bone specimen mold for preparing cement specimens. The mold was made using dimension elite 3D printer. (b) Prepared ASTM 399 standard CBC, nMgO-CBC and μMgO-CBC dog bone cement samples for the mechanical characterization of the samples.

EXPERIMENT

In situ FESEM Tensile test was performed on SENT specimen using a Microtensile tester (Gatan 2KN SEM microtensile tensile test stage) and FEI Quanta 600 FESEM. The deformation process for each specimen was recorded using Veeco Multimode V Scanning Probe Microscope (AFM+STM). All specimens were loaded with a loading rate equal to 1.1 μm/sec until the complete failure of the specimen. This environmental FESEM is equipped with an extremely bright electron source (FEG) that is capable of nano-scale imaging of hydrated samples at atmospheric pressure. Images of the crack path and extension length, Δa, will be obtained simultaneously in backscattering mode at 15 kv and a pressure of 35 Pa by the energy dispersive spectrometer (EDS) and an electron backscattered diffraction system. The images were further analyzed to measure the crack extension length using Nikon NIS element BR image processing software. Figure 3 shows the experimental setup used in this study.

(a)

(b)

Figure 3 Dog bone specimen clamped in the Evex tensile stage. The specimen were gold platted before the experiment on field emission scanning electron microscope at the OSU Microscopy Lab, Stillwater, and (b) Evex tensile stage is assembled in the FESEM chamber to conduct the *R*-curve experiment.

DATA ANALYSIS

The stress intensity factors at the tip of a kink crack of length *l* as shown in Figure 1 can be written for $l \ll a_0$ as:

$$K_I(\alpha,a) = \sigma F_I(\alpha)\sqrt{a\pi} \tag{1}$$

$$K_{II}(\alpha,a) = \sigma F_{II}(\alpha)\sqrt{a\pi} \tag{2}$$

where *a* is the crack tip length, Δa is the crack extension length, σ is the applied stress, α is the ratio between crack tip length, a and width of the specimen, W, F_I and F_{II} are geometric functions calculation using the following equation:

$$F_I(\alpha) = F_t(\alpha)\cos^3\left(\varphi/2\right) \tag{3}$$

$$F_{II}(\alpha) = F_t(\alpha)\sin\left(\varphi/2\right)\cos^2\left(\varphi/2\right) \tag{4}$$

where φ is the kink crack angle for a crack extension length Δa. In Eqs. (3) and (4), $F_t(\alpha)$ can be found from the following equation

$$F_t(\alpha) = \frac{1.1215}{(1-\alpha)^{3/2}}\left[1-0.23566(1-\alpha)+\frac{1}{150}(1-\alpha)^7+0.229\exp\left(-7.52\frac{\alpha}{1-\alpha}\right)\right] \tag{5}$$

The effective stress intensity for the deflected crack can be found from the following equation:

$$K_{eff} = \sqrt{K_I^2 + K_{II}^2} \tag{6}$$

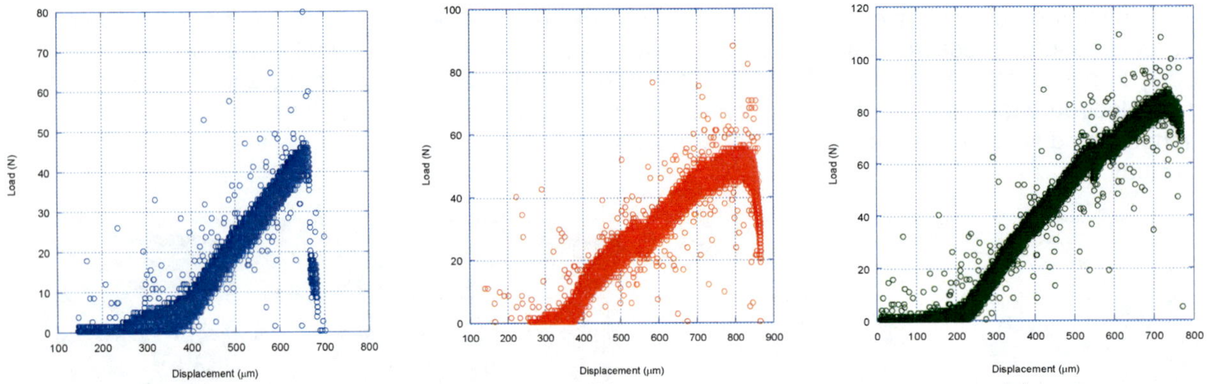

(a) (b) (c)

Figure 4 The graph represents the typical load applied versus the displacement of (a) CBC, (b) µCBC, and (c) nCBC specimens tested in the FESEM.

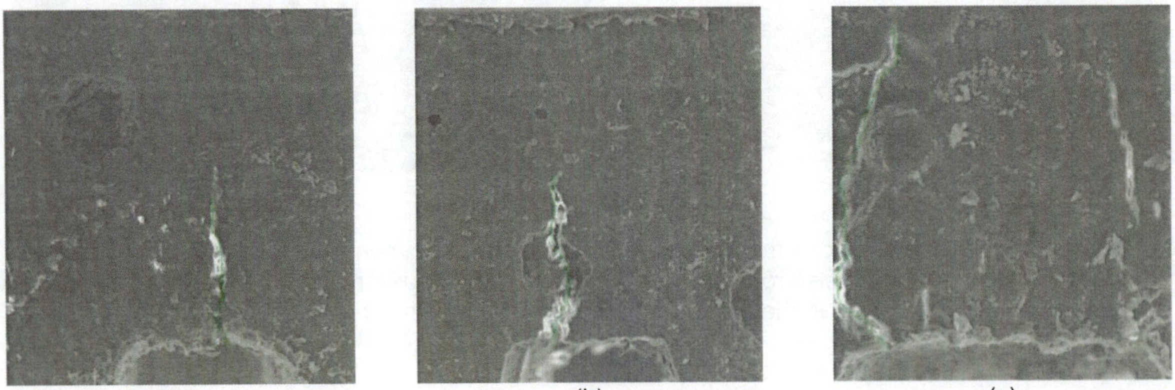

(a) (b) (c)

Figure 5 FESME images reveal the real time crack propagation of the specimens (a) CBC, (b) µCBC and (c) nCBC specimen. The green line in the images represents the crack propagation paths.

RESULTS AND DISCUSSION

Objective 1: Evaluate the fracture properties of composite cements using in-situ field emission scanning electron microscope (FESEM): Our studies showed a clear difference of the load-displacement curve among different kinds of bone cement specimens (Figure 4). The result shows the bone cement with no MgO additives has higher stiffness compared to cement with MgO additives. The result also shows that bone cement incorporated with micro and nano MgO additives has the ability to store higher energy before reaching a breaking point than bone cement without MgO additives. Comparing the properties of different kinds of CBC specimens, we found improved brittleness of CBC due to the inclusion of MgO micro/nano additives to CBC. But we did not find any noticeable improvement of integrity, mineralization, and amount of energy before breaking for CBC specimen due to the inclusion of MgO micro/nano additives to CBC. The values of the fracture strength of CBC found lower compare to MgO incorporated CBC specimen. Benzyl peroxide monomer penetrates deep inside the PMMA beads in case CBC. The large molecular weight (M_W) of the PMMA beads (M_W = 100000) made it difficult for the PMMA beads to mix uniformly with the monomer. It creates voids, which are the weak points in the cement. The weak points in the cement lead to stress concentration under external loading. Inclusion of MgO additives to CBC reduces its elastic properties compare to CBC due to the fact that particles with higher MW like PMMA dissolves in monomer slower than lower MW particles like MgO (M_W = 40.3, provided by the manufacturer). This decreases the amount of monomer available for wetting the PMMA powder during mixing, which leads to more weak points in the μMgO-CBC specimens. The detrimental effect on fracture properties of μMgO-CBC and nMgO-CBC has been associated with the lack of adhesion between the MgO filler materials and the polymeric matrix. Fracture originates at the grain boundary between the MgO additives particles and the polymeric matrix and propagates along the boundary of the polymeric matrix as shown in Figure 5. The non-catastrophic failure behavior of the μMgO-CBC and nMgO-CBC specimens (Figure 5) indicates the presence of fracture toughening mechanism. Extrinsic toughening factors, like particle bridging, debonding at the poles of particle/matrix interface, crack trapping, and crack deflection around the particles, can be the cause for this improved fracture toughness.

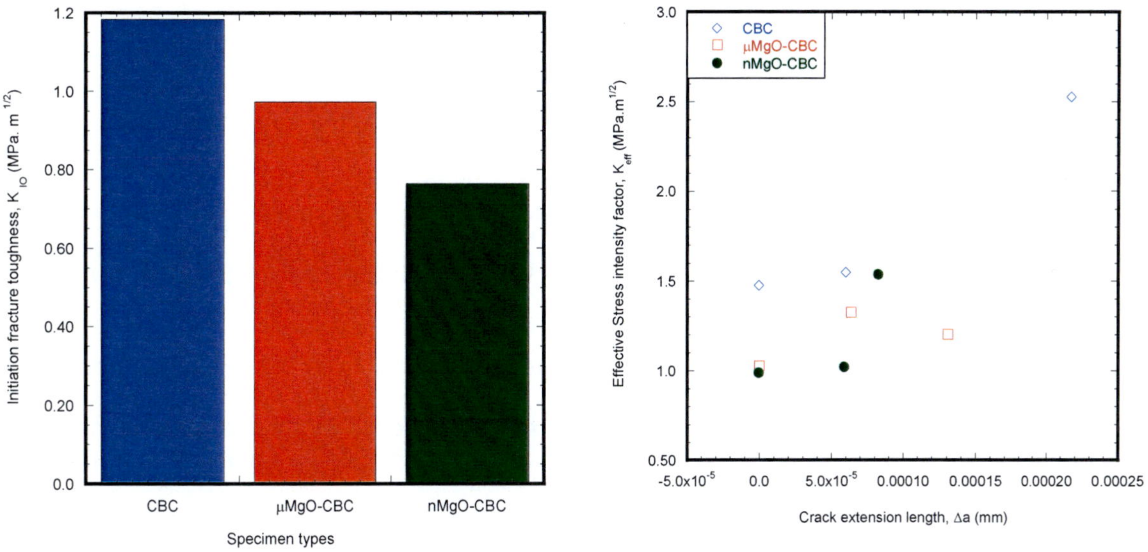

Figure 6. (a) The initiation fracture toughness of various kinds of bone cement tested during this study. (b) Increase of effective Stress intensity factor with the increase of crack extension length for various kinds of bone cement.

Objective 2: Measurement of R-curve affects of composite bone cements: Figure 6(a) shows the variation of initiation fracture toughness calculated from the load-displacement curve of three groups of bone cement tested during this study. The results show that the initiation fracture toughness of the CBC specimen was higher compared to MgO additives incorporated CBC specimen. All specimen shows extrinsic toughening behavior or R curve behavior. R-curve characterizes the resistance to fracture of a propagating crack. SEM images of the crack propagation path reveled distinct R-curve for different kinds of cements (Figure 5). Higher energy is required for CBC specimen compared to the composite CBC specimens to propagate crack up to failure. MgO particles contribute to the origin of voids in the cement structure which affect the structural integrity MgO incorporated PMMA bone cement.

CONCLUSIONS AND FUTURE WORKS

This research concludes the followings: (1) inclusion of 10% (w/w) micro/nano sizes MgO particles in commercial Cobalt™ HV orthopedic bone cement (CBC) has negative effect on the fracture properties of CBC; and (2) inclusion of MgO addition to bone cement increase the fracture resistance though the fracture resistance is higher for CBC specimen compare to MgO incorporated CBC specimen. These findings conclude that incorporating 10% (w/w) micro/nano sizes MgO additives into CBC negatively affect on the fracture properties of PMMA cement, but positively affect the fracture resistance behavior. The improvement of fusion between bone and MgO included PMMA is currently under investigation. The detrimental effect of addition of MgO in to PMMA cement may supersede the benefit of increase of fracture resistance when MgO added to PMMA as well as the increase of osteoblast function of MgO incorporated bone cement compare to PMMA cement only. Further study is required to evaluate whether addition of MgO filler materials adversely affect the longevity of the cemented prosthetic joint. Also the modeling the fracture behavior of the bone-composite cement is currently ongoing.

ACKNOWLEDGEMENTS

This publication was made possible by Grant Number P2PRR016478 from the National Center for Research Resources (NCRR), a component of the National Institutes of Health (NIH) and on-campus research grants support from University of Central Oklahoma (UCO). Its contents are solely the responsibility of the authors and do not necessarily represent the official views of NCRR or NIH or UCO.

REFERENCES

1. Moreo, P., Perez M.A., Garcia-Amar J.M., Doblare M., "Modelling the mixed-mode failure of cement-bone interfaces," Engineering Fracture Mechanics, **73**(10), 1379-1395 (2006).
2. Lucksanasombool, P., Higgs W.A.J., Higgs R., Swain M.V., "Interfacial fracture toughness between bovine cortical bone and cements," Biomaterials, **24**(7), 1159-1166 (2003).
3. An, Y.H., Draughn R.A., Mechanical testing of bone and the bone-implant interface. CRC (2000).
4. Wang, X.D., Agrawal C.M., "A mixed mode fracture toughness test of bone-biomaterial interfaces," Journal of Biomedical Materials Research, **53**(6), 664-672 (2000).
5. Mann, K.A., Damron L.A., "Predicting the failure response of cement-bone constructs using a non-linear fracture mechanics approach," Journal of Biomechanical Engineering-Transactions of the Asme, **124**(4), 462-470 (2002).
6. Ichim, I., Li Q., Li W., Swain M.V., Kieser J., "Modelling of fracture behaviour in biomaterials," Biomaterials, **28**(7), 1317-1326 (2007).
7. Kopperdahl, D.L., Roberts A.D., Keaveny T.M., "Localized damage in vertebral bone is most detrimental in regions of high strain energy density," Journal of Biomechanical Engineering-Transactions of the Asme, **121**(6), 622-628 (1999).
8. Shi, Z., Neoh K.G., Kang E.T., Wang W., "Antibacterial and mechanical properties of bone cement impregnated with chitosan nanoparticles," Biomaterials, **27**(11), 2440-2449 (2006).
9. Heo, S.J., Park S.A., Shin H.J., Lee Y.J., Yoon T.R., Seo H.Y., Ahn K.C., Kim S.E., Shin J.W., "Evaluation of bonding stress for the newly suggested bone cement: Comparison with currently used pmma through animal studies," Key Engineering Materials, **342-342**, 373-6 (2007).
10. Liu, H., Webster T.J., "Nanomedicine for implants: A review of studies and necessary experimental tools," Biomaterials, **28**(2), 354-369 (2007).
11. Ricker, A., Liu-Snyder P., Webster T.J., "The influence of nano mgo and baso4 particle size additives on properties of pmma bone cement," International Journal of Nanomedicine, **3**(1), 125-1 (2008).
12. Sirinrath, S., Thomas J.W., "Multiwalled carbon nanotubes enhance electrochemical properties of titanium to determine in situ bone formation," Nanotechnology, **19**(29), 95101-95101 (2008).
13. Lewis, G., "Alternative acrylic bone cement formulations for cemented arthroplasties: Present status, key issues, and future prospects," Journal of Biomedical Materials Research - Part B Applied Biomaterials, **84**(2), 301-319 (2008).
14. Khang, D., Kim S.Y., Liu-Snyder P., Palmore G.T.R., Durbin S.M., Webster T.J., "Enhanced fibronectin adsorption on carbon nanotube/poly(carbonate) urethane: Independent role of surface nano-roughness and associated surface energy," Biomaterials, **28**(32), 4756-4768 (2007).

15. Sirinrath, S., Chang Y., Xingcheng X., Brian W.S., Thomas J.W., "Greater osteoblast functions on multiwalled carbon nanotubes grown from anodized nanotubular titanium for orthopedic applications," Nanotechnology, **18**(36), 65102-65102 (2007).
16. ASTM, "Annual book of astm standards. Section 8," in D790-03 Standard Test Methods for Flexural Properties of Unreinforced and Reinforced Plastics and Electrical Insulating Materials. Philadelphia, PA, 2006.
17. Graham, J., Ries M., Pruitt L., "Effect of bone porosity on the mechanical integrity of the bone-cement interface," Journal of Bone and Joint Surgery-American Volume, **85A**(10), 1901-1908 (2003).
18. Cowin, S.C., Bone mechanics handbook, (2001).
19. Lucksanasombool, P., Higgs W.A.J., Higgs R., Swain M.V., "Fracture toughness of bovine bone: Influence of orientation and storage media," Biomaterials, **22**(23), 3127-3132 (2001).
20. Zou, L., Huang Y., Wang C.-a., "The characterization and measurement of interfacial toughness for si3n4/bn composites by the four-point bend test," Journal of the European Ceramic Society, **24**(9), 2861-2868 (2004).

Impact Resistance of Antibiotic-Impregnated Orthopedic Bone Cement

Samaneh Choupani, BSc, Ata Hashemi, PhD
Biomedical Engineering Department, Amirkabir University of Technology, Hafez Ave., Tehran, Iran

ABSTRACT

The usess of antibiotic--impregnated bone cement in total joint replacement (TJR) is a common practice but remains extremely variable. The benefit of adding antibiotics to bone cement as a method of infection prophylaxis in TJR needs to be weighed against a number of drawbacks primarily a drop in the mechanical properties of the bone cement. The present study utilized the pendulum impact strength testing to report on the properties of antibiotic-impregnated cement. All testing was conducted in accordance with ASTM D256-06. Test groups included: PMMA mixed with vancomycin, gentamicin, erythromycin or no antibiotic (control). Also, the effect of aging samples in air versus saline on the impact strength was studied. The effect of antibiotic addition to cement on the impact strength was found statistically significant ($p < 0.05$). The Tukey's HSD test indicated that the liquid Gentamicin impregnated samples aged in air had significantly lower mean impact strength than all other groups ($p < 0.05$). The mean impact strength of the vancomycin added samples aged in saline was lower than other samples aged in saline but is not statistically significant ($p > 0.05$).

INTRODUCTION

Bone cements have been used to anchor artificial joints (hip, knee, shoulder and elbow joints) since 1950 due to its excellent tissue compatibility. Bone cements are provided as two-component materials, a powder (polymethyl-methacrylate or PMMA, initiator) and a liquid (methyl-methacrylate or MMA monomer, inhibitor). As the two components mix, its' viscosity changes over time from a runny liquid into dough like state and finally hardens into solid hardened material. The set time can be customized to help the physician safely apply the cement into the bone to either anchor metal implant to bone or to treat osteoporotic compression fractures in the spine.

A common practice in total joint replacement (TJR) surgery, particularly with re-implantation of components following a previously infected total joint arthroplasty is to employ the antibiotic-impregnated cement for component fixation. The combination of several factors such as, microbial sensitivity, acceptable mechanical integrity, low resistance profiles, and preparation methods could affect the choice of antibiotic-cement combination [1]. In 1995, a survey showed that the majority of surgeons in the continental United States preferred tobramycin, followed by gentamicin, vancomycin, cephalosporins, and rarely combinations thereof [2]. Also, the survey indicated that about 11% of surgeons were still using liquid antibiotic in PMMA [2].

The long-term survival of TJR would require an accurate determination of the properties of surgical bone cement. Many biomechanical studies have investigated the effects of adding various types of antibiotics in bone cements on its mechanical properties [3–7]. Cement testing studies have incorporated different testing methodologies and conditions resulting in a wide range of reported findings. The cement strength studies have been based on uniaxial, biaxial and fatigue testing methodologies using tension, flexural, or compressive testing [6–20]. Among the reported properties, the impact strength of bone cement has received very little attention; with the only reports are those by Ungethulum and Hinterberger [21], Ege [22] and Lewis and Mladsi [23]. The latter study developed a bench-top in-house impact tester to determine the impact strength of four different bone cements, i.e., Surgical Simplex® P (Stryker Howmedica Osteonics Corp., Rutherford, NJ) and three forms of Palacos® R (Smith & Nephew Inc., Memphis, TN), one in which the powder constituents had been sterilized by the manufacturer and the other two had been sterilized with different dozes of gamma radiation. Based on the correlation found between the fracture toughness and impact strength of the four different bone cements investigated, it was suggested to estimate the relative change in the fracture toughness of bone cement from the measured relative change in its impact strength [23].

Although the effect of factors such as sterilization method, test specimen configuration and the type of cement on the impact strength was reported [21-23], to our knowledge the effects of adding different antibiotics to cement has not specifically been

T. Proulx (ed.), *Mechanics of Biological Systems and Materials, Volume 2*, Conference Proceedings of the Society for Experimental Mechanics Series 9999, DOI 10.1007/978-1-4614-0219-0_20,

investigated. Therefore, the objective of the present work was set to accurately measure the change in the impact strength of several commonly used antibiotic-cement configurations. Also, the effect of aging samples in air versus saline on the impact strength was studied. The pre-testing hypothesis was that the addition of powder or liquid antibiotics to the cement may lower the impact strength. It was also thought that liquid antibiotics are likely to produce a stronger effect.

MATERIAL & METHODS

Preparation of Specimens

The materials used in the present work are listed in Table I. The proposed concentrations of individual antibiotics reflect those generally used in clinical practice for component re-implantation following a previously infected arthroplasty. Four different test groups have been specified using antibiotics outlined in Table I. A control test group with no antibiotics has been also used.

A total of 32 samples were prepared for testing purposes. Simplex® P bone cement components were stored at ambient temperature (23 ± 2°C, relative humidity of 50 ± 5%) for at least 24 h before hand-mixing at about 1 Hz. First the powder component poured into a bowl and antibiotics were added (erythromycin antibiotic were premixed by the manufacturer) then the liquid monomer were added and manually mixed for about 1-2 minutes until a homogeneous mixture was obtained. This was then quickly transferred into a casting. The casting was made of carbon steel and measured 136.8 mm in length, 86 mm in width and 42.7 mm in depth (Figure 1). Following polymerization of the cement, the hardened cement sample was taken out from the casting and kept in the ambient temperature to cure prior to the testing. All preparations were performed in similar fashion.

A light spray of food grade silicone was applied to the casting surface to facilitate the removal of test specimens after they solidified. Pilot data confirmed that the silicone spray did not change the impact strength of the test specimens, as reported in other works [7]. The final configuration of each sample resembled a rectangular block measuring 63.5x12.7x12.7 mm. Figure 2 presents the schematic of test specimen. The casting was fabricated to allow the manufacturing of four molds per casting at a time. Tow samples were discarded because of gross molding flaws due to a partial filling of the final casting slot. All samples groups were duplicated with one half of the samples stored in room air and the second half were stored in normal saline. Since the mechanical properties of bone cement have been previously reported to peak one week after preparation [11], all test groups were maintained in their respective storage environments for two weeks before subsequent testing.

Experimental Procedures

All testing was conducted in accordance with ASTM D256-06, the standard test methods for determining the pendulum Impact resistance of notched specimens of plastics [24]. To carry out the test, individual samples were first placed on the sample platform and centered. An impact tester with the capability of delivering energy of 3J is used to perform the tests. The specimens were carefully examined to ensure that the nominal values are in the allowable range. Also, only samples with no surface defects were considered for testing. The tests were carried out on a bench-top impact testing machine (Ueshima IM-1100, Tokyo, Japan).

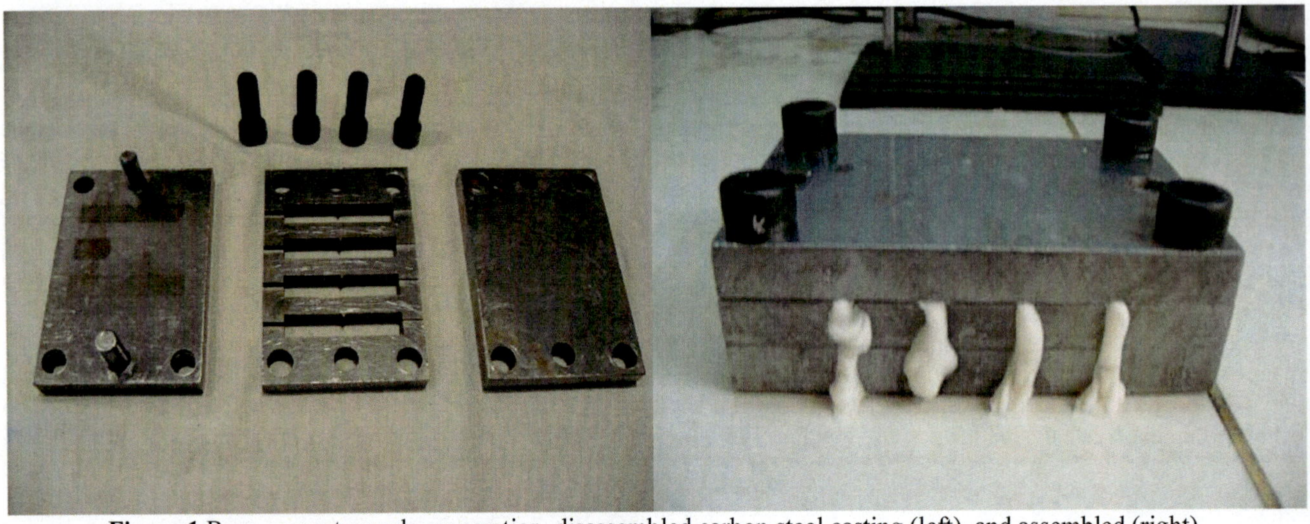

Figure.1 Bone cement sample preparation, disassembled carbon steel casting (left), and assembled (right)

Table.1Materials used in this study

Material	Composition		Mixing ratio
	Powder	Liquid	
Simplex	Methyl methacrylate-styrene copolymer, Barium sulphate and Polymethyl-methacrylate	Methyl methacrylate (monomer), N,N-dimethyl para toluidine, and hydroquinone, USP	40 g/20 mL (powder/liquid)
Vancomycin	1.0 g	-	1.0 g/40 g cement
Gentamicin	-	2.0 mL	2.0 mL/40 g cement
Erythromycin	0.5 g	-	0.5 g/40 g cement (pre-mixed)

Statistical Analysis

A computer based statistical package is used to perform statistical analysis. Data was initially analyzed with a 2-way analysis of variance (ANOVA). The impact strength was the measured or dependent variable. The independent variables were the antibiotic management and the storage condition. A Tukey-Kramer honesty significant difference (HSD) *post hoc* test was used to determine significant differences among the results in each test group. A p-value of <0.05 was taken as significant. It was predicted that the statistical analysis would answer if the addition of antibiotics to bone cement changes the impact strength and also, if the cement storage condition affects the impact strength.

(a)

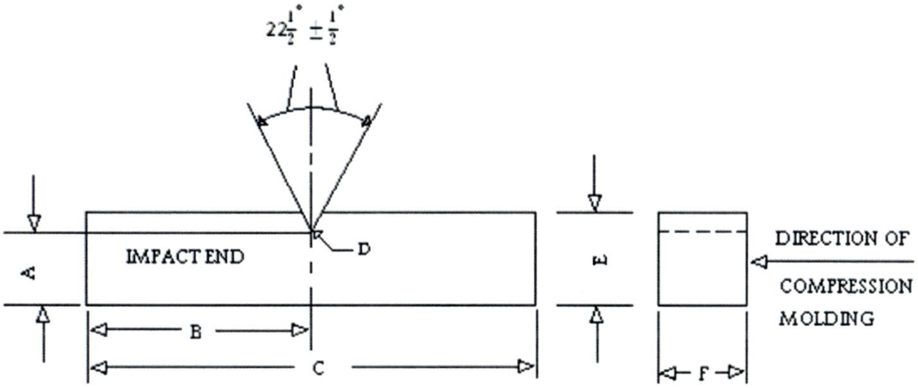

	mm	in
A	10.16 ± 0.05	0.400 ± 0.002
B	31.8 ± 1.0	1.25 ± 0.04
C	63.5 ± 2.0	2.50 ± 0.08
D	0.25R ± 0.05	0.010R ± 0.002
E	12.70 ± 0.20	0.500 ± 0.008
F	12.70 ± 0.20	0.500 ± 0.008

(b)

Figure.2 Photograph of the bone cement sample (a), dimensions of the bone cement sample according to ASTM D256-06 (b)

RESULTS

At failure, samples broke into two pieces and impact strength was recorded for each individual cement samples. The specimens were then individually examined under a microscope to quantify the size and frequency of air pockets or voids (Figure 3).

Voids in the bone cement specimens at fractured surface were present mainly due to fact that manual mixing method was adopted. However, the presence of the voids has been also reported using the vacuum and shaker table mixing techniques [7]. Visual inspection (10x magnification) of fractured surfaces, except those discarded for molding flaws, demonstrated that the liquid Gentamicin samples contained mostly smaller voids (<1 mm) and fewer larger voids (>3 mm) in comparison to the powdered antibiotic test groups. All test groups showed a decrease in the frequency of voids as the void size increased. It was also noticed that only one sample had pores directly at the notch of the sample. Since this is where the failure first initiated, it could indicate that the pores have no or a negligible effect on the impact strength of samples. The visual inspection of the fractured surface of the samples also demonstrated a brittle failure.

Figure 4 shows the measured impact strength of antibiotic-impregnated bone cement samples aged in air or saline. The impact strength of the control group aged in air and saline was found to be 1.6 ± 0.06 and 2.05 ± 0.25 KJ/m^2, respectively. For liquid Gentamicin impregnated group aged in air, it was 1.27 ± 0.09 and in Saline was 2.02 ± 0.17 KJ/m^2. The impact strength of the vancomycin added bone cement samples was 1.54 ± 0.18 and 1.92 ± 0.15 KJ/m^2, respectively, for specimens aged in air and in saline. And for erythromycin it was 1.92 ± 0.07 in air and 2.14 ± 0.44 in Saline. Two-way ANOVA showed a significant dependence on storage condition ($p < 0.05$). Similarly, the effect of antibiotic addition to cement on the impact strength was also found statistically significant ($p < 0.05$). The Tukey's HSD test indicated that the liquid Gentamicin impregnated samples aged in air had significantly lower mean impact strength than all other groups ($p < 0.05$) as shown in

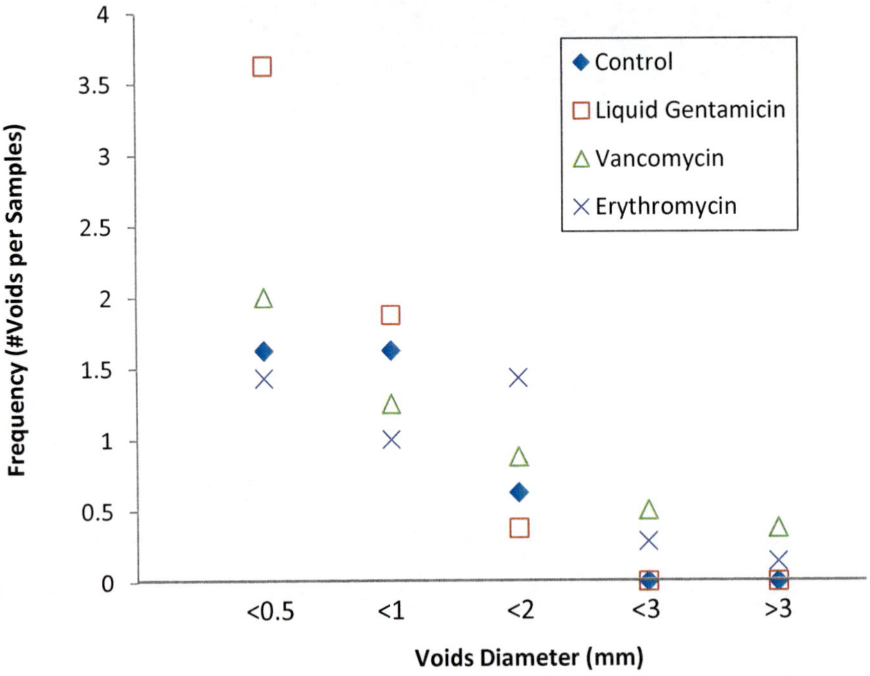

Figure.3 Frequency and size of voids observed on the fractured surfaces

Figure 4. The erythromycin added samples aged in air had higher mean impact strength than all other groups aged in air ($p < 0.05$) but that is not statistically significant with other groups aged in saline ($p > 0.05$). Figure 5 presents the proportionate change in the impact strength of each sample group as a comparison to the control (bone cement without antibiotic) following the addition of antibiotic.

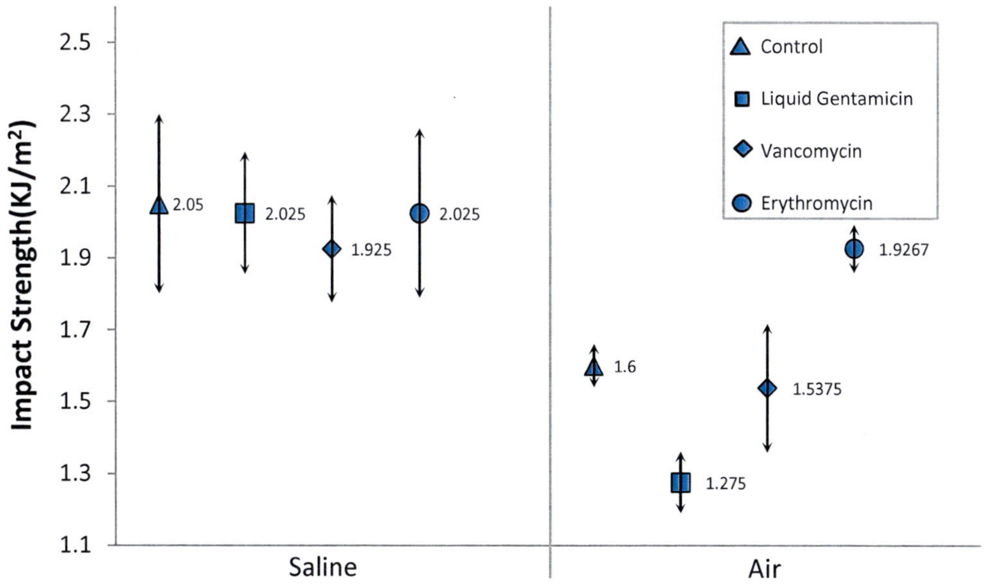

Figure.4 Impact strength of antibiotic-impregnated bone cement samples aged in air or saline.

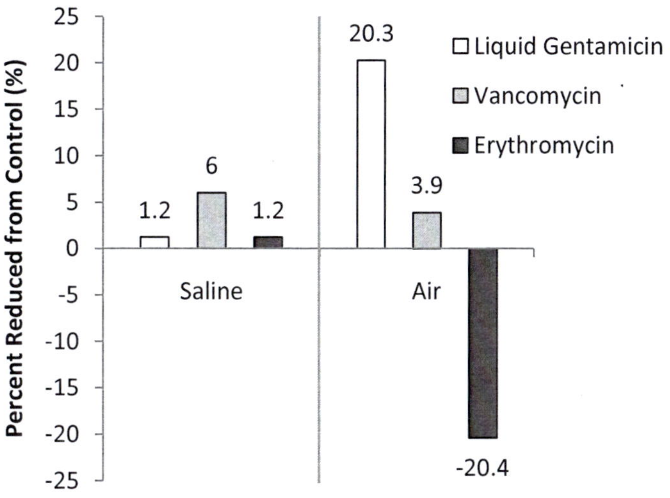

Figure. 5 Percent reduction in the impact strength (compared to control bone cement) due to the addition of antibiotics, aging media is Air or Saline.

DISCUSSIONS

The use of antibiotic-impregnated bone cement in total joint arthroplasty remains extremely variable.[2] The main shortcomings of adding antibiotics to bone cement could be a change in the mechanical integrity of the bone-cement interface, the emergence of antibiotic-resistant bacteria, and the occurrence of an allergic or toxic reaction to the antibiotic. [7,25,26] Therefore, the use of antibiotic-cement combination may only be proposed for cases such as re-implantation of components or primary joint replacement in patients with an increased risk of joint infection. The uni-axial, bi-axial and flexural testing methods have been employed to report the effect of antibiotic addition on bone cement mechanical properties. [3, 5-7] Furthermore, a few study reports the impact strength of bone cement but no data is currently available on the impact strength of antibiotic impregnated bone cement. [21, 22, 23] The focus of the current biomechanical study was, hence, set to determine the change in impact strength of commonly used antibiotic-cement configurations. For this purpose, the pendulum impact resistance testing method, as described in ASTM D256-06, was accurately employed to measure the impact strength of different bone cement configurations. In addition, the effect of storage condition (air vs. saline) on the impact strength was determined. It was reported that a change in the mechanical properties of orthopedic bone cement greater than 10% is considered unacceptable for use in TJR fixation [1].

Two-way ANOVA indicated that the addition of the antibiotic to bone cement lowers the impact strength ($p<0.05$). A similar significance was also noticed when samples aged in air vs saline ($p<0.05$). Others reported the dependence of impact strength on sample size, method of sterlization, cement type and radiation doze. [22, 23] Despite the efforts spent on mixing and proper transferring the cement to the casting, the results indicate slightly higher void per sample than that reported by vacuum and shaker table mixing techniques [7]. It could be proposed that mixing method mainly increases the number of small and mid size voids per sample (2 mm and smaller) while has almost no effect on the number of larger voids (3mm and larger). Visual inspection confirmed that the liquid gentamicin samples contained mostly smaller voids (<1 mm) and fewer larger voids (>3 mm) in comparison to the powdered antibiotic test groups, in agreement with the work of Leone et al. [7] The number of larger size voids was also less than that of smaller voids.

The impact strength of cement has been reported to vary from 2.4 to above 4 KJ/m^2. Also, higher numbers was reported for the impact strength of Palacos bone cement than that of Simplex cement, considered in this study. The impact strength range was measured to be 2.8-4.05 KJ/m^2 and 2.42-3.08 KJ/m^2 for Palacos and Simplex bone cements, respectively. [23] The higher impact strength of Palacos R is proposed to be due to its higher molecular weight. The impact strength of the control samples (with no antibiotics) was measured to be 1.6 ± 0.06 and 2.05 ± 0.25 KJ/m^2 which is below the avialable literature data. The previous studies used smaller sample size than that of the present work. The cross-sectional area of samples used at the present study was about 129 mm^2 while that reported in literature from one source was 18 mm^2, which used an in house impact tester machine, and another source reporting a cross sectional area of 12-45 mm^2. [23, 22] Both studies had considerably smaller cross-sectional area than that of this work which may explain the higher impact strength numbers measured.

Briefly, this was the first study investigating the impact strength of antibiotic-cement combination. The largest drop in the impact strength was found for the samples containing liquid gentamicin aged in the air, about 20% (Figure 5). Based purely on the present test results, powdered erythromycin at clinically acceptable doses showed better biomechanical effects on cement impact strength in comparison to the other antibiotic impregnated cement configurations tested. All samples aged in air showed lower impact strength than that of counterpart aged in saline.

REFERENCES

[1]. Joseph TN, Chen AL, Di Cesare PE. Use of antibiotic-impregnated cement in total joint arthroplasty. J Am Acad Orthop Surg, 11:38–47, 2003.

[2]. Heck D, Rosenberg A, Schink-Ascani M, Garbus S, Kiewett T. Use of antibiotic-impregnated cement during hip and knee arthroplasty in the United States. J Arthroplasty, 10:470–475, 1995.

[3]. He Y, Trotignon JP, Loty B, Tcharkhtchi A, Verdu J. Effects of antibiotics on the properties of poly(methylmethacrylate)-based bone cement. J Biomed Mater Res, 63:800–806, 2002.

[4]. Ger E, Dall D, Miles T, Forder A. Bone cements and antiobiotics. S Afr Med J, 51:276–279, 1977.

[5]. Nelson RC, Hoffman RO, Burton TA. Effect of antibiotic additions on the mechanical properties of acrylic cement. J Biomed Mater Res, 12:473–490, 1978.

[6]. Askew MJ, Kuel MF, Fleissner PR, Gradisar IA, Salstrom S, Tan JS. Effect of vacuum mixing on the mechanical properties of antibiotic-impregnated polymethylmethacrylate bone cement. J Biomed Mater Res, 24:573–580, 1990.

[7]. Leone J, Johnson A, Ziada S, Hashemi A, Adili A,de Beer J, Biaxial Flexural Modulus of Antibiotic-Impregnated Orthopedic Bone Cement. J Biomed Mater Res Part B: Appl Biomater, 83B: 97–104, 2007.

[8]. Krause W, Mathis RS. Fatigue properties of acrylic bone cements: Review of the literature. J Biomed Mater Res, 22(A1):37–53, 1998.

[9]. Krause W, Grimes L, Mathis RS. Fatigue testing of acrylic bone cements: Statistical concepts and proposed test methodology. J Biomed Mater Res, 22(A2):179–190, 1988.

[10]. Lewis G, Nyman JS. Toward standardization of methods of determination of fracture properties of acrylic bone cement and statistical analysis of test results. J Biomed Mater Res, 53:748–768, 2000.

[11]. Lee AJC, Ling RSM, Vangala SS. The mechanical properties of bone cements. J Med Eng Technol, 1:137–140, 1977.

[12]. Beaumont PWR, Young RJ. Slow crack growth in acrylic bone cement. J Biomed Mater Res, 9:423–439, 1975.
[13]. Lewis G. Effect of mixing method and storage temperature of cement constituents on the fatigue and porosity of acrylic bone cement. J Biomed Mater Res, 48:143–149, 1999.
[14]. Johnson JA, Provan JW, Krygier JJ, Chan KH, Miller J. Fatigue of acrylic bone cement-effect of frequency and environment. J Biomed Mater Res, 23:819–831, 1989.
[15]. Freitag TA, Cannon SL. Fracture characteristics of acrylic bone cements. II. Fatigue. J Biomed Mater Res, 11:609–624, 1997.
[16]. Davies JP, Burke DW, O'Connor DO, Harris WH. Comparison of the fatigue characteristics of centrifuged and uncentrifuged Simplex P bone cement. J Orthop Res, 5:366–371, 1987.
[17]. Pittet C, Lemaitre J. Mechanical characterization of brushite cements: A mohr's circle approach. J BiomedMater Res, 53:769–780, 2000.
[18]. Vila MM, Ginebra MP, Gil FJ, Planeli JA. Effect of porosity and environment on the mechanical behavior of acrylic bone cement modified with acrylontrile-butadiene styrene particles. II. Fatigue crack propagation. J Biomed Mater Res, 48:128–134, 1999.
[19]. Vila MM, Ginebra MP, Gil FJ, Planeli JA. Effect of porosity and environment on the mechanical behavior of acrylic bone cement modified with acrylonitrile-butadiene styrene particles. I. Fracture toughness. J Biomed Mater Res, 48:121–127, 1999.
[20]. Kim HY, Yasuda HK. Improvement of fatigue properties of poly (methyl methacrylate) bone cement by means of plasma surface treatment of fillers. J Biomed Mater Res, 48:135–142, 1999.
[21] Ungethulum M, Hinterberger J. Die normung von implantatwer to !en ambeispiel &knockenzemente'. Z Orthopaed, 116:303-311, 1978.
[22] Ege W. Material properties of PMMA bone cements. In: Bucchorn TH, Willert HG, editors. Technical principles: design and safety of joint implants. Seattle, WA: Hogrefe and Huber, 49-52, 1994.
[23]. Lewis G, Mladsi S, Correlation between impact strength and fracture toughness of PMMA-based bone cements, *Biomaterials*. 21:775-781, 2000.
[24] American Society for Testing and Materials (ASTM), Standard D 256-97, Standard test methods for determining the pendulum impact resistance of notched specimens of plastics, 1999 Annual Book of ASTM Standards, vol. 08.01, West Conshohocken, PA, USA: ASTM, 1-20, 1997.
[25]. Garvin KL, Hinrichs SH, Urban JA. Emerging antibiotic-resistant bacteria. Their treatment in total joint arthroplasty. Clin Orthop Relat Res, 369:110–123, 1999.
[26]. Hanssen AD, Osmon DR. The use of prophylactic antimicrobial agents during and after hip arthroplasty. Clin Orthop Relat Res, 369:124–138, 1999.

Use of Nanoindentation for Investigating the Nanostructure of Dentin Tissue

Bo-Hsiung Wu[1], Chung-Jen Chung[1], Chang-Fu Han[2], Terry Yuan-Fang Chen[1,2], Shu-Fen Chuang[3], Wang-Long Li[4], Jen-Fin Lin[2]

[1] Center for Micro/Nano Science and Technology, National Cheng Kung University, Tainan, Taiwan.
[2] Department of Mechanical Engineering, National Cheng Kung University, Tainan, Taiwan.
[3] Institute of Oral Medicine at the Medical College, National Cheng Kung University, Tainan, Taiwan.
[4] Institute of Nanotechnology and Microsystems Engineering, National Cheng Kung University, Tainan, Taiwan.

ABSTRACT—Human dentin contains millions of tubules, and the distribution and size of tubule is none-homogeneous, which resulting in various mechanical properties. In this study, we utilize a nanoindentor on the human dentin, and a focused ion-beam (FIB) is adopted to prepare the dentin specimen for transmission electron microscopy (TEM) to evaluate the nano-structure of human dentin after the residual impression of the nanoindents. The nanoindentation results show that the loading–unloading curve is continuous and smooth in both the loading and the unloading steps, which suggests that no cracking occurs. As to the TEM specimen, 50 nm thick specimens are prepared by FIB cross-sectioning. The nano-structural observations reveal that nanoindentation induces an atomic re-orientation, and the indent results in the formation of high-stress deformation regions beneath the indenter. High resolution images via TEM and diffraction patterns indicate the dentin is poly-crystallized and present anisotropic properties.

Keywords: human dentin, tubules, nanoindentor, FIB, TEM

1. Introduction

The instrumented indentation technique has become a popular method to measure the mechanical properties of bulk specimens [1-4] and thin film materials [5-7]. Microstructures of teeth, as well as the mechanical properties, have been well-characterized. Dentin consists of microscopic channels, dentinal tubules, which radiate outward through the dentin from the pulp to the exterior cementum or enamel border [8]. The mechanical properties of teeth have been characterized by micro-indentation techniques. V. Imveni *et. al.*[9] profiled the hardness and toughness of human teeth using a Vickers indentor. The microhardness testing may not be a suitable technique for reliably evaluating the physical properties of nano-scale mechanical properties of dentin, since the results are significantly affected by the substrate's deformation [10]. However, teeth contain nano-sized structures; nano-indentation provides a powerful technique for obtaining extremely accurate information regarding the physical properties of materials at the nano-scale [11], therefore, in this study a nano-indentation technique is adopted to acquire the nano-mechanical properties of individual dental structure using loads of hundred milli-Newtons.

Transmission electron microscopy (TEM) has long been used for microstructural examinations at nanostructural size scales, with numerous studies on dentin being reported [12]. However, the specimen preparation of biological tissues for TEM is quite challenging. The decalcification and milling processes may impede the observation of real biologic morphology. By the assistance of the focused ion beam (FIB) system, which utilizes a finely focused gallium ion beam operated at low-beam currents for imaging and at high-beam currents for site-specific milling, thus, cross-sectional morphologies specimens of dentine could be precisely observed. Accordingly, in the present study, we used a comprehensive approach, which involved nano-indenter to characterize the nanomechanical properties and utilized FIB-based milling for TEM sample preparation of dentin to observe the nanostructure of dentin tissue after the nanoindentation.

T. Proulx (ed.), *Mechanics of Biological Systems and Materials, Volume 2*, Conference Proceedings of the Society for Experimental Mechanics Series 9999, DOI 10.1007/978-1-4614-0219-0_21,

2. Experimental Procedure

2.1 Sample preparation

The teeth specimens were cut into horizontal sections within 1mm thick, as shown in Fig. 1. A fine polish procedure is carried out for the later on nano-indentation and TEM experiments.

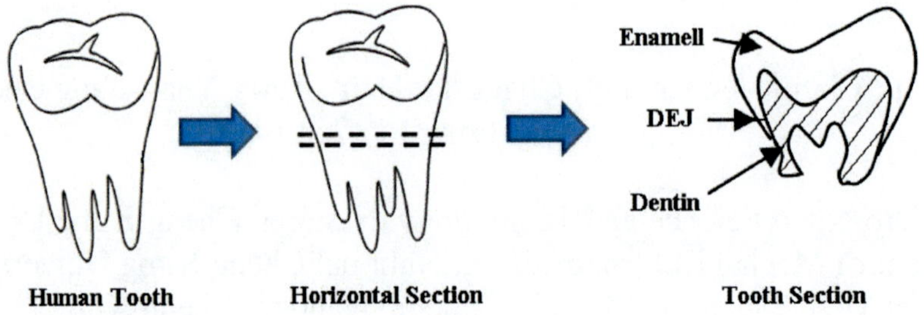

Fig. 1 The schematic figure of human tooth section

2.2. Nanoindentation measurement

The nanomechanical properties of the polished teeth section were investigated using a MTS Nano Indenter® with CSM technique in G200 mode and using a Berkovich indenter. These values are obtained by an indentation depth of 3000 nm with a theoretical displacement resolution less than 0.01 nm in CSM mode. A frequency of 45 Hz was used to avoid the sensitivity to thermal drift [13] and the loading resolution was 50 nN. The loading process was controlled by a constant strain rate of 0.05 s^{-1}. Indentation tests performed on a fused silica substrate before and after the dentin indentation tests, respectively, confirmed that the indenter was well calibrated since the hardness and Young's modulus values of the silica substrate computed using the indentation data were consistent with the theoretical values. Six indentations were done in each sample and excluding abnormal variation of the data then averaged for the results with a deviation. Fig. 2. presents SEM image of the morphology of the residual-indentation mark in dentin surface using a Berkovitch indentor.

Fig. 2 The SEM micrograph of nano-indentation marks on dentin region

Nanoindentation nanohardness is defined as the indentation load divided by the projected contact area of the indentation. It is the mean pressure under the indenter [2]:

$$H = \frac{P_{max}}{A} \quad (1)$$

where P_{max} is the maximum applied force and obtained directly from the force-displacement curve; A is the projected contact area of indenter tip with the material. A geometry independent relation involving contact stiffness, contact area, and reduced elastic modulus can be derived by Sneddon [14] as follows:

$$S = 2\beta\sqrt{\frac{A}{\pi}}E_r \quad (2)$$

where β is a constant which depends on the geometry of the indenter (β = 1.034 for a Berkovich indenter) [2, 15], and E_r is the reduced elastic modulus which accounts for the fact that elastic deformation occurs in both the sample and the indenter. E_r is given by [2]

$$\frac{1}{E_r} = \frac{1-\upsilon_f^{\,2}}{E_f} + \frac{1-\upsilon_{id}^{\,2}}{E_{id}} \quad (3)$$

where E and υ represent the elastic modulus and Poisson ratio, respectively, and the subscript f and id represent the film and the indenter, respectively. The indenter properties used in this study's calculations are E_{id}= 1141 GPa and υ_{id}=0.07 [2], assuming that υ_f=0.25. The relationship between the hardness and Young's modulus values of dentin were investigated in this study.

2.3. Nanostructure observations

The morphologies and nanostructures of the dentin tissue nanoindented specimens were observed using a scanning electron microscope (SEM) equipped with a focused ion beam (FIB) system and a high-resolution transmission electron microscope (TEM), respectively. The high quality TEM specimens are prepared by the dual-beam focused ion beam (FEI Nova-200 NanoLab, USA). To avoid electron and ion charging during the milling process, a platinum thin film is sputtered onto the teeth section in advance. In this study, we fabricate the TEM specimen, 10 um in length, at the indentation mark on the horizontally-cut dentin. Low ion-beam currents are employed to precisely mill the dentin as thin as 50 nm thick. The finished specimen is attached to a carbon-coated copper grid for the TEM (JEOL HR-TEM 2100F Electron Microscope, Japan) examine. The HR-TEM is equipped with a 200kV field-emission gun, thus it provides an atomic resolution in imaging and good energy resolution in spectroscopy. In addition to fundamental imaging, it offers the atomic lattice images with high contrast.

3. Results and Discussion

The nanoindentation tests is the loading–unloading curves produced from the load versus displacement data recorded during the indentation process. Continuous stiffness measurement (CSM) is adopted, and the indentation depth was continuously recorded during both loading and unloading cycles. The loading force and displacement of dentin is analyzed. Fig. 3(a) presents the loading–unloading curve for the current dentin region. Both the loading and the unloading parts of the curve are smooth and continuous, which suggests that no debonding or cracking occurs. Fig. 3(b) illustrates the variation of the Young's modulus value with the nanoindentation depth. It is observed that Young's modulus increases significantly as the tip first penetrates the dentin surface. However, the modulus value then increases linearly with increasing indentation depth, and achieves a final value of approximately 20.2 GPa at the indentation depth of 1000 nm. Fig. 3(c) illustrates the variation of the dentin hardness with the nanoindentation depth. At very low indentation depths, the dentin has an extremely low hardness. As discussed previously for the case of Young's modulus, the low hardness and Young's modulus value reflects the fact that the surface roughness effected under low penetration depths of the indenter. However, as the indentation depth increases, the presence of the dentin causes the hardness to a value of approximately 0.68 GPa. Hence, it is apparent that the dentin is a vital tissue which is harder than bone and weaker than enamel, with a physiological elasticity that helps to prevent tooth fracture [16].

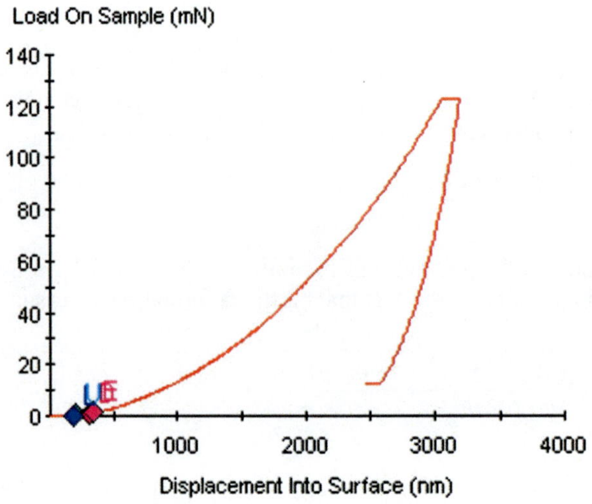

(a)

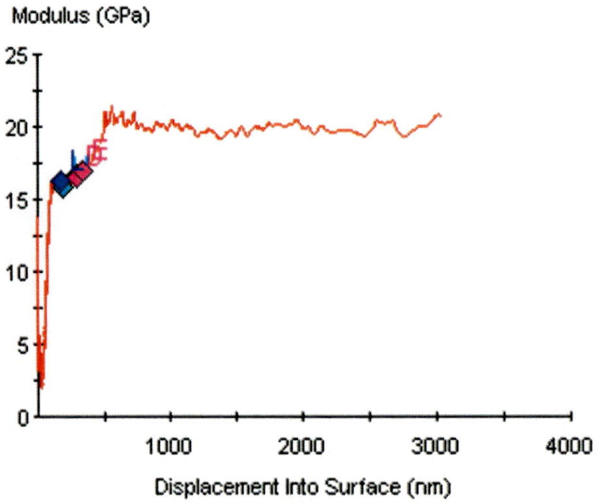

(b)

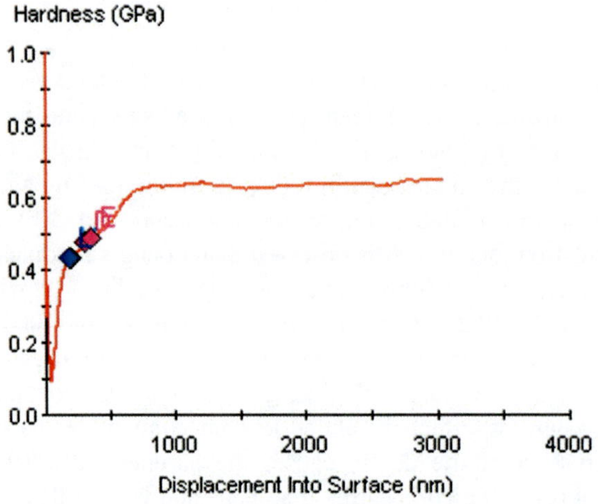

(c)

Fig. 3 (a) The typical load-displacement curves during nanoindentation of dentin, (b) the variation of Young's modulus and (c) hardness of dentin with indentation depth

In the present work, we seek to explore the use of FIB-based milling for examining structure and TEM sample preparation of dentin in this case studies. SEM images equipped with FIB system of dentin and the yellow arrows on representative dentinal tubules, as shown in Fig. 4.

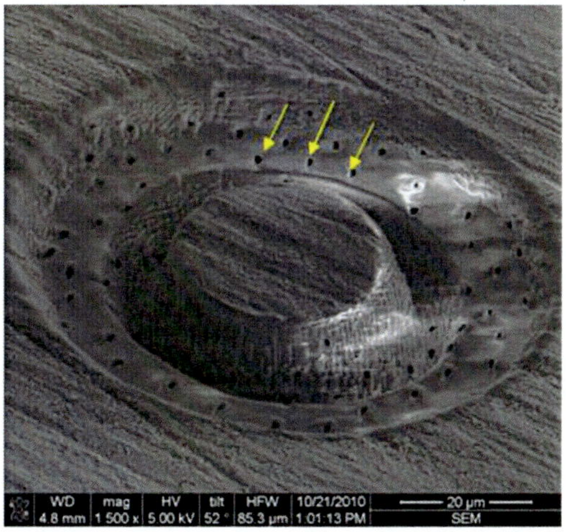

Fig. 4 SEM images of dentin and the yellow arrows on representative dentinal tubules

Fig. 5(a) presents a cross-sectional TEM image of dentin indented specimen. The specimen is progressively thinned until it is thin enough for the electron transparency. Fig. 5(b) shows a high-magnification view of the morphology of the indentation-affected zone in Fig. 5(a). The insets in the lower and upper right corner of Fig. 5(b) shows the TEM diffraction patterns of the dentin regions within the blue arrows (labelled A and B), and confirm that the indentation-affected zone is an anisotropic properties and present poly-crystalline lattices. The polycrystal reorganization which takes place in the indentation-affected area is caused by the pressure induced beneath the indenter during penetration.

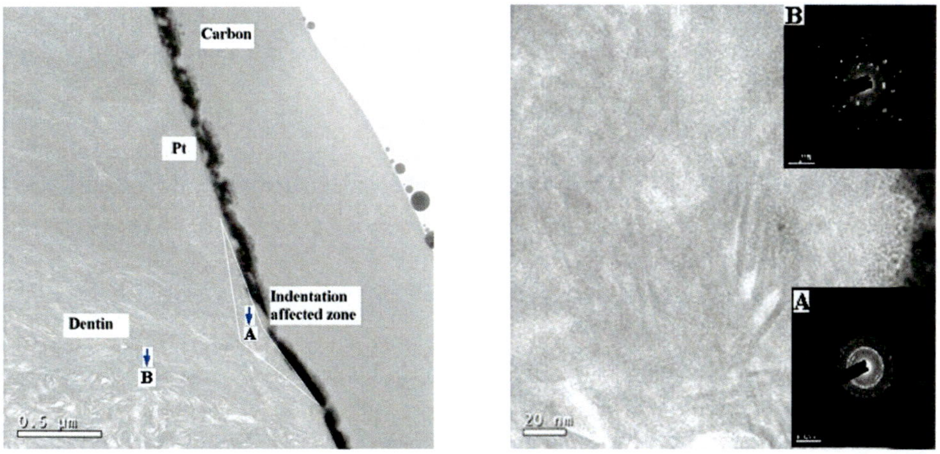

(a) (b)

Fig. 5 (a) A bright field TEM micrograph of dentin indented specimen; (b) a high-magnification micrograph of the indentation-affected zone

4. Conclusion

The nano-mechanical properties and nanostructures of dentin tissue have been investigated by using a Berkovitch nano indentor under continuous stiffness measurement (CSM) mode. The nanostructures of dentin tissue indented specimens a have been examined via FIB and TEM. On the basis of the load– displacement data, the Young's modulus and hardness values of the dentin have been calculated to be 20.2 GPa and 0.68 GPa, respectively. It is apparent that the dentin is a vital tissue which is harder than bone and weaker than enamel properties, with a physiological elasticity that helps to prevent tooth fracture.

As to the TEM specimen, transparent dentin specimen is progressively thinned by FIB using low ion-beam currents, thus, good electron transparency and high resolution TEM can be achieved. The indentation-affected zone of dentin tissue presents anisotropic properties, and poly-crystalline lattices can be clearly observed.

ACKNOWLEDGMENT

I would like to thank the Center for Frontier Materials and Micro/Nano Science and Technology (CFMMNST) for financial support and Mei-Lan Liang, Shih-Wen Tseng, and Shih-Sheng Hung for the assistance on the FIB and TEM.

REFERENCES

[1] A. A. Volinsky, N. R. Moody and W. W. Gerberich, "Interfacial toughness measurements for thin films on substrates," Acta Mater. 50, pp. 441-466, 2002.

[2] W. C. Oliver and G. M. Pharr, "An improved technique for determining hardness and elastic modulus using load and displacement sensing indentation experiments", Journal of Materials Research 7, pp. 1564-1583, 1992.

[3] A. E. Giannakopoulos, S. Suresh, "Determination of elastoplastic properties by instrumented sharp indentation", Scripta Mater., 40, pp. 1191- 1198, 1999.

[4] Taljat, B.; Zacharia, T.; Kosel, F.;, "New Analytical Procedure to Determine StressStrain Curve from Spherical Indentation Data", International , Journal of Solids Structures; 35(33), pp. 4411-4426,1998.

[5] A. C. Fischer-Cripps, Nanoindentation, Springer, New York, 2004.

[6] T. F. Page, G. M. Pharr, J. C. Hay, W. C. Oliver, B. N. Lucas, E. Herbert, L. Riester, "Nanoindentation characterization of coated systems. P:S2 – a new approach using the continuous stiffness technique", Proc. MRS 522, pp. 53-64, 1998.

[7] P. J. Wei and J. F. Lin, "A new method developed to evaluate both the hardness and elastic modulus of a coating–substrate system", Surf. Coat. Tech. 200, pp. 2489-2496, 2005.

[8] M. H. Ross, I. K.Gordon, and P. Wojciech, Histology: a text and atlas: With Cell And Molecular Biology, 4th edition, p. 450, 2003.

[9] V. Imbeni, J. J. Kruzic, G. W. Marshall, S. J. Marshall, and R. O. Ritchie," The dentin-enamel junction and the fracture of human teeth," Nature Materials, vol. 4, pp. 229-232, 2005.

[10] Y. R. Jeng, T. T. Lin, T. Y. Wong, H. J. Chang, and D. B. Shieh, "Nano-mechanical Properties of Fluoride-treated Enamel Surfaces", Journal of Dental Research 87(4), pp.381-385, 2008.

[11] Marshall GW Jr, Balooch M, Gallagher RR, Gansky SA, Marshall SJ, " Mechanical properties of the dentinoenamel junction: AFM studies of nanohardness, elastic modulus, and fracture. J Biomed Mater Res 54, pp.87-95, 2001.

[12] R. K. Nalla, A. E. Porter, C. Daraio, A. M. Minor, V. Radmilovic, E. A. Stach, A. P. Tomsia, R. O. Ritchie, "Ultrastructural examination of dentin using focused ion-beam cross-sectioning and transmission electron microscopy", Micron 36, pp. 672-680, 2005.

[13] G. H. Yang, B. Zhao, Y. Gao and F. Pan, "Investigation of nanoindentation on Co/Mo multilayers by the continuous stiffness measurement technique", Surface & Coatings Technology 191, pp. 127-133, 2005.

[14] Sneddon, I. N., "The relation between load and penetration in the axisymmetric Boussinesq problem for a punch of arbitrary profile," Int. J. Eng. Sci. 3, pp. 47-57, 1965.

[15] X. Li, B. Bhushan, "A review of nanoindentation continuous stiffness measurement technique and its applications", Materials Characterization 48, pp. 11-36, 2002.

[16] Vijay K. Goel, Satish C. Khera, Jeffrey L. Ralston and Kuang H. Chang, "Stresses at the dentinoenamel junction of human teeth—A finite element investigation", The Journal of Prosthetic Dentistry Vol. 66, .pp. 451-459, 1991.

Epinephrine Upregulates Sickle Trait Erythrocyte Adhesion to Laminin and Integrins

Jamie L. Maciaszek, Graduate Assistant, University of Connecticut, Department of Mechanical Engineering, 191 Auditorium Rd., Unit 3139, Storrs, CT 06269

Biree Andemariam, Doctor of Hematology/Oncology, The Carole and Ray Neag Comprehensive Cancer Center, University Connecticut Health Center, 263 Farmington Avenue, Farmington, CT 06030

George Lykotrafitis, Assistant Professor, University of Connecticut, Department of Mechanical Engineering, 191 Auditorium Rd., Unit 3139, Storrs, CT 06269

ABSTRACT

Erythrocytes from sickle cell trait (SCT) patients aggregate during strenuous exercise, resulting in inflammation, blood flow impairment, and vaso-occlusive events. While the exact role of exercise conditions and how they lead to complications is virtually unknown, it likely occurs because of their adhesive interactions with the extracellular matrix (ECM) and endothelial cells (ECs), irregular viscoelastic properties, and morphology. The current study employs single-molecule atomic force microscopy (AFM) experiments to quantitatively measure the frequency and bond strength of SCT erythrocytes with ECM laminin α5 and endothelial αvβ3. Epinephrine, a hormone secreted during stressful conditions, is administered in vitro to determine its direct effect on SCT adhesion. In the presence of epinephrine, we measured an increase in the frequency of adhesion events of BCAM/Lu and LW expressed on the SCT erythrocyte surface with ECM laminin α5 and endothelial αvβ3, respectively. The observed clustering of BCAM/Lu and LW receptors into nanodomains explains the measured amplification in bond strength. In conclusion, we found that epinephrine modulates the BCAM/Lu-laminin α5 and LW-αvβ3 bond interactions. The measured increases in adhesion frequency and bond strength suggest that epinephrine contributes to the pathophysiology of vaso-occlusive sequelae related to sudden death in SCT individuals during strenuous exercise.

INTRODUCTION

Sickle cell trait (SCT, HbAS) is a risk factor for collapse and sudden death in physical training [1-4]. SCT is characterized by the presence of both normal adult hemoglobin (HbA) and abnormal sickle hemoglobin (HbS) in the red blood cell (RBC, erythrocyte). In comparison to the homozygous SCD, the heterozygous SCT is traditionally regarded as a benign condition. However, during strenuous exercise, individuals can develop a syndrome resembling SCD with vaso-occlusive events resulting from changes in the RBC morphology, viscoelasticity, and adhesion [1-2, 4]. Vaso-occlusion occurs in crises in which the arterial circulation is blocked at one or many sites, leading to organ damage and ischemic pain [5]. Endothelial cell activation, cytoadherence, inflammation, and coagulation activation contribute to blood flow obstruction and play a critical role in the pathophysiology of vaso-occlusion [6-7]. While it is known that epinephrine enhances SCD erythrocyte adhesion of Lutheran blood group (Lu) and basal cell adhesion molecule (BCAM), known as BCAM/Lu, and LW glycoprotein, this mechanism has not yet been quantitatively explored in SCT or in SCD [8-9]. Epinephrine is a hormone secreted from the adrenal medulla during periods of stress and strenuous exercise. Increased circulating levels of epinephrine act on the RBC β2-adrenergic receptor, thereby activating $G\alpha_s$ proteins which stimulate adenylyl cyclase (AC). This enzyme catalyzes the conversion of adenosine triphosphate (ATP) to cyclic adenosine monophosphate (cAMP), leading to protein kinase A (PKA) activation, an intermediate step in the up-regulation of BCAM/Lu- and ICAM-4-mediated adhesion [8-10]. The interaction of BCAM/Lu with the α5 chain of laminin (LAMA5) on the ECM, as well as the interaction of LW with the αvβ3 integrin on ECs, may contribute to vaso-occlusive events in SCT during exercise.

METHODS

The study included RBCs from a SCT adult subject (n = 1) and RBCs from healthy adult subjects (n = 2) purchased from Research Blood Components, LLC (Brighton, MA). Cells were immobilized on AFM grade mica (Novascan Technologies, Inc., Ames, IA) coated with 1 mg/ml poly-L-lysine (PLL) solution to increase cell adherence. Silicon nitride cantilevers functionalized with 250 μg/mL anti-Laminin α_5, clone 4C7 (Millipore, Temecula, CA) or 50 μL of 210 μg/mL purified

T. Proulx (ed.), *Mechanics of Biological Systems and Materials, Volume 2*, Conference Proceedings of the Society for Experimental Mechanics Series 9999, DOI 10.1007/978-1-4614-0219-0_22,

human integrin $\alpha_v\beta_3$ (Millipore, Temecula, CA) were employed. Single-molecule atomic force microscopy (AFM) adhesion maps and force-distance curves were obtained using an AsylumMFP-3D-BIO (Asylum Research, Santa Barbara, CA) AFM equipped with a "liquid cell" setup. Baseline measurements were performed in PBS solutions at room temperature. Epinephrine-mediated experiments were performed in a 10 nM epinephrine solution at room temperature. Unless otherwise specified, all force measurements were recorded with a loading rate of 24000 pN/s, calculated by multiplying the tip retraction velocity (nm/s) by the spring constant of the cantilever (pN/nm). The nominal spring constant kc of the employed cantilever was 30 pN/nm, as provided by the manufacturer. Exact values for the cantilever spring constants were obtained via a thermal noise based method implemented by the manufacturer and were used in all calculations. The approach and retraction velocities were held constant at 800 nm/s. Probes had nominal tip radii of 20 nm and nominal angle of 20°, as provided by the manufacturer.

RESULTS

To test the hypothesis that an increase in epinephrine levels results in an increase in the magnitude of the adhesive force associated with BCAM/Lu-laminin and ICAM-4-$\alpha_v\beta_3$, as well as an increase in the frequency of adhesive interactions, we employed an approach in which the AFM probe was functionalized with LAMA5 or $\alpha v\beta 3$ to measure the interaction between the $\alpha 5$ chain of laminin and BCAM/Lu receptor expressed on the tested RBC, or the interaction between integrin $\alpha_v\beta_3$ and ICAM-4 receptor expressed on the tested RBC. Treatment of normal RBCs with 10 nM epinephrine, a physiologically relevant level, induced a significant increase from 6.0% to 10.2% ($p < 0.02$) in the frequency of BCAM/Lu-laminin binding events (n = 6), and an insignificant increase from 5.3% to 7.6% ($p > 0.05$) in the frequency of ICAM-4-$\alpha_v\beta_3$ binding events (n = 7). There was no notable increase in the bond strength of either pair with epinephrine stimulation.

Treatment of sickle trait RBCs (n = 7) with epinephrine induced an increase in the frequency of BCAM/Lu-laminin binding events of from 9.3% to 14.0% ($p < 0.04$), as well as a significant increase ($p < 0.002$) in the magnitude of the BCAM/Lu-laminin interaction. Treatment of sickle trait RBCs (n = 7) with epinephrine induced an increase in the frequency of ICAM-4-$\alpha_v\beta_3$ binding events from 7.9% to 13.1% ($p < 0.05$), as well as an increase ($p < 0.02$) in the magnitude of the ICAM-4-$\alpha_v\beta_3$ interaction. Notably, 500 nm x 500 nm spatially-resolved adhesion maps revealed that the BCAM/Lu-mediated laminin binding activity is not homogeneously distributed over the SCT RBC surface, but concentrated on nanodomains. These nanodomains, often attributed to membrane rafts, are only observed on SCT RBCs (n = 7) in the presence of epinephrine

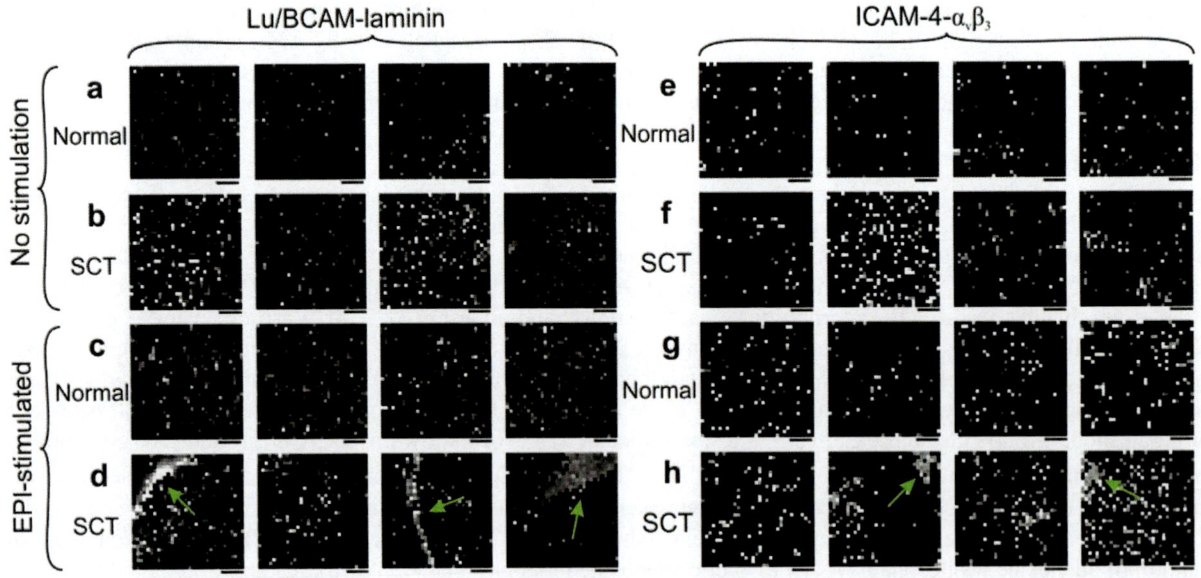

Figure 1. Distribution of BCAM/Lu and ICAM-4 on the surface of normal RBCs and sickle cell trait RBCs using different tips and cell preparations. Adhesion force maps were created via a MATLAB script from data obtained from force-distance curves obtained via AFM. The figure was then compiled using CorelDRAW X4. (A,B) Adhesion force maps (gray scale 0-100 pN) recorded with LAMA5 tips over normal RBCs (A) and sickle cell trait RBCs (B) in PBS. (C,D) Adhesion force maps recorded with LAMA5 tips over normal RBCs (C) and sickle cell trait RBCs (D) in 10 nM epinephrine. (E,F) Adhesion force maps recorded with $\alpha_v\beta_3$ tips over normal RBCs (E) and sickle cell trait RBCs (F) in PBS. (G,H) Adhesion force maps recorded with $\alpha_v\beta_3$ tips over normal RBCs (G) and sickle cell trait RBCs (H) in 10 nM epinephrine. Green arrows indicate nanodomains. Scale bars, 100 nm.

(Fig. 1D). By contrast, normal RBCs (n = 6) stimulated with epinephrine (Fig. 1C) demonstrate a nearly homogeneous distribution of BCAM/Lu receptors. Furthermore, ICAM-4 receptor nanodomains are also observed, to a lesser degree, on the SCT RBC (n = 7) surface in the presence of epinephrine (Fig. 1H), with a heterogeneous receptor distribution on the normal RBC (n = 7) stimulated with epinephrine (Fig. 1G). In the non-stimulated environment, both normal (Fig. 1A, 1E) and sickle cell trait (Fig. 1B, 1F) RBCs maintain a homogenous distribution of BCAM/Lu and ICAM-4 receptors at the local scale.

DISCUSSION AND CONCLUSIONS

Interactions of receptor molecule cytoplasmic domains with the cytoskeleton can play critical roles in adjusting receptor function. It has been determined that BCAM/Lu has a high degree of connectivity to the erythrocyte membrane cytoskeleton via erythroid spectrin [11-12]. Importantly, disruption of the BCAM/Lu-spectrin linkage is accompanied by enhanced cell adhesion to laminin [12]. A major finding of the current study is that BCAM/Lu adhesion events in SCT erythrocytes increased ($p < 0.05$) as compared to the wild-type. It is known that Hb interacts with the spectrin network via band 3 [13-14] and the introduction of HbS has recently been found to have an effect on the mechanical properties of SCT erythrocytes [15]. We conjecture that HbS partially disrupts the spectrin network via its interaction with band 3, indirectly disrupting the BCAM/Lu-spectrin linkage. This results in increased diffusion and clustering and thus significantly enhanced sickle trait RBC adhesion to ECM laminin.

Epinephrine, acting through the β2-adrenergic receptor, increases Lu-laminin α5 sickle mediated red cell binding via a cAMP and PKA-dependent signaling pathway [8]. Further, the Lu cytoplasmic tail is phosphorylated in epinephrine-stimulated sickle cells, which may be a critical factor in sickle RBC's enhanced adhesion to laminin [16]. The BCAM cytoplasmic tail, which lacks the last 40 amino acids of the sequence, does not undergo phosphorylation. It has been demonstrated that PKA-mediated phosphorylation of the Lu glycoprotein at serine 621 positively regulates the adhesion function of Lu under flow conditions. While the molecular basis for the resulting increased adhesion is not yet understood, it has been hypothesized that phosphorylation of the Lu cytoplasmic tail weakens its interaction with spectrin, enabling the freely-floating transmembrane Lu molecules to cluster and thereby generate a larger adhesive force [16]. A major finding of our current study is the mapping of Lu nanodomains on epinephrine-stimulated SCT erythrocytes, accompanied by an increase in the magnitude of the adhesive force ($p < 0.002$) and in the frequency of binding events ($p < 0.04$) from the baseline SCT erythrocytes. The observed increase in the adhesive force shown in Fig. 5B can be attributed to the clustering of membrane receptors, a principle that has been demonstrated in nonerythroid cells [17]. The significant broadening of the bond magnitude distribution agrees with the clustering detected in Fig. 1D. Another significant finding is the dramatic increase of adhesion events on the wild-type RBCs in the stress-induced epinephrine environment, clearly demonstrating the effect of the phosphorylated Lu glycoproteins on the enhanced BCAM/Lu-laminin adhesion. While the frequency of adhesive interactions increases, receptor clustering is not observed (see Fig. 1C) and the bond magnitude remains the same. This observation strengthens our conjecture that HbS disrupts the spectrin network resulting in clustering of BCAM/Lu receptors in SCT RBCs. It is noted that, clustering is not detected in the case of SCT RBCs without epinephrine (see Fig. 1B) probably because of the lower receptor expression.

ICAM-4 mediated binding to endothelial cells and leukocytes is also activated by epinephrine via stimulation of the β2-adrenergic receptor which triggers a cAMP and PKA-dependent signaling pathway [18]. In addition, ICAM-4 on epinephrine-stimulated sickle cells undergo a significant increase in serine phosphorylation, which may be a critical factor in sickle RBC's enhanced adhesion to endothelial integrin $\alpha_v\beta_3$[9]. It has been hypothesized that phosphorylation-induced changes in the ICAM-4 cytoplasmic domain cause a conformational change in its extracellular domain, thus generating a larger adhesive force. We observed an increase in the frequency of binding events ($p < 0.05$) from the baseline SCT erythrocytes which accompanied ICAM-4 nanodomains on epinephrine-stimulated SCT erythrocytes (see Fig. 1H). The increase in the magnitude of the adhesive force ($p < 0.02$) could possibly be attributed to the clustering of membrane receptors [17]. However, the bond magnitude distribution is not broadened as would be expected in this phenomenon. To the contrary, we observed significant narrowing of the distribution, leading to us to infer that the bond magnitude did in fact increase upon epinephrine stimulation. Another important finding of the present data is the quantification of the insignificant increase ($p > 0.05$) of ICAM-4-$\alpha_v\beta_3$ adhesion events on the wild-type RBCs in the stress-induced epinephrine environment. This is consistent with the known fact that there is minimal ICAM-4 phosphorylation in normal RBCs, due to an age-related reduction in the cell's ability to produce cAMP and thereby enhance PKA activity in response to stimulation [8, 19-20]. While the current study tests the effect of epinephrine on single molecule adhesion interactions thought to contribute to vascular blockage in sickle cell vaso-occlusion, epinephrine has also been found to work via increasing RBC trapping in the spleen and liver [18, 21].

REFERENCES

1. Jones SR, B.R., Donowho EM, *Sudden Death in Sickle Cell Trait.* N Engl J Med, 1970. **282**: p. 323-325.

2. Kark, J.A., D.M. Posey, and H.R. Schumacher, *Sickle-Cell Trait as a Risk Factor for Sudden Death in Physical Training.* N Engl J Med, 1987. **317**: p. 781-787.
3. Tsaras, G., et al., *Complications Associated with Sickle Cell Trait: A Brief Narrative Review.* The American Journal of Medicine, 2009. **122**(6): p. 507-512.
4. Kark, J.A. and F.T. Ward, *Exercise and Hemoglobin-S.* Seminars in Hematology, 1994. **31**(3): p. 181-225.
5. Stuart, M.J. and R.L. Nagel, *Sickle-cell disease.* Lancet, 2004. **364**(9442): p. 1343-1360.
6. Ballas, S.K. and N. Mohandas, *Pathophysiology of vaso-occlusion.* Hematology-Oncology Clinics of North America, 1996. **10**(6): p. 1221-&.
7. Conran, N., C.F. Franco-Penteado, and F.F. Costa, *Newer Aspects of the Pathophysiology of Sickle Cell Disease Vaso-Occlusion.* Hemoglobin, 2009. **33**(1): p. 1-16.
8. Hines, P.C., et al., *Novel epinephrine and cyclic AMP-mediated activation of BCAM/Lu-dependent sickle (SS) RBC adhesion.* Blood, 2003. **101**(8): p. 3281-3287.
9. Zennadi, R., et al., *Epinephrine acts through erythroid signaling pathways to activate sickle cell adhesion to endothelium via LW-alpha v beta 3 interactions.* Blood, 2004. **104**(12): p. 3774-3781.
10. Kaul, D.K., *"Stress" and sickle red cell adhesion.* Blood, 2004. **104**(12): p. 3425-3426.
11. Parsons, S.F., et al., *Lutheran blood group glycoprotein and its newly characterized mouse homologue specifically bind alpha 5 chain-containing human laminin with high affinity.* Blood, 2001. **97**(1): p. 312-320.
12. An, X.L., et al., *Adhesive activity of Lu glycoproteins is regulated by interaction with spectrin.* Blood, 2008. **112**(13): p. 5212-5218.
13. Fortier, N., et al., *THE RELATIONSHIP BETWEEN INVIVO GENERATED HEMOGLOBIN SKELETAL PROTEIN COMPLEX AND INCREASED RED-CELL MEMBRANE RIGIDITY.* Blood, 1988. **71**(5): p. 1427-1431.
14. Snyder, L.M., et al., *IRREVERSIBLE SPECTRIN HEMOGLOBIN CROSSLINKING INVIVO - A MARKER FOR RED-CELL SENESCENCE.* British Journal of Haematology, 1983. **53**(3): p. 379-384.
15. Maciaszek, J.L. and G. Lykotrafitis, *Sickle cell trait human erythrocytes are significantly stiffer than normal.* Journal of Biomechanics. **In Press, Corrected Proof**.
16. Gauthier, E., et al., *Protein kinase A-dependent phosphorylation of lutheran/basal cell adhesion molecule glycoprotein regulates cell adhesion to laminin alpha 5.* Journal of Biological Chemistry, 2005. **280**(34): p. 30055-30062.
17. Schneider, H., et al., *CTLA-4 up-regulation of lymphocyte function-associated antigen 1 adhesion and clustering as an alternate basis for coreceptor function.* Proceedings of the National Academy of Sciences of the United States of America, 2005. **102**(36): p. 12861-12866.
18. Zennadi, R., et al., *Epinephrine-induced activation of LW-mediated sickle cell adhesion and vaso-occlusion in vivo.* Blood, 2007. **110**: p. 2708-2717.
19. Porzig, H., et al., *G-protein-coupled receptors in normal human erythroid progenitor cells.* Naunyn-Schmiedebergs Archives of Pharmacology, 1995. **353**(1): p. 11-20.
20. Baumann, R., et al., *Ontogeny of catecholamine and adenosine receptor-mediated cAMP signaling of embryonic red blood cells: Role of cGMP-inhibited phosphodiesterase 3 and hemoglobin.* Blood, 1999. **94**(12): p. 4314-4320.
21. Ebert, E.C., M. Nagar, and K.D. Hagspiel, *Gastrointestinal and Hepatic Complications of Sickle Cell Disease.* Clinical Gastroenterology and Hepatology, 2010. **8**(6): p. 483-489.

Organ culture modeling of distraction osteogenesis

Marnie M Saunders[1], Van Sickels J[2], Heil B[3], Gurley K[3]

[1]Department of Biomedical Engineering, The University of Akron, Akron, OH 44325
[2]College of Dentistry, University of Kentucky, Lexington, KY 40506
[3]Center for Biomedical Engineering, University of Kentucky, Lexington, KY 40506
E-mail: mms129@uakron.edu

ABSTRACT
Bone cell mechanotransduction involves the process by which bone cells sense and coordinate their activity in response to mechanical loading. *In vitro* and *in vivo* models are commonly used but may overly simplify (*in vitro*) or complicate (*in vivo*) the response making the effects of the load difficult to discern or of questionable clinical relevance. The author previously proposed the use of an organ culture system for mechanotransduction studies. In contrast to previous organ culture research addressing accelerated resorption effects, the goal was to determine if a whole bone organ culture could remain viable in culture for a period of time sufficient to study the short-term response of physiologic loading-induced maintenance/osteogenesis. If successful, the organ culture system would provide more of a biomimetic environment simplifying the systemic response seen *in vivo* while increasing the biological relevance over *in vitro* systems. Here we continue with this work. That is, to be useful as a mechanotransduction model, the organ culture system needs to be able to correctly simulate relevant, clinical conditions. In the current paper, the applicability of an organ culture approach to simulate distraction osteogenesis is evaluated and initial effects on bone viability and mechanical performance are presented.

In distraction osteogenesis (DO), mechanical forces are applied to generate new bone. These procedures are conducted in both orthopaedic and craniofacial indications and can range from devices incorporating simple linear to multiplanar vectors. DO devices can rely on internal or external fixation and while internal systems can be limited in use given design dictates within a confined space (eg. craniofacial applications), external systems can be inaccurate given the distance from the distraction mechanism to the bone.

In the distraction procedure, a pseudo growth plate is created and in essence the body is 'tricked' into the osteogenic potential of immature bone. In these cases an osteotomy or corticotomy is created and the distractor is placed to span the fracture site. The distraction procedure encompasses three phases, the latency phase, the distraction phase and the consolidation phase. In dealing with craniofacial distraction in cases involving neonates, such as mandibular distraction to treat airway obstruction, the neonatal bone is highly osteogenic and the corticotomy/osteotomy and latency phase are not necessary. During the phase of active distraction, the system is manually elongated in rates generally on the order of 1 mm/day. Once the desired length is reached, the distractor is locked into place to enable consolidation. The consolidation phase enables the bone to stiffen and ends with the removal of the distractor. With the relative novelty of this technique, limited use and the many variables that contribute to the success of the procedure, much of this work reduces to trial and error. Given the importance of facial symmetry in aesthetics, craniofacial surgeons can be disappointed with results that are inaccurate to fractions of a millimeter. Therefore, systems that could help to isolate the effects of the load and enable the study of the mechanisms and pathways involved in distraction would prove useful to enhance outcome predictability and improve device development.

In previous work, Saunders, et al. developed an *ex vivo*, or organ culture model in a neonatal rat long bone that had as the goal to be used as a model for mechanotransduction research [1]. That is, a model was developed that could be used to study the short-term mechanisms by which bone cells respond to brief bouts of mechanical stimulation in a biomimetic environment incorporating a native matrix and the cells in the appropriate ratios and 3D architecture. In the current work, we employ this model to investigate if the organ culture model is responsive to loading protocols simulating linear distraction for the purpose of extending this model to study DO mechanisms. To accomplish this, we first demonstrate culture viability with microCT analysis, a much more powerful and accurate technique than many of the rudimentary techniques initially utilized [1]. We then employed two loading regimes: one, a single distraction loading bout; and, the second, an equal distraction bout

T. Proulx (ed.), *Mechanics of Biological Systems and Materials, Volume 2*, Conference Proceedings of the Society for Experimental Mechanics Series 9999, DOI 10.1007/978-1-4614-0219-0_23,

repeated three times over the course of one week. At the end of the week, the effects of the two loading regimes on bone properties were characterized and compared to each other and their no-load contralateral controls.

MICROCT ANALYSIS OF ORGAN CULTURE VIABILITY

To more accurately assess viability of the organ culture time period and extend the model analysis out to include an assessment of a 1 month time period, twenty-eight femoral pairs were harvested from 5-day-old neonatal Wistar rat pups and maintained in culture for 1 (n=9), 2 (n=9) or 4 (n=10) wk, then analyzed using a MicroCT-40 computed tomography system (Scanco Medical, Basserdorf, Switzerland) and compared to their day 1 contralateral controls as well as normalized to contralateral controls and compared across culture time periods. Wistar rat femurs were harvested and maintained in BGJb culture medium (Gibco, Invitrogen) supplemented with 15% fetal bovine serum (FBS) and 2% penicillin/streptomycin (P/S) for 1, 2 or 4 wk. Bones were scanned using source settings of 55 kV, 145 μA and high resolution [2]. Each scan produced 50, 2048 x 2048 2D axial slices in the midshaft of the bone (0.4 mm). Inertial, areal, volumetric and density properties were analyzed.

The 1 wk results, as shown in Figure 1, are in agreement with previous model findings from rudimentary techniques supporting the use of the organ culture model in studies out to 1 wk of culture. Specifically, with respect to inertial properties, maximum and minimum moments of inertia increased 67.6% and 60.2%, respectively, while polar moment of inertia increased 64.3% over the 1 wk culture period. All differences were statistically significant ($p<0.001$). With respect to minimum and maximum loading resistance (section moduli), Imax/Cmax and Imin/Cmin increased 69.0% and 67% over the 1 wk culture period, respectively. Differences were statistically significant ($p<0.001$). With respect to density, 1 wk in culture significantly ($p<0.001$) increased bone density 86.6%. With respect to areal properties, bone area and total area increased 69.7% and 36.6%, respectively, while medullary area decreased 70.9% over the 1 wk culture period. All differences were statistically significant ($p<0.001$). With respect to volumetric properties, bone volume and total volume increased 61.3% and 36.3% over the 1 wk culture, respectively. Differences were statistically significant ($p<0.001$).

Comparing results across time points by normalizing microCT time point values to day 1 contralateral controls revealed a significant loss in organ culture viability associated with the 1 month culture period. For example, resistance to bending decreased 25.9% (Imax/Cmax) ($p<0.05$) and 24.0% (Imin/Cmin) ($p<0.05$) over the 2 wk culture period and 39.3%

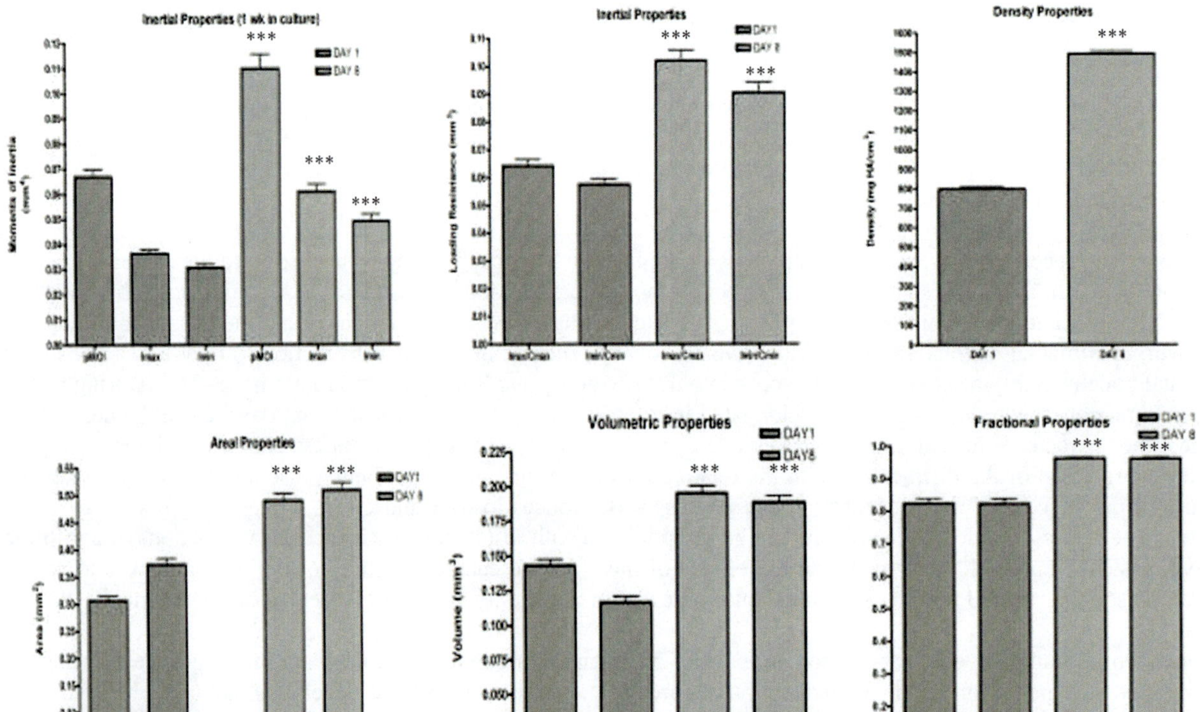

Figure 1. MicroCT results from neonatal bones as a result of 1 wk maintenance in culture. As demonstrated, bones remained viable and continued to experience an increase in density, inertial, areal, and volumetric properties. Results are plotted as mean ± SEM.

(Imax/Cmax) (p<0.001) and 38.6% (Imin/Cmin) (p<0.001) over the 4 wk culture period. Density decreased 41.6% and 53.0% (p<0.001) over the 2 and 4 wk culture periods, respectively. These findings, not surprisingly, suggest that the organ culture models, as maintained here (passive diffusion), are most effective if used for short-term studies (≤ 1 wk) and that while the bones at 2 wk of culture are more dense and stronger than their counterparts at 24 hr post-harvest, their loss of viability is significant given the comparison to the bones from the 1 wk time point. Four week results indicate a relatively dead culture that in many cases was approximately equivalent to the 24 hr controls.

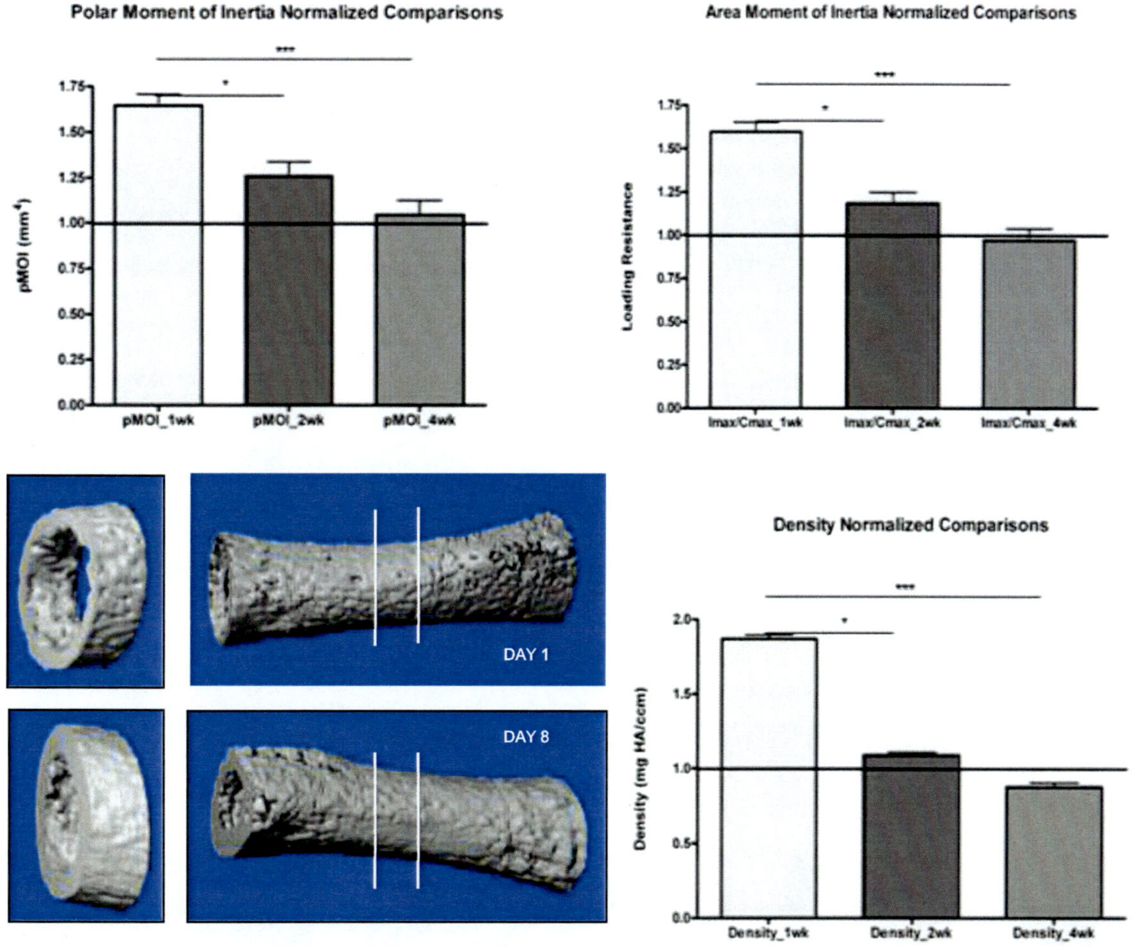

Figure 2. Normalized microCT results show the effects of a 1 month culture period on the organ culture model. Normalized results greater than 1 indicate an increase in culture time point parameter (eg, density) in comparison to the contralateral control at 24 hr post-harvest. Normalized results less than 1 indicate a decrease in culture time point parameter in comparison to contralateral controls at 24 hr. Furthermore, as demonstrated, increases that were observed at 1 wk of culture were significantly reduced by 2 wk of culture and completely lost by 4 wk of culture. These results support the short-term (≤ 1 wk) use of the organ culture system. Results are plotted as mean ± SEM.

EVALUATION OF THE ORGAN CULTURE SYSTEM UNDER DISTRACTION

To determine if the organ culture approach could be an effective model for DO studies, femurs from 5-day-old Wistar rats were utilized. Unless otherwise noted, all femurs were maintained in culture for 1 week (37°C and 5% CO_2). Bones were maintained in culture in 12 well plates on stainless steel mesh inserts to maintain the bones at a liquid/gas interface and organ culture medium (BGJb, Gibco) supplemented with 15% FBS and 2% P/S was added daily to counteract evaporation. Prior to any experimentation, bones were allowed to equilibrate in culture for 24 hr. Following observation of the 24 hr acclimation period, all bones were replenished with fresh medium to minimize any harvest–induced insult culture effects [3].

Group I (GRP I) consisted of 23 bone pairs; Group II (GRP II) consisted of 21 bone pairs. From GRP I, all 23 pair were assessed for non-destructive dimension and mass changes; from GRP II, all 21 pair were assessed for non-destructive dimension and mass changes. For each group, 3 pair were assigned to viability studies and 9 pair were assigned to microCT and destructive massing. For GRP I, 11 pair were assessed for mechanical properties and for GRP II, 9 pair were assessed for mechanical properties. One femur from each pair was assigned to distraction and the contralateral control was maintained in culture for 1 wk without distraction. Given the pilot study nature of this work, two mechanical distraction protocols were applied, Figure 3. Bones assigned to GRP I were subjected to 2% shaft elongation 24 hr post-harvest for a total of 2 hr; bones assigned to GRP II were subjected to 2% shaft elongation 24, 72 and 120 hr post-harvest for 2 hr each time point for a total of 6 hr. The initial waveforms were selected to address the effectiveness of multiple loading bouts (GRP II) while minimizing the risks of contamination in comparison to daily and twice daily manipulation of bones, which more closely mimic clinical distraction protocols but also greatly increase contamination risks. In addition, they were used to address the ability to manipulate the relatively fragile neonatal bones on multiple days while maintaining an intact system. On day 8, following the 1 wk culture period, all bone analyses were completed and the effects of distraction on dimensional and mechanical properties were assessed.

Figure 3. Bones in Group I (GRP I) were distracted 24 hr post-harvest for 2 hr and 2% bone shaft elongation; bones in Group II (GRP II) were distracted 24, 72, 120 hr post-harvest for 2 hr and 2% bone shaft elongation at each time point. All analyses were completed on day 8 following 1 wk in culture.

An in-house device originally developed for liquid metal strain gauge (LMSG) calibration was modified and used as a simple 'distractor'. The device centered around a micrometer head (resolution of 0.0254 mm) affixed to an aluminum plate that was used to dial-in the 2% shaft elongation and a large thumbscrew pushing against a brass bar was used to lock in the distraction, Figure 4. Bone shaft elongation was verified optically, Figure 5. Suture-lassoed diaphyseal bone ends were connected to the distractor on one end via a sliding block and the other end via a fixed block such that the dialed in displacement represents the bone shaft distraction plus suture elongation. Suture strands

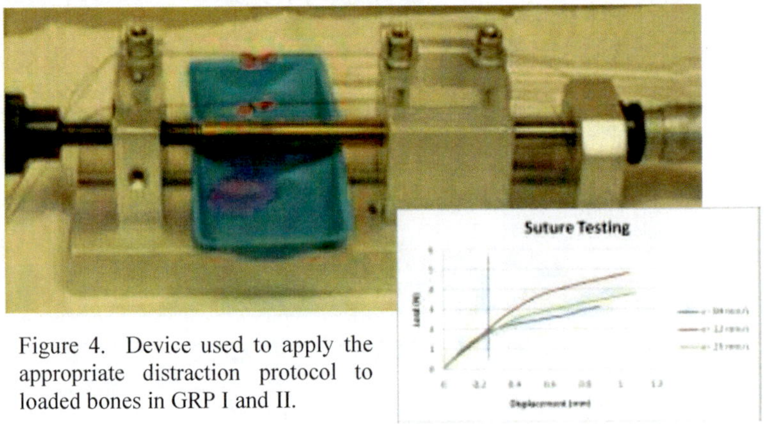

Figure 4. Device used to apply the appropriate distraction protocol to loaded bones in GRP I and II.

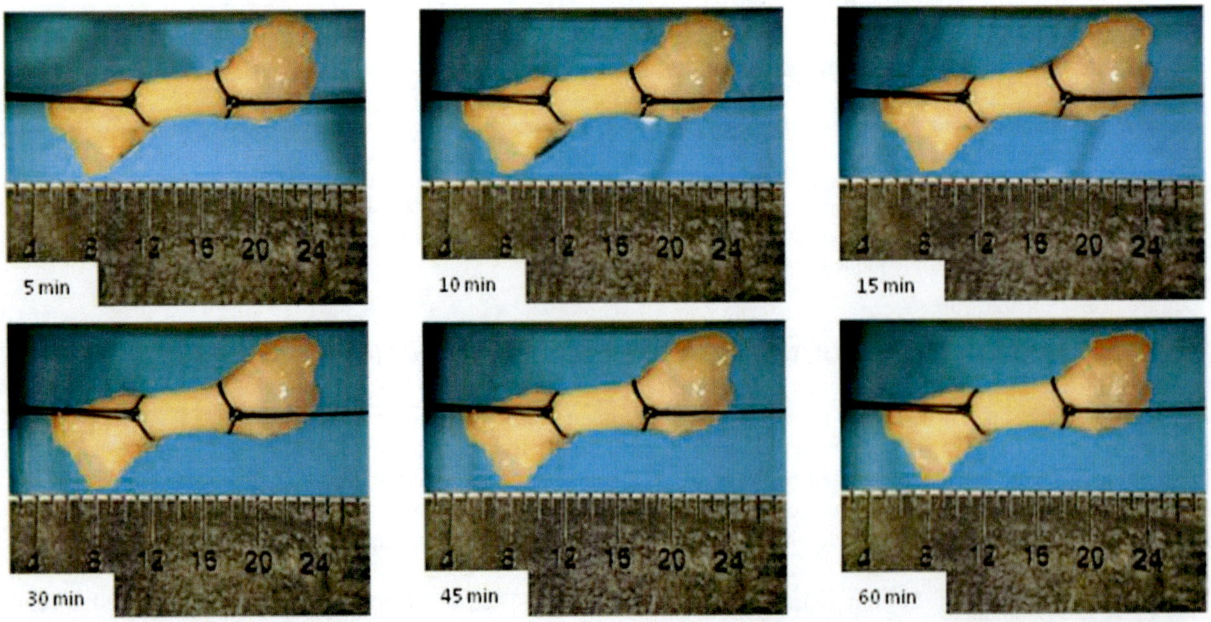

Figure 5. Time lapse photographs used to optically verify bone shaft elongation of 2% during the active loading phase (distraction) of GRP I and II. Lassoed bone elongation measured from the micrometer head on the distractor incorporated both bone shaft elongation and suture stretch; optical measurements enabled separation of the shaft from total elongation.

were sandwiched between the aluminum blocks and polycarbonate plates and held in place with socket head cap screws. A thumbscrew was then engaged against the micrometer rod to minimize relaxation effects. Optical techniques were used to verify the 2% bone shaft elongation independent of suture stretch and studies on suture strength were completed to verify that suture loading was in the elastic region and the 2% shaft elongation load was not of a magnitude to prematurely break the suture (braided, 5-0). During the distraction period, the bones were hydrated and controls were subjected to the same alterations in environment with the exception of loading application. Following the 2 hr distraction, bones were returned to culture to await the next loading bout, or the end of the 1 wk culture period.

DISTRACTION EFFECTS ON CELL VIABILITY

Lactate dehydrogenase (LDH) is an enzyme indicative of energy conversion [4,5]. If a cell is actively metabolizing, LDH will be present in the cell while it is absent from metabolically inactive cells. As such, whereas hematoxylin and eosin (H&E) stains may be used to show the presence of the cell in the bone tissue, the technique does not establish the viability of the cell. In contrast LDH staining effectively assesses not only the presence of the cell but if the cell is viable (metabolically active). A protocol for assessing osteocyte viability using LDH staining was developed from existing techniques [4,5] and successfully used to quantify osteocyte viability in the neonatal organ culture model as a function of time in culture [1]. Briefly, bones were washed and placed in a stock base solution containing Polypep, Gly-Gly and Nictinamide Adenine Dinucleotide in lactic acid. A pH of 8.0 was established and nitroblue tetrazolium was added. After 4 hr of incubation, the reaction chemicals were removed and the bones were rinsed with deionized water, fixed with 4% paraformaldehyde and refrigerated for 24 hr prior to decalcification in Formical-2000 (Decal Chemical, Tallman, NY), with calcium sediment measurement used as a decalcification endpoint. Longitudinal cross-sections (6 per bone) were analyzed under a Nikon Eclipse E600 light microscope (Nikon Instruments Inc., Melville, NY) for both control and distracted bones; pictures of the cross-sections were taken using a Nikon DN100 camera for manual counting. As shown in the Figure 6, osteocytes lying within the bone matrix of the medial and lateral cortices of the diaphyseal bone region were included in the counts. Following manual counts of viability for live (LDH +), dead (LDH-) and absent (empty lacunae) osteocytes, it was determined that GRP I displayed a 6.21% increase in LDH+ osteocytes and a 21.39% decrease in LDH- osteocytes, both of which were significantly different ($p<.05$) in contrast to their non-distracted contralateral controls. GRP II displayed a 13.02% increase in LDH+ osteocytes and a 22.04% decrease in LDH- osteocytes, both of which were significantly different ($p<.01$) in contrast to their non-distracted contralateral counterparts. Furthermore, GRP II exhibited a significant change ($p<.05$) in the number of empty lacunae between control and distracted bones with a 21.85% decrease in the distracted bones. Results are in agreement with the distraction having a stimulatory effect on the bone and a greater stimulatory effect was observed in GRP II with the repetitive loading in comparison to the single bout of loading of the bones in GRP I.

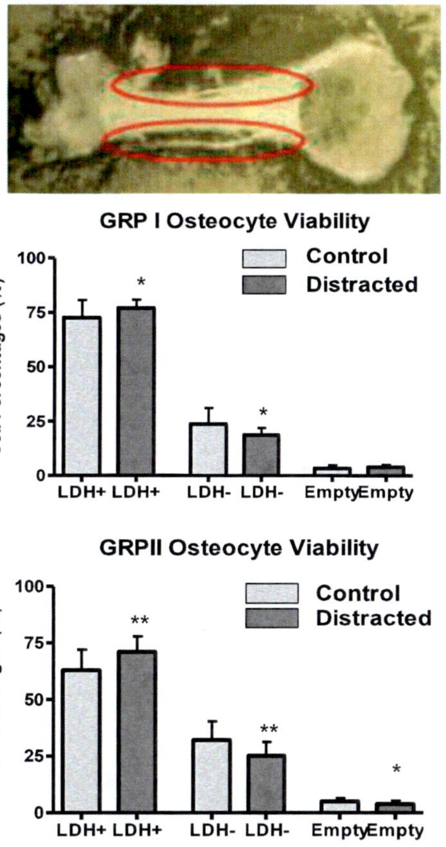

Figure 6. Distraction effects on osteocyte viability for GRP I and GRP II, (mean ± SD). Distracted bones exhibited a significant increase in cell viability in comparison to the no-load contralateral controls. Furthermore, the multiple loading bouts of bones in GRP II were more stimulatory than the single bout of GRP I.

DISTRACTION EFFECTS ON DIMENSIONS

To assess the effects of distraction on dimensional changes, all bones were subjected to nondestructive dimensional analysis and destructive mass analysis (dry and ash massing) was completed on the 3 pair of bones per group assigned to microCT analysis following microCT scanning, Figure 7. Bone dimensions were measured using the Nikon dissecting microscope outfitted with the Nikon digital camera and analyzed with ImageJ software (NIH). Femur lengths, shaft lengths and shaft diameters were determined from the averaging of three independent measurements. Weight changes were determined using a Sartorius (CP-64) balance. Bones were weighed 'wet', following drying off on an absorbent towel; 'dry', following defatting in acetone (72 hr) and air drying (24 hr); and, 'ash', following burning for 6 hr in a muffle furnace (600°C). Percent ash fraction was determined using previously established techniques [6].

Comparing the changes of the distracted bones with their contralateral controls revealed that while there were no significant differences for any of the dimensional results, there were trends that were seen in both groups for all three dimensional measurements (lengths and diameter). Of particular interest are the bone shaft lengths which increased 1.41% for GRP I and 2.09% for GRP II in comparison to their non-distracted contralateral controls. Assessing changes in weight as a function of distraction, differences between the distracted and contralateral control bones followed similar trends. Of particular interest, ash weight increased 1.60% for GRP I and 4.63% for GRP II in comparison to their non-distracted contralateral controls. Ash fraction increased 2.88% for GRP I and 5.23% for GRP II in comparison to their contralateral controls. These trends in dimensional increases of the distracted bones of GRP I and II in comparison to their non-distracted controls in conjunction with the increased trends being higher in GRP II with the multiple distraction bouts support an increase in shaft length with distraction that is qualitatively consistent with the clinical condition. Furthermore, the trends in ash weight increases of the distracted bones of GRP I and II in comparison to their non-distracted controls in conjunction with the increased trends being higher in GRP II with the multiple distraction bouts support an increase in mineral content as a result of distraction. This again is qualitatively consistent with the clinical condition and an increase in new bone formation (osteogenesis) with the distraction, supporting the validity of an organ culture model of DO.

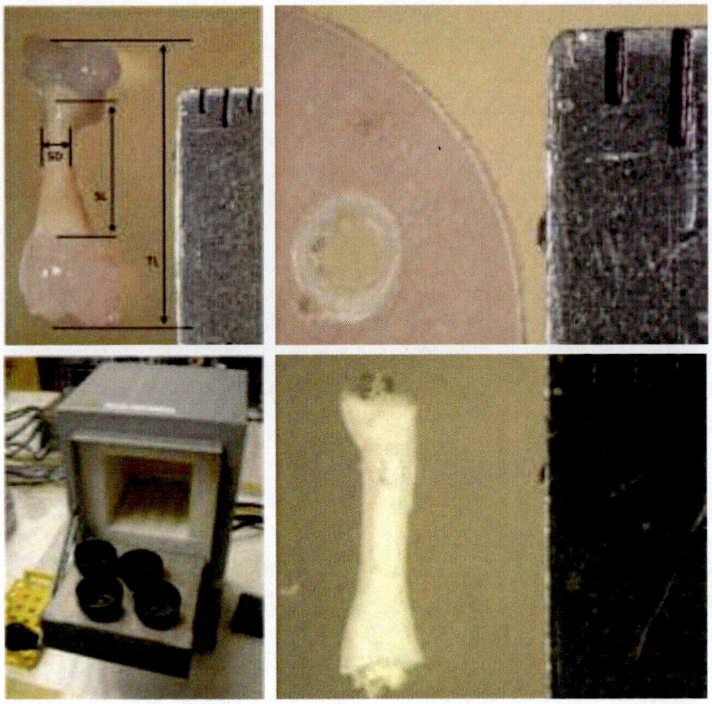

Figure 7. Bone dimensions and weight changes as a function of distraction were evaluated. Clockwise (a) bone shaft dimensions, (b) cross-sectional dimensions and (c) weight changes including ash weight/mineral content following burning in a muffle furnace (d) were determined.

DISTRACTION EFFECTS ON MECHANICAL PROPERTIES

To assess the effects of distraction on mechanical properties, bones were subjected to destructive three-point bend testing using a custom built small-scale loading platform developed in-house [7,8]. Bones were placed on the fixtures such that the anterior cortex of the bone shaft was in tension and the posterior cortex was in compression. Bones were tested to failure under displacement control using a linear ramp waveform at a rate of .4 mm/sec and a data collection rate of 10 Hz. For data collection, the platform was outfitted with a 22.5 N load cell and a 25 mm displacement sensor.

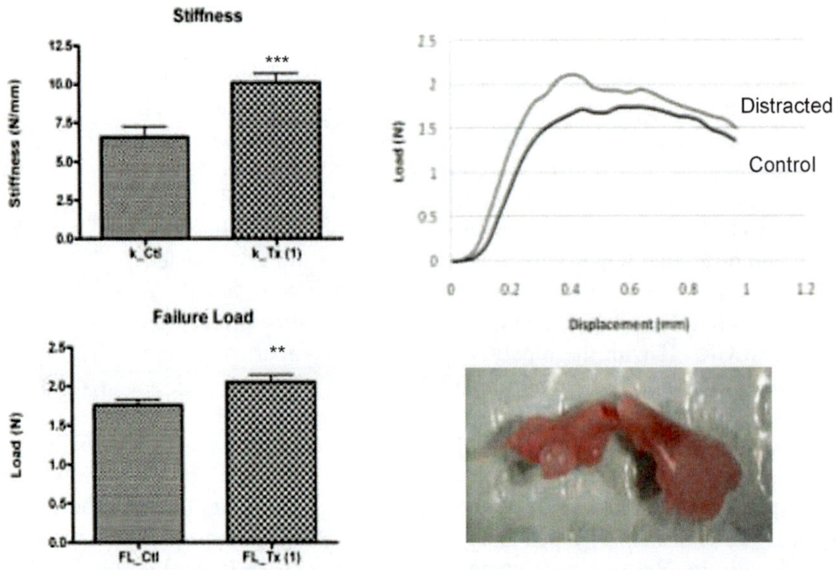

Figure 8. Distraction in organ culture increased bone strength. Failure load (FL) and stiffness (k) were significantly increased in distracted bones in comparison to contralateral counterparts maintained in culture for 1 wk. Results shown are from GRP I.

From analysis of structural properties failure load increased 9.15% for GRP I and 18.85% for GRP II in comparison to non-distracted contralateral controls. Failure displacement decreased 1.45% for

GRP I and 11.49% for GRP II in comparison to contralateral controls. Stiffness increased 31.18% for GRP I and 53.12% for GRP II in comparison to non-distracted contralateral controls. Results for stiffness and failure load were statistically significant with the results of GRP I shown in Figure 8. Taking the mechanical results into consideration, the combination of an increase in failure load (and moment), decrease in failure displacement and increase in stiffness are consistent with an increase in strength as a result of distraction. Moreover the larger percent changes in GRP II distracted bones in comparison to GRP I suggest that the multiple loading bouts are more stimulatory than the single bout. Furthermore, they also support the use of the organ culture model in DO studies suggesting that these systems are capable of surviving multiple loading cycles on multiple days of culture. Had this not been the case, the use of such a system to model a clinical condition that is continuously loaded over several days would certainly bring into question the usefulness of an organ culture system for simulating clinical DO.

DISCUSSION

Understanding the process by which bone cells respond to mechanical loading is critical to eradicating metabolic bone diseases, optimizing implant design, improving fracture healing outcomes, developing bioreactors and improving the success of functional bone tissue engineering. Models that enable the study of the mechanisms and pathways by which bone cells respond to mechanical stimulation are pivotal to this mission. Furthermore, models that enable the isolated cellular response to the load to be measured at the macroscopic scale are useful, particularly if they can accurately simulate the clinical loading condition. In the current work the whole bone organ culture model was assessed for its ability to be used as a model of DO. While this work only lays the foundation for this work, the results are promising and indicate that the model remains viable and is responsive to DO loading protocols. While this work addressed the simplest case of a linear distraction system and did not incorporate a latency period, qualitative and quantitative changes observed from both loading regimes are in agreement with events observed in clinical DO.

ACKNOWLEDGEMENTS

Partial support for this work was provided by a National Science Foundation Career Award.

REFERENCES

1. Saunders MM, Simmerman LA, Reed GL, Sharkey NA, Taylor AF. Biomimetic bone mechanotransduction modeling in neonatal rat femur organ cultures: Structural verification of proof of concept. Biomech Model Mechanobiol 9:539-550, 2010.

2. Bagi CM, Hanson N. The use of micro-CT to evaluate cortical bone geometry and strength in nude rats: Correlation with mechanical testing, pQCT and DXA. Bone 38(1): 136-144, 2006.

3. Garrett R. Assessing bone formation using mouse calvarial organ cultures. In:Helfrich MH, Ralston SH (eds) Bone research protocols, chap 14, Humana Press, Totowa, 2003.

4. Mann V, Huber C. The influence of mechanical stimulation on osteocyte apoptosis and bone viability in human trabecular bone. J Musculoskelet Neuronal Interact 6(4): 408-17, 2006.

5. Wong SY, Dunstan CR, Evans RA, Hills E. The determination of bone viability: a histochemical method for identification of lactate dehydrogenase activity in osteocytes in fresh calcified and decalcified sections of human bone. Pathology 14(4):439-442, 1982.

6. Mikic B, Battaglia TC, Taylor EA, Clark RT. The effect of growth/differentiation factor-5 deficiency on femoral composition and mechanical behavior in mice. Bone 30(5):733-737, 2002.

7. Saunders MM, Donahue HJ. Development of a cost-effective loading machine for biomechanical evaluation of mouse transgenic models. Med Eng Phys 26:595-603, 2004.

8. Saunders MM, Taylor AF, Du C, Zhou Z, Pellegrini VD Jr, Donahue HJ. Mechanical stimulation effects on functional end effectors in osteoblastic MG-63 cells. J Biomech 39(8):1419-1427, 2006.

Mechanical Properties of Tooth Enamel: Microstructural Modeling and Characterization

Toshio Nakamura, Cunyou Lu, and Chad S. Korach
Department of Mechanical Engineering, State University of New York at Stony Brook
Stony Brook, NY, 11794-2300 USA

ABSTRACT
Human tooth enamel possesses unique morphology characterized by repeated cell arrangement. Due to its complex structure, various investigators have reported diverse mechanical models and properties in their experimental and numerical studies. In this paper, the proper behavior described by the monoclinic model is reported and the effects of hydroxyapatite fibers and prism rods on the effective properties of tooth enamel are presented. The results are obtained from 3D finite element analysis with a novel procedure to construct periodic cell models and impose boundary conditions. This specialized approach allows determinations of 13 independent material constants needed for the monoclinic model. These constants may be used to construct homogenized models to study the mechanical behavior of entire tooth under abrasion, erosion, wear and fracture. In addition, a large scale 3D analysis was also performed to simulate instrumented micro-indentations of tooth enamel. The computed results are compared with experimentally obtained load-displacement measurements to verify the proposed model for the tooth enamel.

1. Introduction

A tooth is the hardest substance in the human body and mechanically structured to withstand high stresses. During mastication process, it may experience large physical forces, as much as 300N [1] with contact stresses reaching up to 2.5GPa [2]. The outer layer of tooth is hard enamel, which is supported by dentin, a more resilient but less mineralized connective tissue, and the inner pulp. The highly mineralized enamel functions as a shield with hardness and toughness to withstand masticating forces, as well as protector for the dentin and the nerves from mechanical and chemical impact. Its cross-sectional thickness varies from less than 1mm (in the crown-root junction) to around 2.5mm (in the crown surface). Highly calcified enamel consists approximately 96% in weight by inorganic crystalline hydroxyapatite (HAp) and the rest with organic non-collagenous proteins [3]. The near total absence of protein matrix leads to two major issues of enamel: first, brittleness albeit with high hardness and incapability of self-regeneration with lack of proteins.

The primary structural unit of enamel is a prism rod with variable internal HAp crystal orientations and outer sheath which is mainly organic. The acid etched surfaces of enamel are shown in Fig. 1 for the occlusal and transverse views of prisms. Generally the prism grows perpendicular to the surface of dentin thus they can bear extremely large normal loads during masticating process by transferring high loads into the softer dentin tissue. The size of a single prism is approximately 7×5μm. The cross-section of the prisms is shaped like a keyhole with a large circular head and a tail, periodically repeating in the cross-sectional plane. The prism consists of highly organized and packed hydroxyapatite (HAp) with specific crystal orientations and an organic matrix of proteins

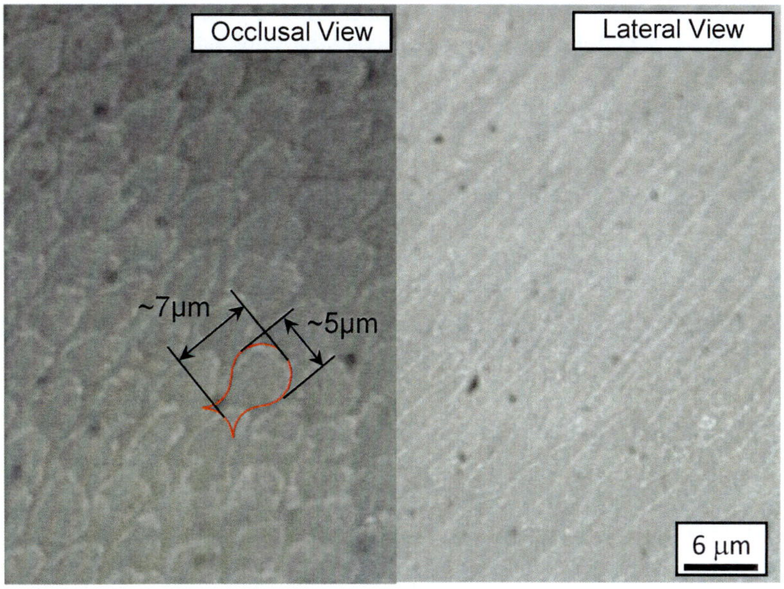

Fig. 1. Optical images of prism rods after etched with acid. A typical shape and size of prism rod are noted.

T. Proulx (ed.), *Mechanics of Biological Systems and Materials, Volume 2*, Conference Proceedings of the Society for Experimental Mechanics Series 9999, DOI 10.1007/978-1-4614-0219-0_24,

surrounding the HAp crystals with thickness of 2nm [4].

The characteristic microstructures at various size scales are illustrated in Fig. 2. At the low scale level, HAp fibers that are 30~70nm wide and over 100μm in length are bonded with organic matrix of proteins, and they make up the prism rod. The calcium phosphate that comprises HAp contains a hexagonal crystal structure, and results in a hexagonal shaped cross-sectional profile of the crystal, though this becomes distorted due to compaction [5]. Young's modulus of HAp has been reported as 110~135GPa [6-9]. The main components of the proteins surrounding HAp crystals are noncollagenous proteins such as amelogenins, enamelin and ameloblastin [5]. In the modeling of enamel, there have been different values of the protein modulus used in past work. Spears used keratin's 4.3GPa [10] as the elastic modulus of the enamel proteins and assumed an isotropic material behavior. Xie's group has utilized a much smaller value of 0.1GPa in their work [11]. Collectively, the effective behavior HAp fibers and protein matrix is similar to those of fiber-reinforced composites. Hence the suitable anisotropic material to effective stress-strain relation is "transversely isotropic". The long axes of the HAp crystal composites run in the general direction of the longitudinal axis of the prism in the head part and rotate in a perpendicular direction to the prism in the tail region as illustrated in Fig. 3.

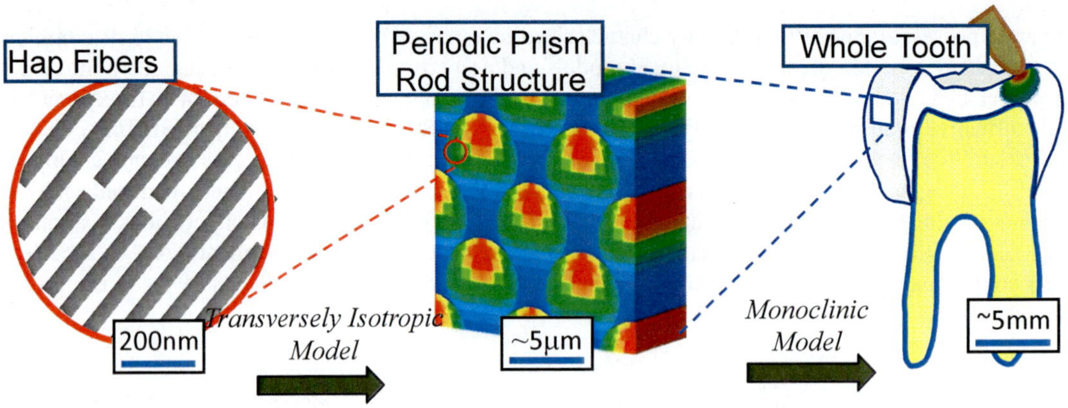

Fig. 2. Scale hierarchy of tooth enamel involves two levels of homogenization with two separate anisotropic models; transversely isotropic model from HAp fibers to prism rod, and monoclinic model from periodically arranged prism rods to a tooth.

The study of mechanical properties of enamel based on its microstructure can offer valuable insights into many aspects, such as mechanisms of wear processes, effects of erosion, and the development of restorative materials that imitate the structure and properties of natural enamel. The elastic mechanical properties of human tooth enamel have been studied through experiments and finite element simulations, because of the complicated microstructure, which contains variations in crystal orientations and non-homogeneous properties. The orientation was taken into consideration in several microscopic tests such as nano-indentation, but appropriate conditions for introducing crystal orientations within enamel rods and the property variations between rods and interrods are yet to be considered. The investigation of the micromechanical response of enamel rods through finite element modeling using a periodic model is presented in this work to determine the effective mechanical properties for enamel as a monoclinic material, and to determine the effects of microstructure on the stress field. Simulation of a spherical indentation in the longitudinal direction of an enamel bulk model is also investigated to study the microstructural response during indentation. Spherical indentation simulates the natural loading of enamel, thus provides a response related to its normal function.

2. Modeling of Prism Rods and Computational Procedure

The objective here is to utilize a proper constitutive relationship of enamel consisted of repeated prism rods as shown in Fig. 4. Within a prism rod, the property can be assumed to be constant along the axis of rod (z-direction). However, due to the various orientations of Hap fibers, the local property is a function of the in-plane coordinates (x and y) with the symmetry plane along the center as indicated. Since the geometry and material property possess only single plane of symmetry, the stress-strain relation must be represented as "monoclinic" model with 13 independent constants. Note that the orientations of HAp fibers (Fig. 3) inherently require 3D modeling of the structure.

2.1. 3D Representative Unit Cell

In order to determine the relationship between local (transversely isotropic) and enamel level (monoclinic) properties, finite element analysis is used to model prism rods. Similar to many composite analyses, a representative unit cell (RUC) or representative volume element (RVE) approach for the periodic nature of prism cells is utilized to reduce the size of modeling.

A RUC is a small material region that can structurally represent the entire material by repeating itself infinitely. Here the smallest such a model is a single prism rod. However, imposing loading conditions would be difficult along non-planar surfaces of key-holed geometry. Thus we have chosen a brick or cuboid geometry as the unit cell as shown in Fig. 4. Essentially this cell contains two equivalent prism rods in the model.

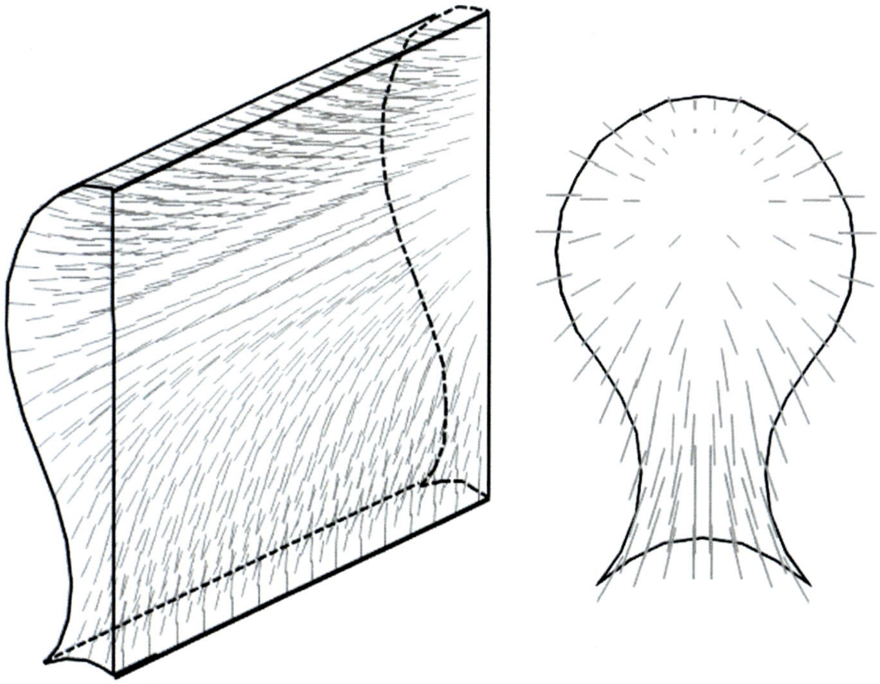

Fig. 3. Schematics of HAp fiber orientations within prism rod shown in two separate perspectives.

A great care was taken to construct 3D mesh of the brick RUC model. First since there is no material variation through thickness, one layer element is only needed along the z-axis. On the x-y plane, since the periodic boundary condition must be imposed on the outer boundaries, nodes on the opposing sides must match exactly. Furthermore some of the interior nodes must follow the physical interfaces of prism rods. Other interior nodes must be located so that the symmetry and conjugate elements in two prism rods are identical. The constructed mesh shown in Fig. 4 contains 240 8noded brick elements and 546 nodes. Here the dimensions are chosen as 7×7×7μm although the dimensional scales of RUC model do not enter in determining the effective properties.

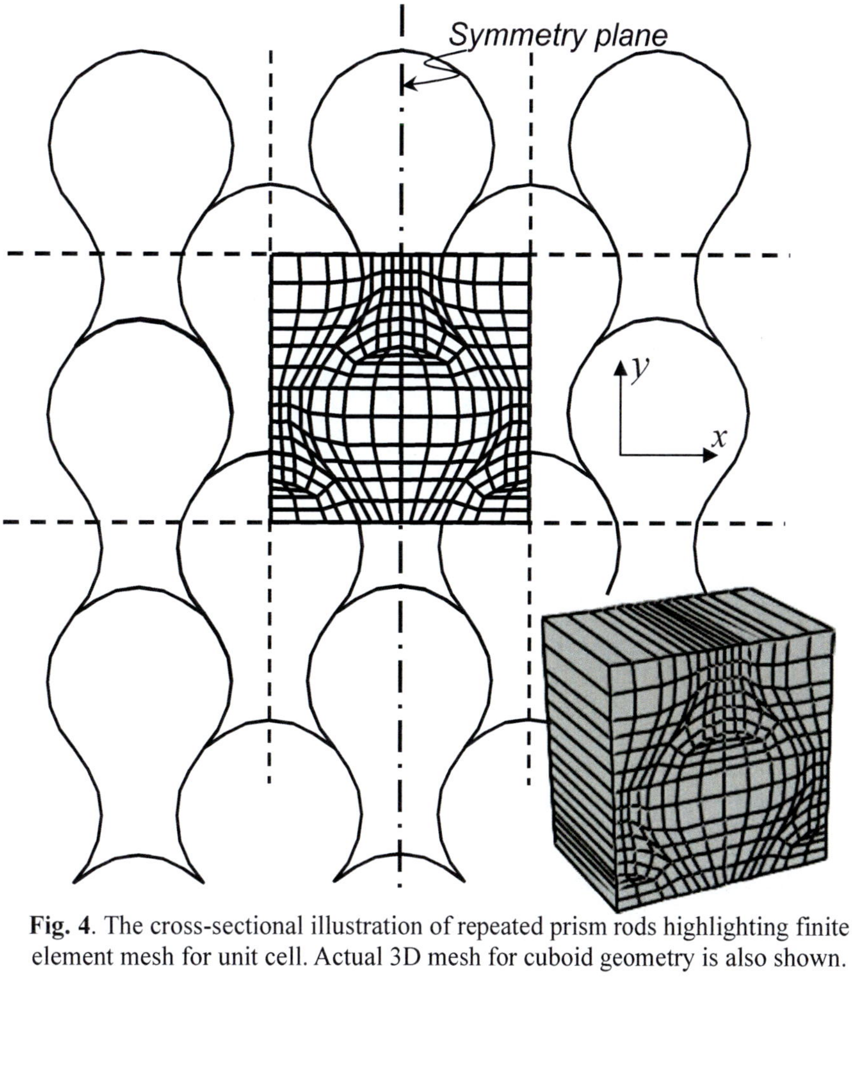

Fig. 4. The cross-sectional illustration of repeated prism rods highlighting finite element mesh for unit cell. Actual 3D mesh for cuboid geometry is also shown.

2.2. Local Properties within Prism Rod

Locally, the enamel prisms are composed of inorganic mineral-hydroxyapatite and organic matrix-proteins with specific orientations. The structure of the hydroxyapatite-protein matrix is similar to that of a fiber reinforced composite, which is axisymmetric in the long direction as shown in Fig. 3. It is suitable to treat the local properties as transversely isotropic. The local properties of the elements are modeled as transversely isotropic (with 5 material constants) and variable orientations of the HAp crystal composite with the elements. The transversely isotropic material contains 5 constants as E_L, E_T, ν_{LT}, ν_{TT} and μ_{LT}. Although some values were reported for the elastic moduli of HAp and protein, there is no measurement available for the effective properties for HAp fibers with protein matrix structures. Thus we have taken the following operational approach to assign the material values. First since the ratio of Young's moduli with respect to the fiber (longitudinal) and transverse directions are generally most critical parameter, we chose E_T / E_L as the primary variable. Although computed results can be normalized by either of the Young's moduli, we have assigned E_L = 100GPa, which should be in the close range of an actual value. The rest of constants, μ_{LT}, ν_{LT} and ν_{TT} are set to follow certain rules.

As shown in Fig. 3, the orientations of HAp fibers or the axis of transversely isotropic model varies with the prism rod. Here two angles are needed to define the orientation due to the 3D geometry. They are the *x-z* plane angle θ with respect to the global *x*-axis, and the *y-z* plane angle ϕ with respect to the global *y*-axis. Within the inner head region of the model, the angle θ ranges from 135° to 45° from left to right horizontally while ϕ ranges from 80° to 180° from top to bottom vertically. These angles were assigned suitably to the elements in the model.

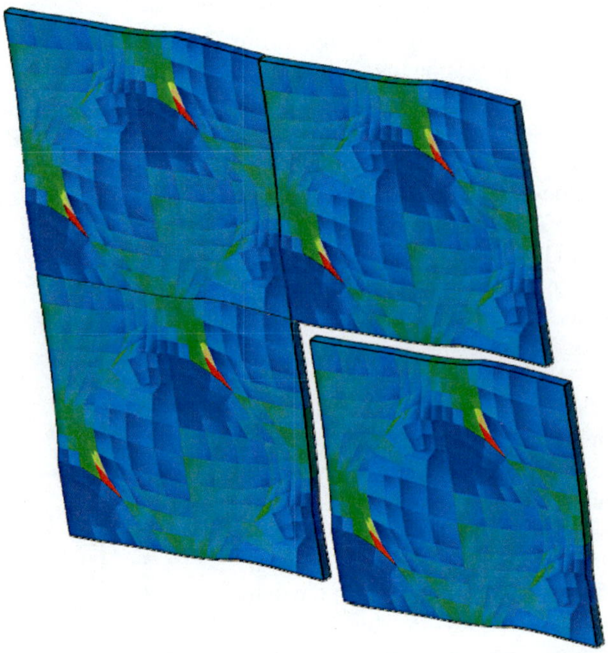

Fig. 5. Boundary displacements are properly constrained to match deformation on boundary and to represent periodic deformation of unit cells.

2.3. Periodic Boundary Conditions

Periodic boundary conditions must be prescribed on the entire perimeter of the model so that top/bottom, left/right and front/back surfaces must deform in the same pattern under any loading conditions. Essentially each pair of deformed surfaces must match exactly to fit as shown in Fig. 5. Unlike unusual symmetric boundary conditions, prescriptions of proper constraints are not simple. Particular degrees of freedom at nodes must be controlled to displace according to those of conjugate nodes while other degrees of freedom are left unconstrained. We have developed a procedure to constrain a vector formed by a node and a reference node to achieve the periodic boundary condition. This method ensures equivalent vectors to remain identical which maintains two conjugate element surfaces to have the same areas and the normal directions. As shown in Fig. 5, the unit cells repeat themselves without any gaps or overlaps.

2.4. Effective Properties of Enamel – Monoclinic

As described earlier, the single plane of symmetry requires the enamel, made up of many repeated prism rods, to be modeled as monoclinic. The macroscopic effective constitutive relation is more appropriately modeled as monoclinic (13 independent elastic constants) as shown in the equation:

$$\begin{pmatrix} \sigma_{11} \\ \sigma_{22} \\ \sigma_{33} \\ \tau_{23} \\ \tau_{31} \\ \tau_{12} \end{pmatrix} = \begin{pmatrix} C_{11} & C_{12} & C_{13} & C_{14} & 0 & 0 \\ C_{12} & C_{22} & C_{23} & C_{24} & 0 & 0 \\ C_{13} & C_{14} & C_{33} & C_{34} & 0 & 0 \\ C_{14} & C_{24} & C_{34} & C_{44} & 0 & 0 \\ 0 & 0 & 0 & 0 & C_{55} & C_{56} \\ 0 & 0 & 0 & 0 & C_{56} & C_{66} \end{pmatrix} \begin{pmatrix} \varepsilon_{11} \\ \varepsilon_{22} \\ \varepsilon_{33} \\ \gamma_{23} \\ \gamma_{31} \\ \gamma_{12} \end{pmatrix} \quad (1).$$

As can be seen in the above equation, unlike isotropic or even orthotropic materials, normal and shear stress-strain relations are not decoupled (e.g., purely normal loading causes shear deformation). Such a behavior makes determinations of the constants more difficult. Furthermore, determinations of the 13 constants in the stiffness tensor require several separate analyses with different loading conditions as described next.

3. Computational Analysis of Representative Unit Cell

Many separate finite element calculations are carried out to determine the effective monoclinic constants under various ratios of local properties (E_T / E_L ratio).

3.1. Determination of Constants under Various Loading Conditions

The model is analyzed under six different loading conditions to determine the monoclinic stiffness constants. They are executed in six directions on six corner nodes as shown in Fig. 6. Under each condition, average stress fields are generated by

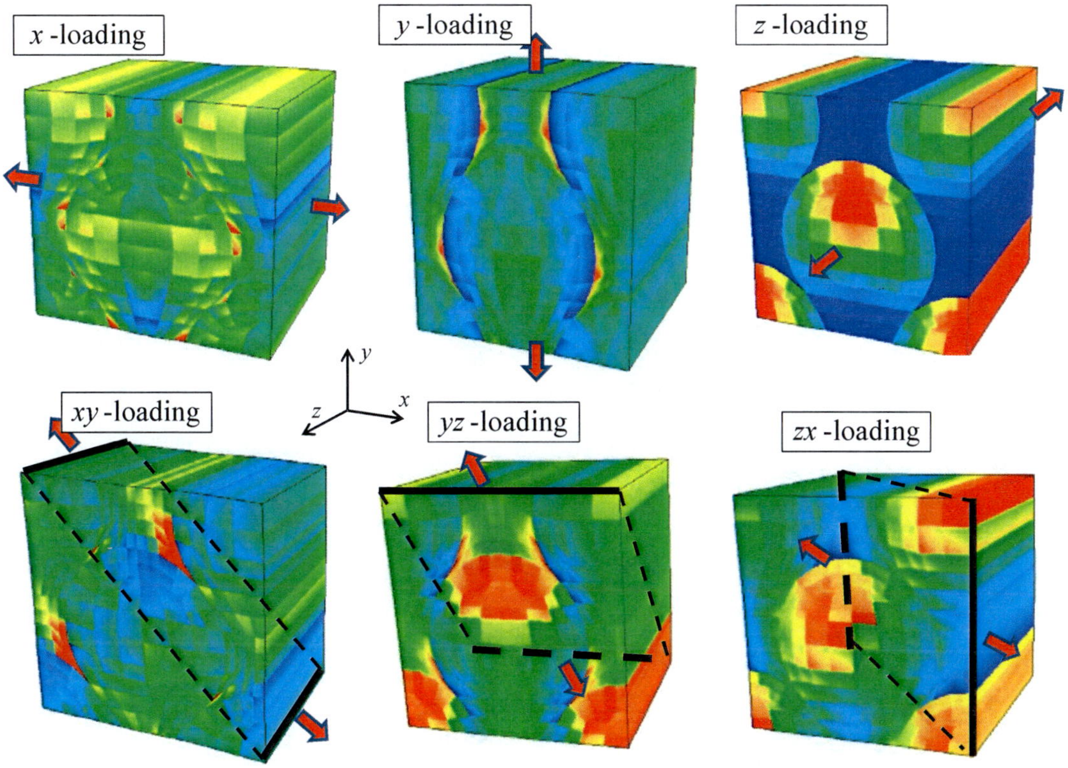

Fig. 6. Six separate loading conditions are applied to determine unknown monoclinic material constants. Contours of effective stress are shown.

dividing the reaction forces of the loading nodes over the nominal area *A* where the load was applied. To determine the average strain fields, the displacements of corner nodes are divided by appropriate lengths of the models. Thus, under each loading condition, six average components of stress and size average components of strain are obtained.

In each loading case, one can establish 6 equations for stress and strains using (1). Suppose under αth loading case, they are shown as $\boldsymbol{\sigma}_\alpha = \boldsymbol{C}\,\boldsymbol{\varepsilon}_\alpha$ where $\boldsymbol{\sigma}_\alpha$ and $\boldsymbol{\varepsilon}_\alpha$ are the stress and strain vectors containing 6 components shown in (1). With six separate loading conditions, one can obtain 36 equations although many of them are linearly dependent. To obtain the best estimates for the 13 material constants, the singular value decomposition (SVD) method is used. Here the equations are rearranged as $\boldsymbol{\sigma}_\alpha = \boldsymbol{\varepsilon}^*_\alpha\,\boldsymbol{C}^*$ where

$$\boldsymbol{\varepsilon}^{*\,i}_{\alpha} = \begin{bmatrix} \varepsilon_{11}^\alpha & \varepsilon_{22}^\alpha & \varepsilon_{33}^\alpha & \gamma_{23}^\alpha & 0 & 0 & 0 & 0 & 0 & 0 & 0 & 0 & 0 \\ 0 & \varepsilon_{11}^\alpha & 0 & 0 & \varepsilon_{22}^\alpha & \varepsilon_{33}^\alpha & \gamma_{23}^\alpha & 0 & 0 & 0 & 0 & 0 & 0 \\ 0 & 0 & \varepsilon_{11}^\alpha & 0 & 0 & \varepsilon_{22}^\alpha & 0 & \varepsilon_{33}^\alpha & \gamma_{23}^\alpha & 0 & 0 & 0 & 0 \\ 0 & 0 & 0 & \varepsilon_{11}^\alpha & 0 & 0 & \varepsilon_{22}^\alpha & 0 & \varepsilon_{33}^\alpha & \gamma_{23}^\alpha & 0 & 0 & 0 \\ 0 & 0 & 0 & 0 & 0 & 0 & 0 & 0 & 0 & 0 & \gamma_{13}^\alpha & \gamma_{12}^\alpha & 0 \\ 0 & 0 & 0 & 0 & 0 & 0 & 0 & 0 & 0 & 0 & 0 & \gamma_{13}^i & \gamma_{12}^\alpha \end{bmatrix} \quad (2),$$

and

$$\boldsymbol{C}^* = \begin{bmatrix} C_{11} & C_{12} & C_{13} & C_{14} & C_{22} & C_{23} & C_{24} & C_{33} & C_{34} & C_{44} & C_{55} & C_{56} & C_{66} \end{bmatrix}^T \quad (3).$$

By combining all six loading cases, one may also treat $\boldsymbol{\varepsilon}^*_\alpha$ as 36 by 13 component matrix $\boldsymbol{E}$, and then solve for the components of $\boldsymbol{C}^*$ vector or the elastic constants. Since $\boldsymbol{E}$ is not a square matrix, there may be multiple solutions of $\boldsymbol{C}^*$. To find the best estimates for the components of $\boldsymbol{C}^*$, the SVD method can be used the transformed strain matrix $\boldsymbol{E}$ [12,13]. It is an effective technique to solve for the unknown parameters based on the linear least-square fit approach when matrices are singular or non-square. Here $\boldsymbol{E}$ can be decomposed as $\boldsymbol{E}=\boldsymbol{USV}^T$, where $\boldsymbol{U}$ is a column-orthogonal matrix, $\boldsymbol{S}$ is a square diagonal matrix, and $\boldsymbol{V}$ is a square orthogonal matrix. Once $\boldsymbol{E}$ is decomposed, the generalized inverse form of $\boldsymbol{E}$ can be expressed as $\boldsymbol{VS}^{-1}\boldsymbol{U}^T$, so $\boldsymbol{C}^*$ can be calculated by $\boldsymbol{C}^*=\boldsymbol{VS}^{-1}\boldsymbol{U}^T\boldsymbol{\sigma}_\alpha$.

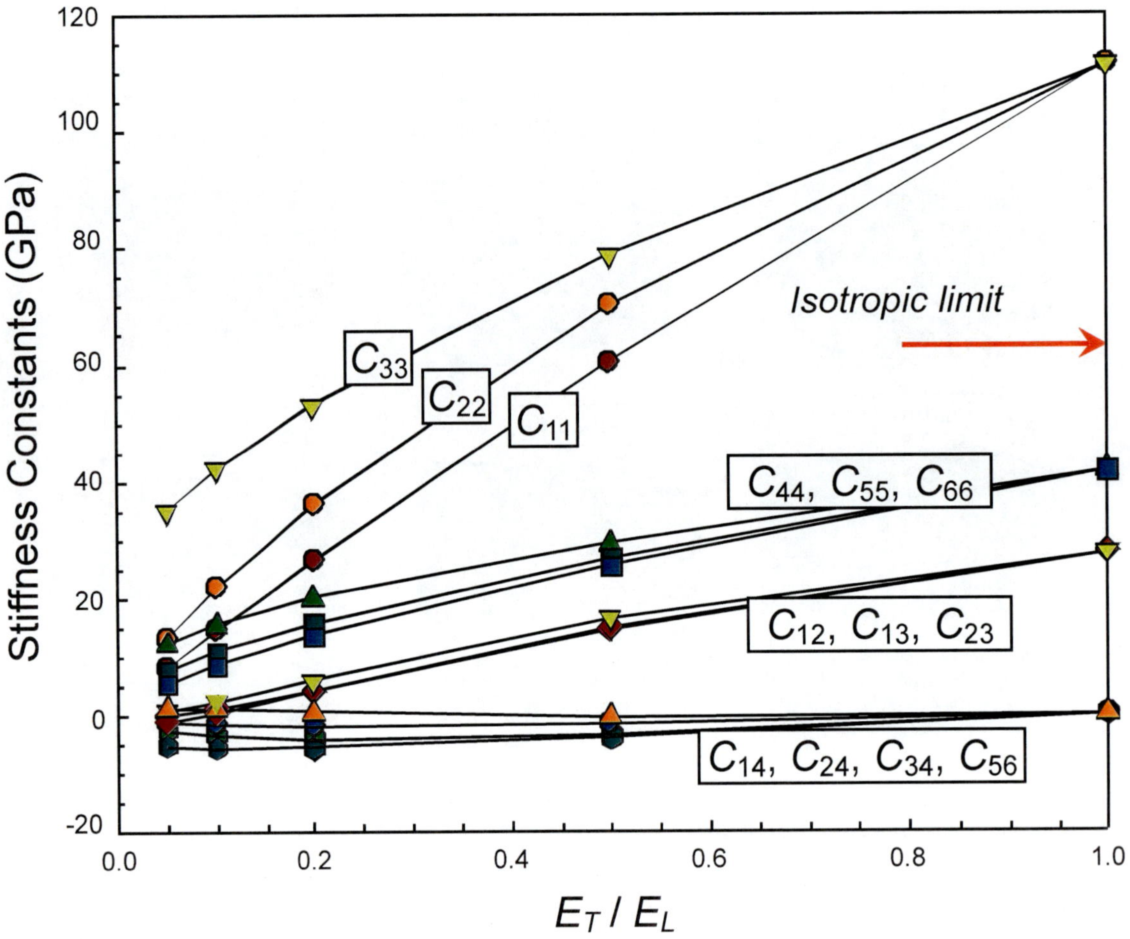

Fig. 7. Effects local modulus ratio (E_T/E_L) assigned for local transversely isotropic model on the 13 elastic constants of effective monoclinic model.

3.2. Computed Elastic Constants

The effective mechanical properties determined from the six loading cases for various E_T / E_L ratios are determined. The effective stress contours of the local property ratio of $E_T / E_L = 0.2$ shown in Fig. 6 indicate large stresses along the boundaries of enamel rods. This may imply an enamel rod boundary is the most likely location where micro-crack would occur under critical or repeated loading due to locally high strains which will disrupt the HAp crystal-protein composite.

Computed elastic constants (1) for five different ratios of E_T / E_L are shown in Fig. 7. Here E_L is fixed at 100GPa. As E_T changes, large variations on normal stress-strain constants C_{11}, C_{22} and C_{33} are observed. In fact, in-plane components C_{11} and C_{22} drop to very small values as E_T decreases to zero. These behaviors are expected since the HAp fibers are oriented more along the *z*-axis and the stiffest direction is always along the prism rod axis. This has implications to understanding the structure of the enamel rods in the tooth crown, where the rod heads point in the direction of the tooth cusp around the entire circumference of the tooth. The data indicates that the higher stiffness in the direction of the head-tail may allow enamel to absorb compressive stresses that are generated due to occlusal contact conditions (such as mastication with the primary molars). Other components also change as the ratio of E_T / E_L changes.

In order to verify the accuracy of the present procedure, a homogenized model with these monoclinic constants is separately analyzed. Using the same finite element mesh, the model was then reloaded in similar ways and then the resulting displacements are compared with the original model with inhomogeneous model (i.e., elements with transversely isotropic properties with variable fiber orientations). The differences among them are less than 0.01% for all of the components in averaged strain and stress fields for every E_T / E_L ratio.

4. Instrumented Indentation

One possible method to determine accurate values for the local transversely isotropic model (e.g., E_T and E_L) is to utilize

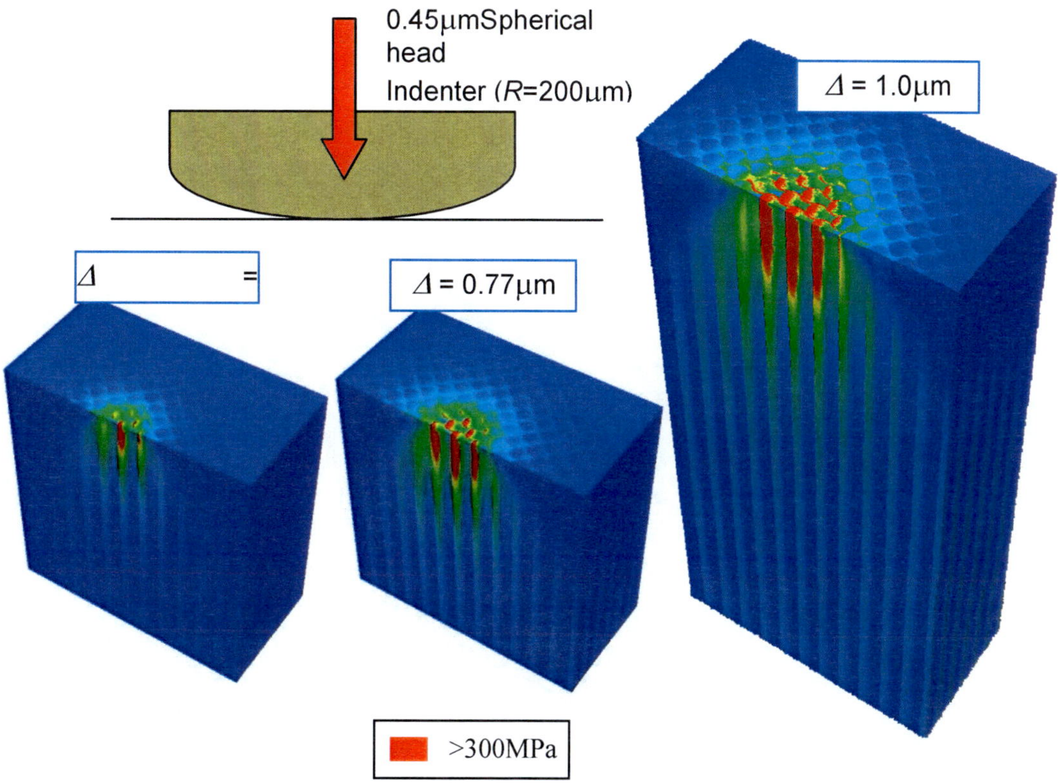

Fig. 8. Effective stress contours during the indentation for the heterogeneous model for E_T / E_L = 0.2.

instrumented indentation. In the present analysis, simulated indentation results are compared with experimental record.

4.1. Experimental Indentation Measurements

Experimental measurement of a spherical-head micro-indentation on the occlusal and transverse surfaces of an enamel sample with a radius of 200μm was conducted. The modulus of occlusal surface was measured as E_{norm}=56.5 ± 2.3GPa and the modulus of the transverse surface was measured as E_{tran}=34.5 ± 3.8GPa using Oliver and Pharr method. The ratio of E_{tran} / E_{norm} = 0.61 is close to E_{tran} / E_{norm} = 0.57 reported by He's group [14] where for shallower indentations (contact radius of ~3μm) was used. Measurements on the local elastic moduli of an enamel sample using atomic force acoustic microscopy exhibited 40~50% difference between the moduli of local properties which approximately correspond to the local property E_T and E_L in the model [15].

4.2. Finite Element Model

Using the unit cell in as the primary geometry, large meshes are constructed for the indentation simulations. Since the model possesses only single symmetry, half geometry must be modeled. After testing several model sizes (must be large enough to avoid boundary effects), we have chosen a model contains 6×12×22 = 866 unit cells with 42×84×168μm. Note that the length along the out-of-plane or the prism rod axis is adjusted for each element layers while the in-plane dimensions are kept the same. The total number of elements is 207,360. Due to the monoclinic nature of material properties, the large depth was needed to exclude the bottom boundary effects.

4.3. Indentation Results

Displacement controlled indentation was simulated for different local property ratios (E_T / E_L = 0.1, 0.2, 0.5) up to the maximum depth of 1μm The effective stress contours are shown in Fig. 8 for three different load levels. Unlike homogenous materials, they clearly show discontinuous stress fields as the material properties are heterogeneous. The large stress shown in red indicates stiffer response of head region of key-hole shaped prism rods. Note also that the large stresses appear to propagate to the bottom instead of remain near the indentation. Again these are the results of unique tooth enamel microstructure.

The indented load-displacement curves are shown in Fig. 9 for three different ratios of E_T / E_L. As the ratio decreases, the response becomes more compliant. For a comparison, the experimentally obtained data is also shown in the figure. Here the loading portion appears to agree with the E_T/ E_L = 0.2 simulation results. However the unloading curve (also shown at a

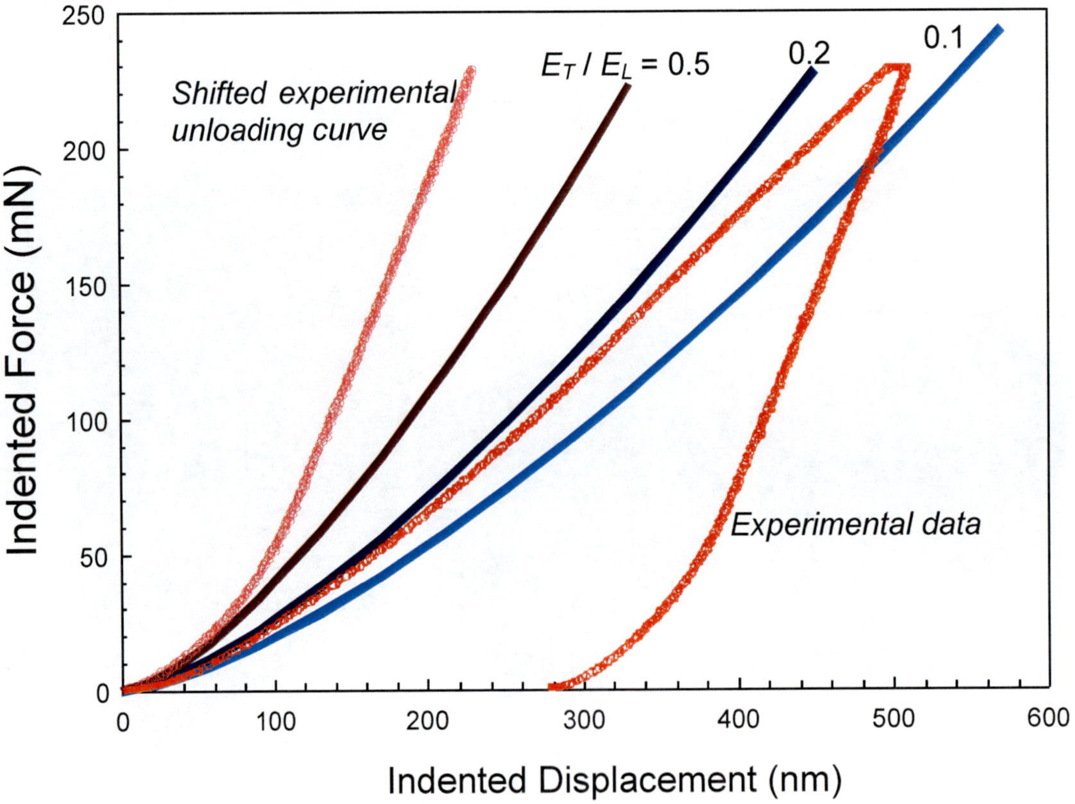

Fig. 9. Simulated indented force-displacements for various modulus ratios of transversely isotropic model. Experimental data is shown for comparison.

shifted position for clarity) indicates a much stiffer response. In fact is stiffer than E_T / E_L = 0.5. The large difference in loading and unloading portions of load-displacement cures is probably due to crushing of actual tooth enamel and compacting upon unloading. A further test is needed to elucidate the results and make comparisons with the simulated results.

5. Conclusions

This work studies mechanical properties of human tooth enamel based on the complex microstructure through a periodic finite element model. In the analysis, a representative unit-cell model was utilized effectively to study the connection between microstructure and the macroscopic mechanical behavior. Also a procedure was developed to prescribe 3D periodic boundary conditions. The present analysis revealed the accurate monoclinic nature of the enamel microstructure and the suitable constitutive law was used to determine 13 independent material constants. The model was further developed for the simulations of micro-indentation which can be used to compare with experimental results. The indentation analysis also revealed that the maximum effective stress occurred at the depth of 8.5μm beneath the surface for the heterogeneous model.

Acknowledgements:

The authors would like to extend appreciation to NSF award CMMI-0626025 in part for support of this work.

References:

1. Fernandes, C. P., P. O. J. Glantz, et al. (2003). "A novel sensor for bite force determinations." Dental Materials 19(2): 118-126.
2. Hayasaki, H., A. Okamoto, et al. (2004). "Occlusal contact area of mandibular teeth during lateral excursion." Int J Prosthodont 17(1): 72-76.
3. Avery, J. K., P. F. Steele, et al. (2002). Oral development and histology. Thieme Stuttgart, New York.
4. Jeng, Y. R., T. T. Lin, et al. (2009). "Nanotribological characterization of tooth enamel rod affected by surface treatment." J Biomech 42(14): 2249-2254.
5. Nanci, A. and A. R. Ten Cate (2008). Ten Cate's oral histology: development, structure, and function. St. Louis, Mo., Mosby Elsevier.
6. Katz, J. L. and K. Ukraincik (1971). "On the anisotropic elastic properties of hydroxyapatite." J Biomech 4(3): 221-227.

7. Katz, J. L. (1985). "The Biophysical and Biomechanical Properties of Bone, Bone-Mineral and Some Synthetic Bone Biomaterials." Bulletin De La Societe Chimique De France(4): 514-518.
8. Weiner, S. and H. D. Wagner (1998). "The material bone: Structure mechanical function relations." Annual Review of Materials Science 28: 271-298.
9. Viswanath, B., R. Raghavan, et al. (2007). "Mechanical properties and anisotropy in hydroxyapatite single crystals." Scripta Materialia 57(4): 361-364.
10. Spears, I. R. (1997). "A three-dimensional finite element model of prismatic enamel: a re-appraisal of the data on the Young's modulus of enamel." Journal of Dental Research 76(10): 1690-1697.
11. Xie, Z., M. V. Swain, et al. (2009). "Structural Integrity of Enamel: Experimental and Modeling." Journal of Dental Research 88(6): 529-533.
12. Mottershead, J. E. and C. D. Foster (1991). "On the Treatment of Ill-Conditioning in Spatial Parameter-Estimation from Measured Vibration Data." Mechanical Systems and Signal Processing 5(2): 139-154.
13. Ramanujam, N. and T. Nakamura (2007). "Estimating surface damage of composite panels with inverse analysis." Journal of Composite Materials 41(20): 2471-2498.
14. He, L. H., N. Fujisawa, et al. (2006). "Elastic modulus and stress-strain response of human enamel by nano-indentation." Biomaterials 27(24): 4388-4398.
15. Zhao, W., C. Cao, et al. (2010). Measurement of Structural Variations in Enamel Nanomechanical Properties using Quantitative Atomic Force Acoustic Microscopy. Proceedings of the 2010 SEM Annual Conference and Exposition, June 7-10, 2010, Indianapolis, IN.

A Novel Biomimetic Material Duplicating the Structure and Mechanics of Natural Nacre

Deju Zhu[1] and Francois Barthelat[2]

[1] Postdoctoral research fellow, Department of Mechanical Engineering, McGill University, Montreal, QC, Canada, E-mail: deju.zhu@mail.mcgill.ca
[2] Assistant Professor, Ph.D., Department of Mechanical Engineering, McGill University, Montreal, QC, Canada, corresponding author, E-mail: francois.barthelat@mcgill.ca

ABSTRACT

Nacre from mollusk shell is a high-performance natural composite, composed of microscopic mineral tablets bonded by a tough biopolymer. Under tensile stress these tablets slide on another in a highly controlled fashion and over large volumes, and this unique mechanism leads to an overall toughness of nacre three orders of magnitude higher than the brittle mineral. This degree of toughness amplification is currently not matched by any synthetic composite. This article presents the first synthetic material based on the structure of nacre that successfully duplicates the mechanism of tablet sliding. This material was made of millimeter size poly-methyl-methacrylate (PMMA) tablets arranged in columns and held by fasteners. Strain hardening was provided by tablet waviness, leading to strains at failure 3-5 times greater than bulk PMMA. Analytical and finite element models successfully captured the locking mechanisms, enabling a rigorous design and optimization of similar composites based on different materials or at different length scale. This work demonstrates how key features and mechanisms in natural nacre can be successfully harnessed in engineering material, and unveils two new mechanisms, the effect of free surfaces and unzipping. Both mechanisms may be relevant to natural materials such as nacre or bone.

Keywords: nacre; biomimetics; tablet; analytical model; finite element model

1. Introduction

Materials produced by nature exhibit remarkable properties which are attracting the attention of engineers, material scientists, chemist and zoologists in search of inspiration for novel material designs [1]. Often made of materials with relatively poor structural qualities, biological materials achieve their performance through intricate microstructures finely tuned over millions of years of evolution [2-4]. Elucidating the structure-properties relationships in these materials is a challenging task, but is nevertheless an essential step for a successful biomimetic "transfer of technology" from natural to synthetic materials and systems. Nacre from mollusk shells, also known as mother of pearl, is now identified as an excellent model for high-performance materials offering attractive combinations of stiffness, strength and toughness [5]. Nacre is a highly mineralized material made of 95% of the mineral calcium carbonate, which comes in the form of microscopic polygonal tablets closely stacked to form a dense, three-dimensional brick wall [6]. The remaining constituents are proteins and polysaccharides, which form an organic matrix concentrated at the interfaces between the tablets [7]. Nacre is relatively stiff and hard compared to other biological materials, which is easily explained by its high mineral content. The most remarkable and less obvious property in nacre is actually its toughness, which is 3000 times that of the mineral it is made of (in J_{IC} terms) [2]. Much of this toughness is explained by nacre's ability to deform past a "yield point" and to develop large inelastic strains over large volumes around defects and cracks. This mechanism generates tremendous toughness amplification [8], effectively making nacre damage tolerant [9]. Failure strains in nacre exceed 1% [10], which is 100 times the typical failure strain of engineering ceramics. These high strains in nacre are now well explained by two mechanisms. First, the tablets have the ability to "slide" on one another when nacre is subjected to tension [11]. This sliding is mediated by soft organic materials that provide cohesion over large

T. Proulx (ed.), *Mechanics of Biological Systems and Materials, Volume 2*, Conference Proceedings of the Society for Experimental Mechanics Series 9999, DOI 10.1007/978-1-4614-0219-0_25, © The Society for Experimental Mechanics, Inc. 2011

sliding distances [12] and possibly lubrication [9]. In addition nanoscale mineral features at the interface can increase the resistance to sliding [9, 11]. These combined features are necessary for large deformation but they are not sufficient. A hardening mechanism has to operate in order to spread the sliding mechanisms over large volume and to delay strain localization in order to maximize energy dissipation at the interfaces. This second key mechanism has recently been demonstrated to be generated by the waviness of the tablets, which makes the ends of some tablets thicker at the periphery than near the center, effectively providing low angle dovetails [10]. Under tension these dovetails provide progressive locking stresses that impede tablet sliding and generate hardening. The "core" of each column (central region of the tablets) is also reinforced by mineral bridges [13] which probably add stability to the structure and prevents delamination.

The unique architecture and mechanisms of nacre and how they lead to its impressive mechanical performance have motivated the development of numerous "artificial nacres" over the past 20 years. Various fabrication techniques were employed, including innovative and ingenious approaches such as layer by layer deposition [14], bottom-up colloidal assembly [15], molecular scale self-assembly and biomineralization [16], or ice-templated sintering of alumina powders [17]. These materials successfully duplicate several features of natural nacre: unidirectional micron or sub-micron size mineral tablets with high aspect ratio embedded in softer, ductile polymeric matrices. However, while these materials display structures which closely resemble nacre, none could truly duplicate its mechanism of tablet sliding with progressive locking.

An alternative approach to alleviate this major fabrication challenge is to "relax" the length-scale condition and to produce structures with millimeters rather than micrometers length scales. Current fabrication technologies allow a far greater control over geometries and structures in the millimeter scale, which can be harnessed to duplicate specific features in natural nacre. Recent work by Espinosa et al. [18] followed this approach: rapid prototyping was used to produce a two-dimensional tiling resembling nacre, including tablet overlap and waviness in the form of dovetails. The mechanical response showed softening at first followed by some locking and the beginning of spreading of deformation. The strain hardening, progressive locking and massive spreading of inelastic distribution observed in natural nacre could however not be replicated.

In this article a novel composite based on the structure and mechanics of nacre is presented. This composite was based on millimeter size poly-methyl-methacrylate (PMMA) tablets, constructed following the brick and mortar arrangement of natural nacre in a two-dimensional tiling. In place of the mineral bridges present in the core regions of nacre, transverse fasteners were used to hold the tablets together. The tablets duplicated the waviness of natural nacre in order to generate hardening and spread deformations. The first section of the article presents an overview of the structure, mechanics, fabrication and testing of this new material. The mechanics is further explored, where analytical and finite element models are presented and compared to the experiments.

2. Overview of the Composite

The objective of this work was to fabricate a material that incorporated the structural features required to duplicate the tablet sliding mechanism observed in natural nacre. The proposed nacre-like composite material is showed in fig. 1a. Its main features directly borrow the following design features of natural nacre:

1) Stiff and brittle tablets are arranged in a columnar fashion with well defined overlap and core regions. This arrangement is achieved by alternating long tablets (major tablets) with shorter tablets (minor tablets). The tablets were machined from PMMA, which was chosen for its machinability and because it is relatively stiff and brittle compared to other polymers.

2) Interfaces between the tablets that maintain cohesion over long sliding distances: Here no material was actually used at the interfaces, and direct contact and dry friction controlled the interaction between the tablets.

3) Waviness on the tablets in order to generate strain hardening and spread deformations: Here a triangular shaped waviness was implemented on the surface of the tablets, which generated dovetail features at the ends of the major and minor tablets.

4) Reinforcements in the core regions: In natural nacre mineral bridges, concentrated in the center of the tablets, reinforce the core regions and prevent delamination. In the proposed material these bridges were replaced by transverse fasteners that hold the tablets together and reinforce the core.

The anticipated behavior in tension of such composite is that tablets slide on one another and progressive locking spreads inelastic deformations over the entire material, which successfully duplicates the behavior of nacre. As a result the strain at failure and the energy dissipated through friction is expected to be several times that of the tablets themselves. With flat tablets the deformation would inevitably lead to immediate localization of strains. The dovetails are critical to generate progressive locking, and the stresses involved in this mechanism are showed on fig. 1b. In addition to dry friction, the dovetails generate transverse compressive stresses in the overlap regions, which increase with sliding distance and generate hardening. Transverse compression is balanced by transverse tension in the core region, which can become high enough to delaminate the tablets. Delamination immediately shuts the locking mechanism and leads to premature failure. A preliminary finding that lead to this prototype was that no synthetic glue could prevent delamination while at the same time allow for tablet sliding. Fasteners were therefore chosen to hold the tablets together, emulating the mineral bridges observed in the core regions of nacre.

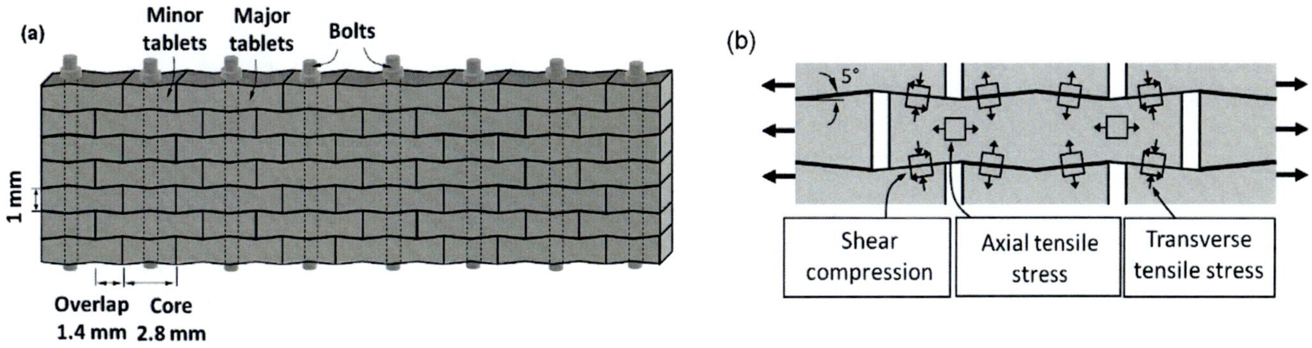

Fig. 1 Overview of the composites: (a) schematic with dimensions; (b) some of the stresses involved in progressive locking

3. Fabrication and Testing

As a first fabrication step, 50 mm by 50 mm PMMA plates were cut from a 1.5 mm thick PMMA sheet. An end mill sharpened at a 5º angle was used on a CNC milling machine to generate the wavy features on the surface of the plates, and a small diameter end mill was then used to machine the deep grooves that defined the ends of the tablets. The PMMA plate was then flipped and the machining process was repeated on the other face. A special fixture was used to ensure that the features on both faces were properly aligned. As a final machining step, an array of holes with a diameter of 1.0 mm was machined through the centers of the tablets. The tablets were then separated manually, and after sonication and cleansing with ethanol they were assembled on 0.8 mm diameter stainless steel rods and nuts to form a nacre-like structure. The resulting material consisted of 7 layers and composed of 8 columns of tablets (fig. 2a). The nuts were either lightly tightened by hand ("snug-tight"), or tightened using a wrench. The amount of tightening was then controlled by counting the fraction of turns imposed on the nuts.

The resulting material was tested in uniaxial tension along the direction of the tablets, using a miniature loading stage (Fullam, NY). The sample was held in place by clamping the end columns of tablets. The setup was then placed under a digital camera equipped with a macro lens. The sample was stretched at a rate of 3.0 mm/min (corresponding to a strain rate of 2.8×10^{-3} s^{-1}), until the specimen failed, and images were captured at regular interval to keep track of deformations and failure. Fig. 2b shows the displacement results from digital image correlation (DIC) in one typical test. Displacement jumps across every potential sliding site are evident. It also indicates that the columns of tablets also rotate during sliding, and that the overlap regions actually progressively fail by "unzipping". This instability is probably triggered by little amount of misalignment of the individual tablets during assembling process or preexisting defects in the assembly. The stress-strain curves resulting from this tensile test are shown in fig. 2c. After a short linear region the material enters a nonlinear regime with significant strain hardening. In this regime the tablets are sliding on one another, the dovetail geometry providing progressive locking and hardening. As the tablets slide and hardening develops, the contact area between tablets diminishes, which decreases the amount of force transferred at the interfaces. This softening phenomenon competes with hardening until the stress reaches its maximum value, after which softening prevails and a strain localization appears: any additional stretch imposed on the sample generates tablet separation in only a single overlap region, the other being "frozen" at the same stage of sliding. Complete separation of the composite occurred at strains of about 12%, which is 3-5 times higher than the strain at failure of PMMA. This nacre-like mechanism can therefore effectively turn a relatively brittle material into a ductile one. Interestingly, tightening the fasteners increased the maximum stress. This is explained by (i) the initial transverse compression between tablets is higher, and therefore the friction force is higher and (ii) higher transverse compression delays delamination between the tablets which sustains the strength of the locking mechanism.

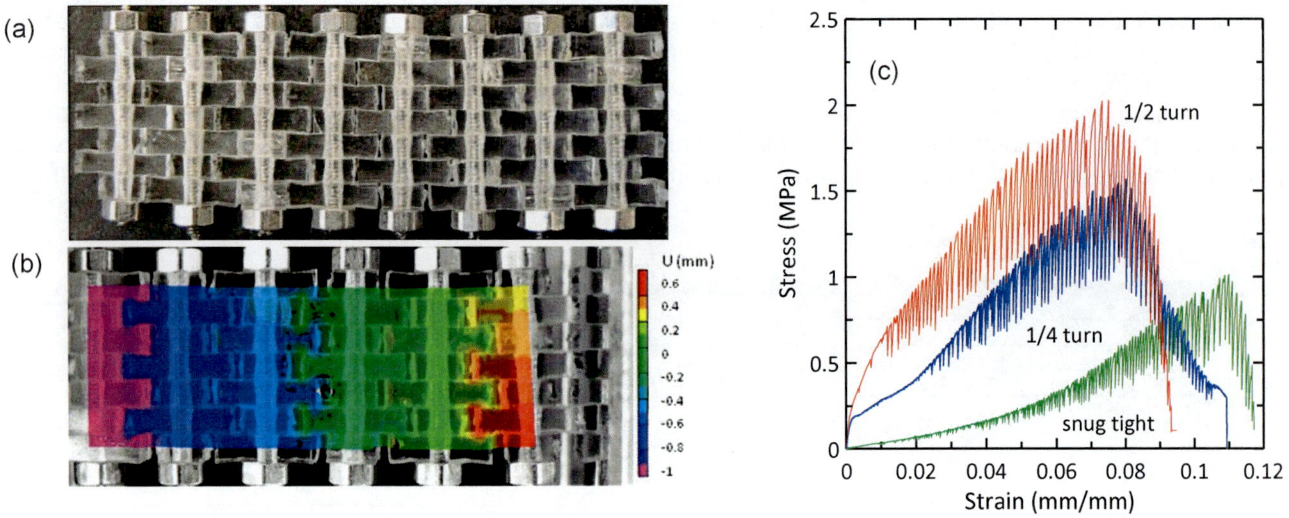

Fig. 2 (a) Actual composite after assembly; (b) displacement using DIC; (c) tensile stress-strain curves from the nacre-like composite for different amounts of tightening on the fasteners

4. Representative Volume Element (RVE) Models

A two-dimensional analytical model of the nacre-like composite was first developed, with the aim of capturing the progressive locking mechanism and to identify key structural parameters. In this first model the structure was assumed to be periodic in the two in-plane directions so that only the representative volume element (RVE) was modeled, as shown in fig. 3a.

The depth of the composite was smaller than the in-plane dimensions plane stress conditions were assumed. The tablets were modeled as linear elastic (modulus E and Poisson's ratio ν). The sliding of the tablets at the interfaces was assumed to be controlled by Coulomb friction, with friction coefficient (f). An analytical model could be derived to take in account the progressive locking of the elastic tablets and the effect of the fasteners. The main result from this model is the tensile stress-strain curve of the composite, given by:

$$\frac{\bar{\sigma}}{E}=\frac{\kappa\phi+(1+\kappa)\alpha\bar{\varepsilon}\theta}{\frac{1}{\alpha(\beta-\bar{\varepsilon})}\left[\frac{1-\bar{\varepsilon}}{1-\beta}+\kappa\right]\frac{1-\theta f}{\theta+f}+\nu[1+2\kappa]} \quad (1)$$

where $\bar{\sigma}$ and $\bar{\varepsilon}$ are the stress and strain in the composite. The tensile stresses could be normalized by the modulus of the tablets E, with the immediate result that the elasticity of the tablets dominates the locking mechanism. Equation (1) is also function of the nondimensional tablet aspect ratio $\alpha=\frac{L}{t}$, of the overlap ratio $\beta=\frac{L_0}{L}$, of the nondimensional bolt stiffness $\kappa=\frac{A_t}{wL}\frac{E_b}{E}\frac{1}{1-\beta}$ (with E_b being the modulus of the bolt, A_t its cross section and w the depth of the RVE), and of the nondimensional bolt tightening $\phi=\frac{Np}{t}$ (where p is the pitch of the thread and N is the fraction of turns given to tighten the fastener).

While this relatively simple analytical expression for the stress-strain curve is convenient for parametric studies, several assumptions were made in its derivation. For this reason the accuracy of that solution was assessed by comparison with a finite element model. Only one quarter of the RVE was model using symmetries, and periodic-symmetric boundary conditions were imposed (fig. 3b). The tablets had the elastic properties of PMMA (E = 2.3

GPa obtained by three-point bending test; $\nu = 0.35$). The fastener was explicitly modeled with a bar element, whose end displacements were tied to the upper and lower edges of the RVE. The fastener had a diameter of 0.8 mm and a modulus $E_b = 200$ GPa. In some cases an initial tightening was imposed on the fastener, by the same amounts as for the experiments. Surface to surface contact elements were inserted at the interfaces with friction coefficient $f = 0.2$. Fig. 3c shows that the stress-strain curves obtained by the analytical and the finite element models are in excellent agreement. However, the experimental stress-strain curves are much lower than these predictions. This indicates that these RVE based models do not properly capture major phenomena in the actual material.

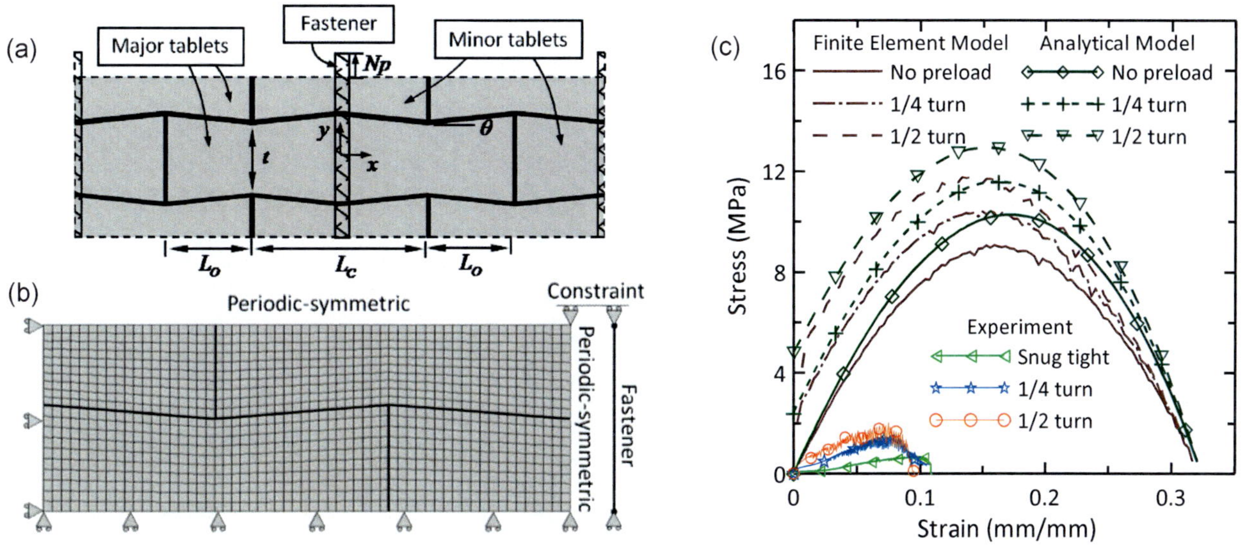

Fig. 3 (a) RVE model; (b) mesh and boundary conditions of the RVE based finite element model; (c) comparison of the stress-strain curves of analytical model, finite element model and experiment

5. Finite Width Model

In order to refine the model to capture the experiments, several features were added to the model: (i) the free surfaces on the side of the composite were considered, (ii) initial gaps between tablets, due to imperfect assembly were included; (iii) an instability was introduced in order to induce the "unzipping" phenomenon observed experimentally.

The progressive locking mechanism requires a build-up of transverse stresses. This build-up is possible for a periodic RVE, which is assumed to be far enough from free surfaces not to feel their effect. In reality the present model material has only 7 layers so the effect of the free surface is likely to be prominent. Therefore a finite width finite element model composed of 7 layers was constructed (fig. 4a). The top and bottom edge were free surfaces while nodes on the left and right edges were constrained to remain on a straight line by using slider elements. Fasteners were modeled with bar elements on the right and left of the model, and their extension or contraction was tied to the upper and lower edges of the material. The resulting stress-strain curve (fig. 4d) shows stress levels less than a third of the stresses from the small RVE FE model. The effect of the free surfaces is therefore significant: transverse stresses cannot fully develop and the efficacy of the locking mechanisms is greatly diminished.

As a second refinement to the model, initial gaps were introduced, following experimental observation (fig. 2a). The presence of these gaps further reduces the tensile strength and the strain at failure (fig. 4d), as they reduce the contact area between tablets. In order to simulate unzipping in the finite width model, a defect was introduced in the model by slightly decreasing the modulus of a few elements in the lower half of the model. The model was then pulled at only 2 points only, located half-way through the thickness. This triggered an "unzipping" type of failure where the overlap areas near the lower surfaces (where defects were introduced) would fail first (fig. 4c). The model then progressively opened ("unzipped") towards the upper side of the model. Incorporating unzipping in the model also had an impact on the stress-stain curves, by decreasing the overall strength by about a third and by reducing the strain at failure (fig. 4d).

Incorporating the free surfaces, the initial gaps and unzipping brought the stresses close to the experimental values. The results of this refined model are compared with the experiments on fig. 4e for the 3 levels of bolt tightening (snug-tight, ¼ turn and ½ turn). The agreement is reasonable, considering the assumptions made in the model and possible imperfections in the material. The tensile strength obtained by the finite width model is reasonably close to the experiments with pre-loads of ¼ turn and ½ turn, but is approximately 30% larger than the experiment under the snug-tight condition. The predicted failure strain is approximately 50% larger than the experimental value, which was expected since the model has only two columns as opposed to eight in the experiment. The model properly captures the following effects: (i) strain hardening up to a maximum stress followed by softening; (ii) the initial stress increases as the bolt pre-load is increased; (iii) the hardening rate is not dependant on tightening; (iv) the maximum stress increases as the bolt pre-load is increased; and (v) the strain at localization decreases as the bolt pre-load is increased.

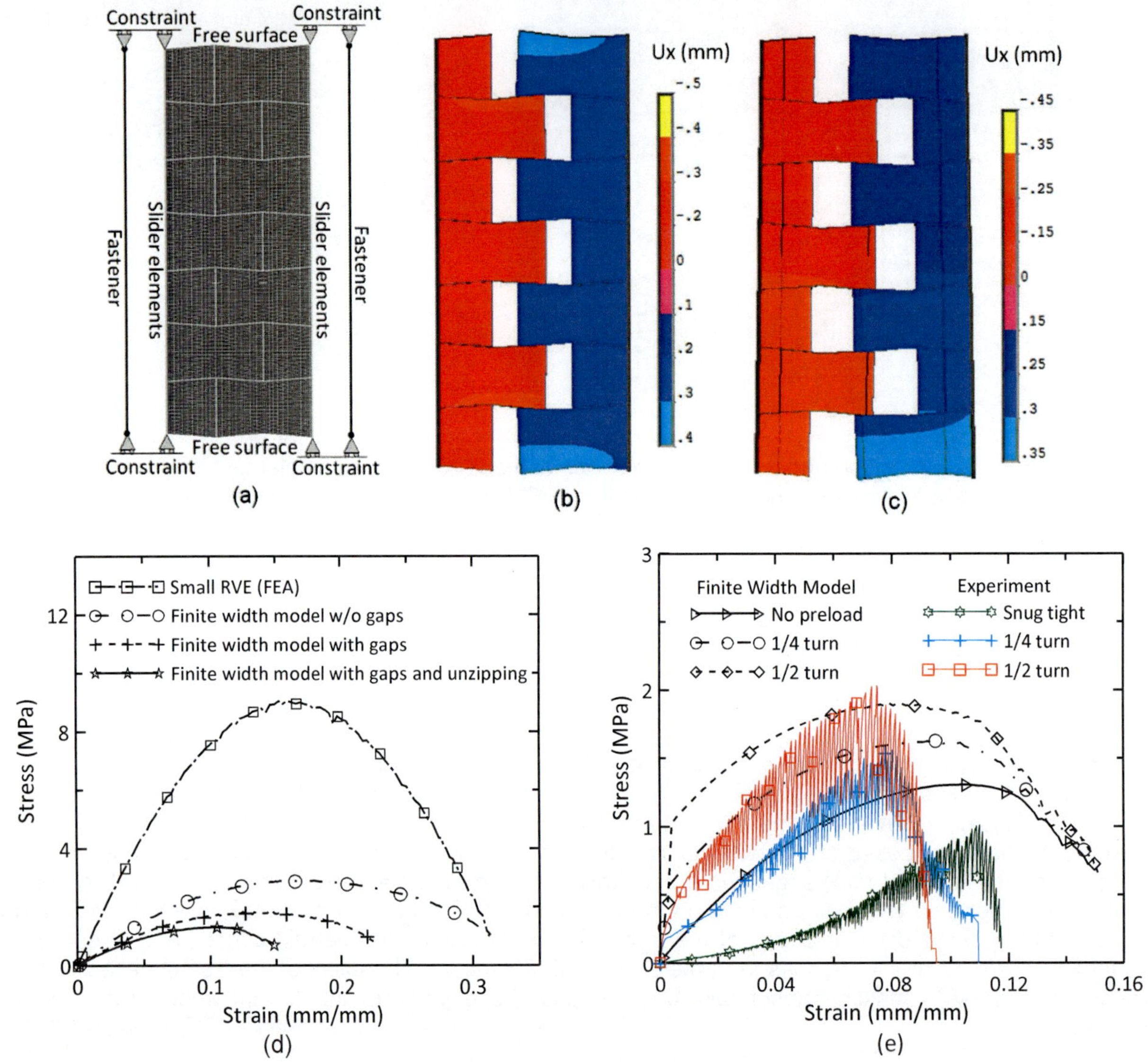

Fig. 4 (a) Mesh and boundary conditions of finite width model; (b) uniform deformation; (c) unzipping; (d) the stress-strain curves of RVE based FEA model and finite width model; (e) the stress-strain curves of finite width model and experiment with different levels of bolt tightening

6. Conclusions

This article presents the first prototype made of PMMA tablets in millimeter size based on the structure of natural nacre which properly duplicates, for the first time, its unique mechanisms. Progressive tablet locking, strain hardening and spreading of large deformations over large volumes were all achieved in this new material. The objectives of this model material were to (i) demonstrate how tablet waviness leads to the properties of nacre and (ii) to validate models that can be used in the design of this type of composites. Transverse fasteners act as a strong transverse reinforcement to the composite, which was needed to generate tablet locking. In natural nacre mineral bridges provide a similar type of reinforcements. Even though the structure of this PMMA material was not optimized, it still showed strain hardening behavior and large strain capacity (9-12%). The material is however very weak in tension making it not suitable for engineering applications. With the proper mechanical model this mechanism can however be translated to similar composites made of stronger and perhaps smaller tablets (metals, ceramics), and the microstructure can be optimized accordingly to achieve new combinations of strength, modulus and toughness that may be of interest for engineering applications. The analytical and finite element models presented here provide useful platforms to design and optimize similar materials based on the same mechanisms, but using different materials or length scales. Interestingly, experiments on the model material revealed new insights on the mechanics of staggered structures that would have been difficult or impossible to gain from studying natural nacre. The effects of the free boundary, the initial defects and the "unzipping" are new factors that should be considered in the modeling and design of artificial nacres.

Reference

1. Barthelat F. Biomimetics for next generation materials. Philosophical Transactions of the Royal Society a-Mathematical Physical and Engineering Sciences 365(1861), 2907-2919, 2007.
2. Wegst UGK and Ashby MF. The mechanical efficiency of natural materials. Philosophical Magazine 84(21), 2167-2181, 2004.
3. Ballarini R, Kayacan R, Ulm FJ, Belytschko T, et al. Biological structures mitigate catastrophic fracture through various strategies. International Journal of Fracture 135(1-4), 187-197, 2005.
4. Meyers MA, Chen PY, Lin AYM and Seki Y. Biological materials: Structure and mechanical properties. Progress in Materials Science 53(1), 1-206, 2008.
5. Barthelat F. Nacre from mollusk shells: a model for high-performance structural materials. Bioinspiration & Biomimetics 5(3), 1-8, 2010.
6. Currey JD. Mechanical-Properties of Mother of Pearl in Tension. Proceedings of the Royal Society of London Series B-Biological Sciences 196(1125), 443-463, 1977.
7. Schaeffer TE, IonescuZanetti C, Proksch R, Fritz M, et al. Does abalone nacre form by heteroepitaxial nucleation or by growth through mineral bridges? Chemistry of Materials 9(8), 1731-1740, 1997.
8. Barthelat F and Rabiei R. Toughness amplification in natural composites. Journal of the Mechanics and Physics of Solids, in press, doi:10.1016/j.jmps.2011.01.001, 2011.
9. Evans AG, Suo Z, Wang RZ, Aksay IA, et al. Model for the robust mechanical behavior of nacre. Journal of Materials Research 16(9), 2475-2484, 2001.
10. Barthelat F, Tang H, Zavattieri PD, Li CM, et al. On the mechanics of mother-of-pearl: A key feature in the material hierarchical structure. Journal of the Mechanics and Physics of Solids 55(2), 306-337, 2007.
11. Wang RZ, Suo Z, Evans AG, Yao N, et al. Deformation mechanisms in nacre. Journal of Materials Research 16(9), 2485-2493, 2001.
12. Smith BL, Schaeffer TE, Viani M, Thompson JB, et al. Molecular mechanistic origin of the toughness of natural adhesives, fibres and composites. Nature (London) 399(6738), 761-763, 1999.
13. Song F and Bai YL. Effects of nanostructures on the fracture strength of the interfaces in nacre. Journal of Materials Research 18(8), 1741-1744, 2003.
14. Tang ZY, Kotov NA, Magonov S and Ozturk B. Nanostructured artificial nacre. Nature Materials 2(6), 413-418, 2003.
15. Bonderer LJ, Studart AR and Gauckler LJ. Bioinspired design and assembly of platelet reinforced polymer films. Science 319(5866), 1069-1073, 2008.
16. Li CM and Kaplan DL. Biomimetic composites via molecular scale self-assembly and biomineralization. Current Opinion in Solid State & Materials Science 7(4-5), 265-271, 2003.
17. Munch E, Launey ME, Alsem DH, Saiz E, et al. Tough, Bio-Inspired Hybrid Materials. Science 322(5907), 1516-1520, 2008.
18. Espinosa HD, Rim JE, Barthelat F and Buehler MJ. Merger of structure and material in nacre and bone - Perspectives on de novo biomimetic materials. Progress in Materials Science 54(8), 1059-1100, 2009.

Adaptive Hybrid Control for Low Resolution Feedback Systems with Application on a Novel Microinjector: Ros-drill

Z. Zhang, J.F. Diaz, N. Olgac
University of Connecticut, Mechanical Engineering Department, CT, USA.
***email: olgac@engr.uconn.edu**

Abstract

A novel hybrid control logic is deployed on a novel cellular microinjector called the Ros-Drill (Rotationally Oscillating Drill). Ros-Drill technology is primarily developed for ICSI (Intra Cytoplasmic Sperm Injection). The new structure offers an inexpensive tool which performs precise micro motion control on an actuator (DC motor), creating high-frequency rotational oscillations at the tip of an injecting pipette, which, in turn, drills the cells. Angular motion control presents no particular difficulties when the application uses high fidelity measurements of the motion being controlled. However, size, costs and accessibility of the needed technology relevant to the hardware components negatively influence the design of the control settings. In this presentation we describe an adaptive-hybrid control structure which is developed around a commonly available microcontroller core. We consider three limiting factors: bounded sampling speed of the controller, and extremely low relative resolution of position measuring sensor, as well as the limited capabilities of generating the desired rotational trajectory. We present a novel methodology to overcome these specific problems utilizing an adaptive, robust and optimal PID (proportional-integral and derivative) control strategy. It is demonstrated that the required fine rotational motion for the cellular injections is obtained with desirable fidelity, despite the stated shortfalls in the components. Field experiments and preliminary biological test results will also be included in the discussions.

ROS-Drill micro-injector description and test results

We review the very recent development of a novel technique in ICSI (intracytoplasmic sperm injection) in this presentation. It is called the Ross-Drill (Rotationally Oscillating Drill). The underlying procedure is one of the broadly used cellular-injection processes. A drawn glass pipette is pushed against the cell, and a series of impulsive pulses are exerted on it axially to achieve the piercing through the membrane (oolemma) and injection, Fig. 1. The most advanced version of this procedure is currently conducted using piezoelectric assistance (so called *piezo-assisted drilling*). It requires a very small mercury column inside a glass pipette in order to stabilize the pipette tip against transverse oscillations. But the toxic mercury is generally prohibited in biological clinics, which makes the technique deployable only in laboratory settings. In a recent study we introduced a novel procedure to remedy the shortfalls of the piezo technology, both from mercury requirement as well as the transverse oscillations aspects [1,2]. This new technology is based on a *rotational oscillatory drilling* thus it is named (**ROS-Drill**).

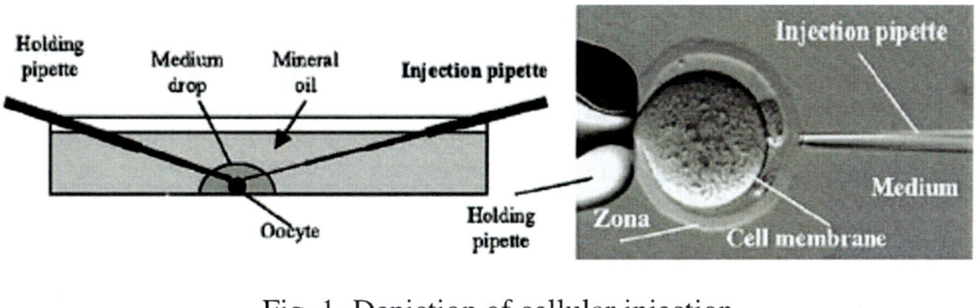

Fig. 1. Depiction of cellular injection

T. Proulx (ed.), *Mechanics of Biological Systems and Materials, Volume 2*, Conference Proceedings of the Society for Experimental Mechanics Series 9999, DOI 10.1007/978-1-4614-0219-0_26,

A prototype of ROS-Drill microinjector is shown in Fig.2. A precision DC-servomotor rotationally oscillates the pipette tip, which is attached to a holder. The entire structure is on a micro-positioner (robot) in order to guide the pipette tip to its proper location, in microscopic scale. This manipulator plays a very important role as it establishes the pre-piercing deformation on the cell membrane. This deformation amount is crucial as it sets the membrane tension just prior to the start of the Ros-Drill action. The main task starts from that point on: one needs to create desired rotational oscillations at the tip [3].

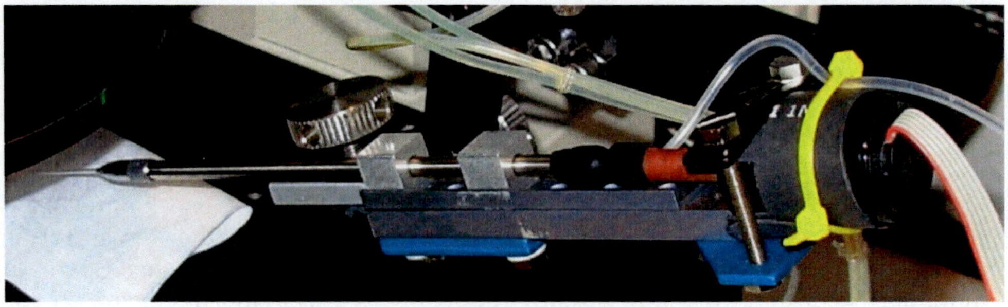

Fig. 2 **Ros-Drill©** prototype

A general description of the proposed control structure of Ros-Drill is shown in Fig. 3. The feedback control signal is generated by a digital controller (Silicon Labs, C8051 microcontroller). The input reference signal for the desired trajectory is harmonic $A\sin(2\pi ft)$,where A is the amplitude of the oscillation (degrees) and f (Hz) is the frequency of the motion. Just to give a point of reference, typical operational ranges of the desired trajectory for mouse oocytes are ***A* (0.2-0.4 degree), *f* (400-1000 Hz)**.

Several observations are noteworthy: A pure harmonic reference trajectory avoids excitation to the natural vibration modes of the drawn pipette tip, therefore it is desired. The range of the oscillation frequency is selected to create sufficient bandwidth differential between the membrane responses to harmonic stimulus and that of the pipette tip. We hypothesize that this differential is the main mechanism to create a clean and self-healing piercing of the membrane.

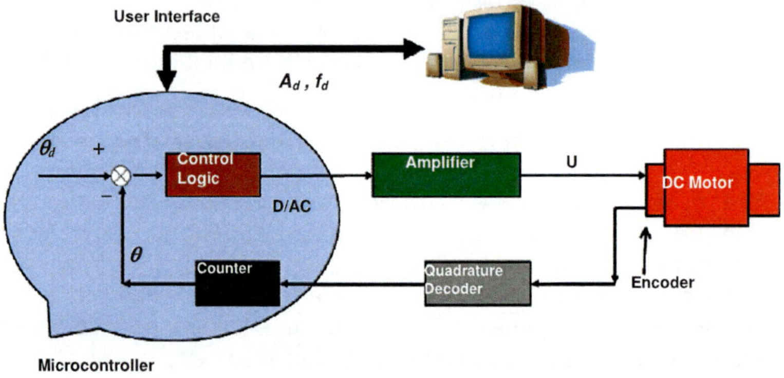

Fig. 3 New Ros-Drill© control structure.

Highlights of the controller: A harmonic trajectory is feasible to implement. In ROS-Drill, PID (Proportion- Integral-Derivative) control logic is selected. The position sensing is provided by an incremental optical encoder which is attached to the shaft of the micro-motor. It has a restrictive 1000 pulses/rev rotational sensing capability. When it is used in quadrature decoding mode it creates a resolution of 0.09 degree. As such, the operating amplitudes generate 4-8 encoder pulses for each peak-to-peak stroke, which represents a very low-resolution sensing capability. This constraint is very stiff for a healthy and reproducible trajectory tracking task. Main difficulty is due to the numerical limitations in creating the appropriate control signals (especially that of the **D**–derivative component). Under these severe restrictions the only practical remedy would be an "**Adaptive hybrid feedback control**" logic, which is currently implemented on Ros-Drill. It offers several interesting features:

(a) tuning of the controller numerically and experimentally by proper selection of feedback control gains,
(b) categorizing these optimum settings in clusters of user-defined amplitude and frequency,
(c) creating a look-up table routine within the microcontroller program to make these selections in-situ.

Fig. 4 shows sample trajectory tracking results which are repeatable and optimum for the given controller setting. It shows the sustained oscillatory behavior of 0.2 degree in amplitude (4 steps of encoder pulses peak-to-peak). These mechanistic capabilities are expected to be transformed into biological end results in the form of higher success rate of piercing, better survival rates of embryos and ultimately the pops.

Having developed the control logic to a satisfactory level, Ros-Drill is presently tested for biological influence. The tests have been very encouraging to date. Some recordings from experiments and preliminary comparison data vis-à-vis the conventional, piezo-assisted ICSI procedures will also be presented during the talk. Interested reader is directed to several publications of the group [4]

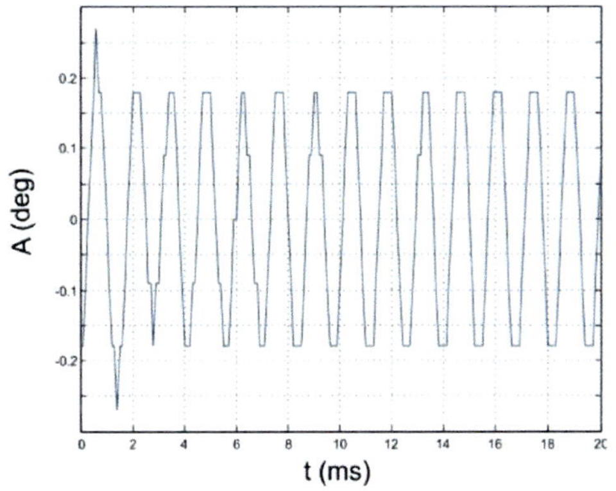

Fig. 4 Rotational oscillations at $f_d = 700$ Hz.

References:

1. K. Ediz, N. Olgac, "Micro-dynamics of the piezo-driven pipettes in ICSI", Biomedical Engineering, IEEE Transactions on ,Volume:51, Issue:7, July 2004 Pages:1262 – 1268.

2. Ediz, K. and Olgac, N., 2005. Effect of mercury column on the microdynamics of the piezo-driven pipettes. J. Biomech. Eng. 127, 531-535.

3. Ergenc, A.F., Li, M., Toner, M., Biggers, J., Kent, J.C. and Olgac, N. 2008. Rotationally Oscillating Drill (Ros-Drill©) for Mouse ICSI without using Mercury. Mol. Reprod. Dev. 1-8.

4. Ergenc, A.F. and Olgac, N., 2007. New Technology for Cellular Piercing: Rotationally Oscillating Micro-injector, Description and Validation Tests. J. Biomedical Microdevices Vol. 9, no. 6, 885-891.

Design of a mechatronic positioner for a holographic otoscope system

I. Dobrev[1], C. Furlong[1,2], J. J. Rosowski[2]
[1] Center for Holographic Studies and Laser micro-mechaTronics - CHSLT
NanoEngineering, Science, and Technology - NEST
Mechanical Engineering Department
Worcester Polytechnic Institute
100 Institute Road, Worcester, MA 01609, USA

[2] Eaton-Peabody Lab, Massachusetts Eye and Ear Infirmary
243 Charles Street, Boston, MA, USA

ABSTRACT

We are developing an advanced computer-controlled digital optoelectronic holographic system (DOEHS) with the ability to measure both shape and acoustically induced deformations of the tympanic membrane of several species, including humans. The DOEHS have been deployed for testing and use in clinical environment.

The DOEHS consists of laser delivery (LD), optical head (OH), sound presentation (SP), computing platform (CP), and a mechatronic otoscope positioner (MOP) subsystems. In this paper, we present advances in our development of the MOP subsystem, which is capable of positioning the OH subsystem near the patient's ear and maintaining of its relative position and orientation during holographic eardrum examinations. Our work focuses on the development and implementation of various approaches for mechanical stabilization of the MOP-OH subsystems, including custom designed packaging of the OH as well as automatic interferometric compensation against measuring disturbances induced by periodic oscillations, such as those produced by heartbeat and breathing of a patient during examination.

We present preliminary results of our investigations of acoustically induced motions on tympanic membranes by measurements with our DOEHS enabled with a structurally stable MOP subsystem.

Keywords: structural stabilization, mechatronics, middle ear mechanics, otology, shape and deformation measurements.

1. INTRODUCTION

Current ear exams are assessing the state of the patient's hearing mainly based on the healthiness of the tympanic membrane (TM). Various examination procedures [1, 2] of the response of the TM to a controlled sound input include the Laser Doppler vibrometry (LDV) measurement. While giving valuable quantitative feedback for the healthiness of the patients hearing , this is limited to providing information for the response of only one point on the membrane leaving the examiner blind to the complex patterns unfolding across the full surface of the membrane.

In order to overcome this limitation we are developing an advanced computer-controlled digital optoelectronic holographic system [3, 4] (DOEHS) that is capable of simultaneously measuring the displacement of the full surface of the tympanic membrane induced by a controlled input. Our current otoscope system can provide near real-time quantitative measurements of the sound-induced motion of the eardrum and is already deployed for clinical testing. The system also allows for lensless focusing that relies on completely numerical reconstruction of the image.

The stability of the OH during measurement is crucial for the quality of the measured data. We have developed a mechatronic positioning system that allows for the otoscope head to be easily positioned near a patient's ear by an examiner and

T. Proulx (ed.), *Mechanics of Biological Systems and Materials, Volume 2*, Conference Proceedings of the Society for Experimental Mechanics Series 9999, DOI 10.1007/978-1-4614-0219-0_27,

autonomously kept stable during duration of holographic measurements. In order to ensure that the positioning system is intrinsically safe a novelty control approach was taken by using only adjustable friction elements [5] together with haptic feedback resulting in a non-actuated admittance control system. This non-actuated yet active system design approach renders a positioning system that is as safe as ordinary static positioner systems. This allows for the positioner to be operated in existent otology clinic settings without the need for expensive and time consuming testing that is required with similar actuated mechanisms.

A proof of concept prototype positioner (PCPP) based on the new control scheme was first developed that allows for full 6 degrees of freedom (DOF) orientation of the otoscope head with respect to the patient's ear. Based on the development and testing of the PCPP and quantification of the required DOF in real clinical examination conditions the first fully functional prototype of the MOP was developed and deployed in a real clinic environment. Initial tests shows promising results for the measurement capabilities and versatility of the system and they also suggest the system to be safe and reliable.

2. DIGITAL OPTOELECTRONIC HOLOGRAPHIC OTOSCOPE SYSTEM

2.1. Realization of DOEHS

The system consists of four physically independent modules where every one of them is easily disconnectable to allow better mobility in the transportation between examination rooms and various otology clinics.

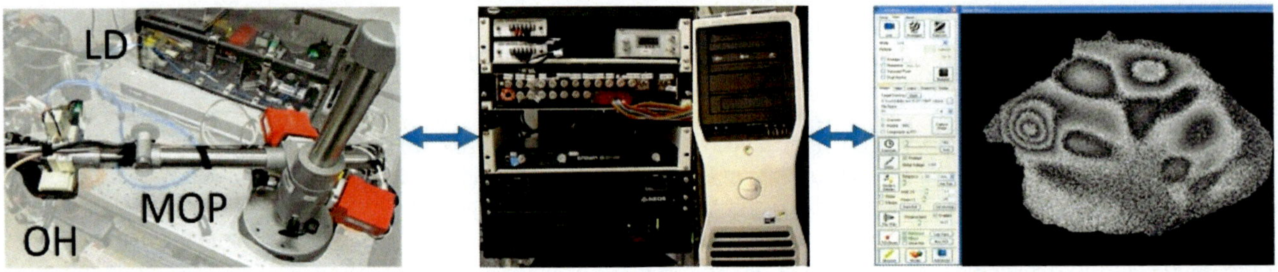

(a) (b) (c)

Fig. 1. Full system overview: (a) Measurement subsystem; (b) Mobile computing and control platform; (c) LaserView [6] image processing software.

The otoscope optical head (OH), the sound presentation (SP) system and the mechatronic otoscope positioner (MOP) are all in one assembly that can be attached close to the patient to an existing clinical examination setup. The laser delivery system (LD) is separate from the rest of the measurement system to reduce the size and increase mobility of the measurement system. The computing platform (CP) consists of two physical parts: a driver and power unit, and PC that controls the whole system through the driver module.

The examiner controls the whole system through a single unified data processing software platform [6] that allows for various measurement modes and numerous live and offline image processing techniques.

2.2. Otoscope optical head system and sound presentation system design and packaging

Reconstructing an image with the DOEHS is based on the four phase stepping digital holography method [3, 4]. Technique allows for the recording of holograms without any optical elements, thus reconstructing purely numerically the object image with the help of a computer.

Without the lenses and the required alignment mechanisms associated with them, the lensless system allows for much better packaging densities. This allows for a mobile and versatile system as it greatly reduces machining and packaging requirements as well as specific size and mass of the system. The implemented design of the OH achieves nearly 60% size reduction as compared to a packaging design based on a traditional holographic setups [7] thus allowing for truly handheld portability.

Figure 2 represents a CAD model and a image of digital holography based OH system implemented in clinical conditions.

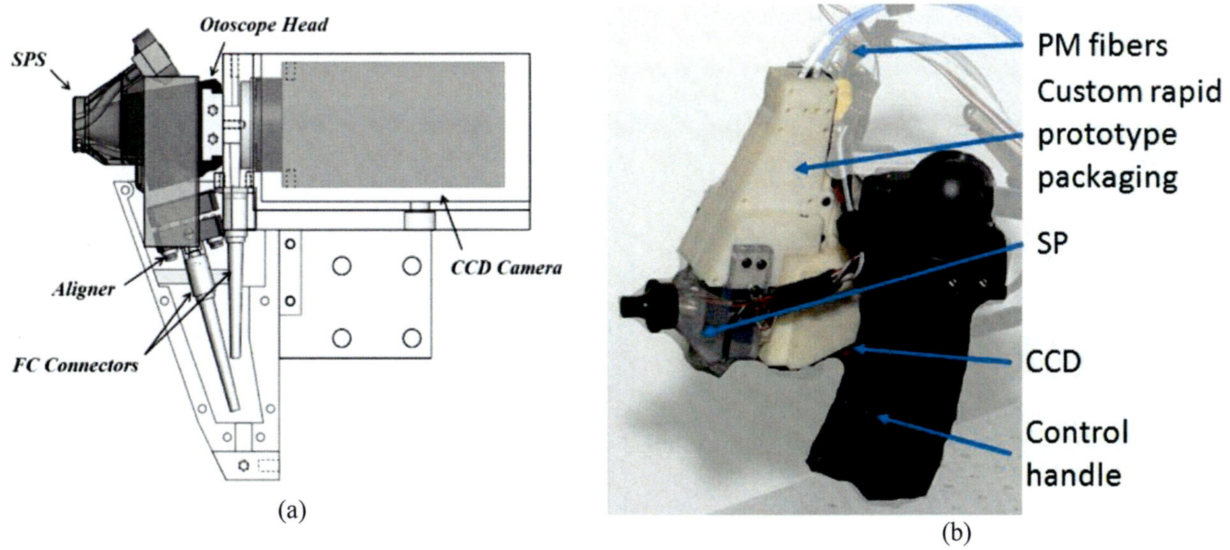

Fig. 2. Otoscope head subsystem configurations: (a) Lensless otoscope system CAD model; (b) and manufactured system. Both systems share the following elements: high speed digital camera (CCD or CMOS); beam splitter (BS); reference beam (RB); object beam (OB).

The OH represents a multi-domain measurement device as it incorporates a sound presentation system (SP) that includes a calibrated microphone for pressure impedance measurements as well as speaker for controlled sound excitation of the human TM.

3. DESCRIPTION OF MECHATRONIC OTOSCOPE POSITIONER

3.1. Proof of concept prototype positioner

An initial proof of concept positioner was designed to explore the plausibility of the concept of using a versatile and mobile positioner for a holographic system as opposed to more conservative exiting designs that would limit the maneuverability of the OH during the delicate positioning operation in a human ear in clinical environments.
The device allows for the examiner to position the OH to the patient's ear by intuitively applying a force to the positioner's control handle. The positioner's controller uses a novel approach [13] that does not actively position the OH, rather only constrains and limits the user's input. This feature makes the device inherently safer than traditional actuated manipulators.

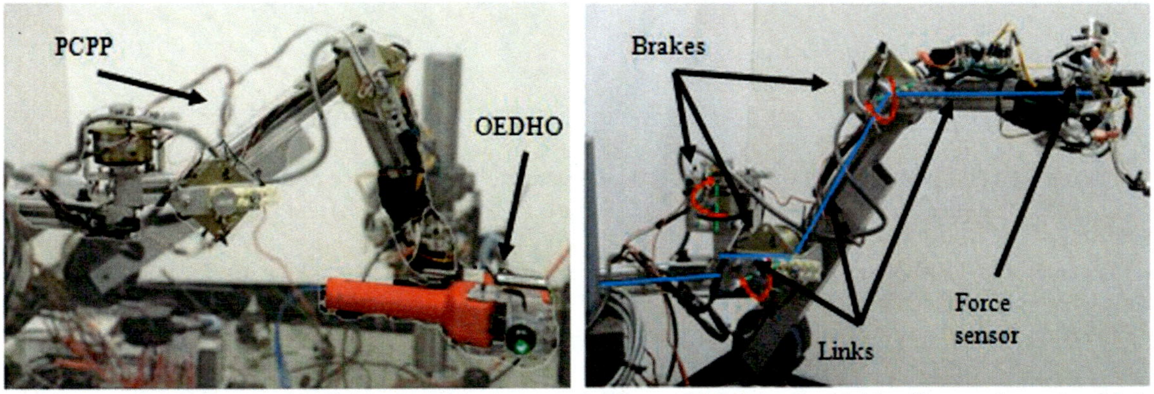

Fig. 3. Full system test setup and individual positioner parts. Links are highlighted with blue lines and rotational joints are marked with green center lines.

Aside from providing ease in the positioning of the OH, the PCPP also provides stable conditions for the recoding of nano-meter displacements of TM. Figure 4 compares stroboscopic measurements of the nano-meter scale vibrations of a 7mm circular copper membrane acquired with the OH, while being supported by an optical post and the PCPP.

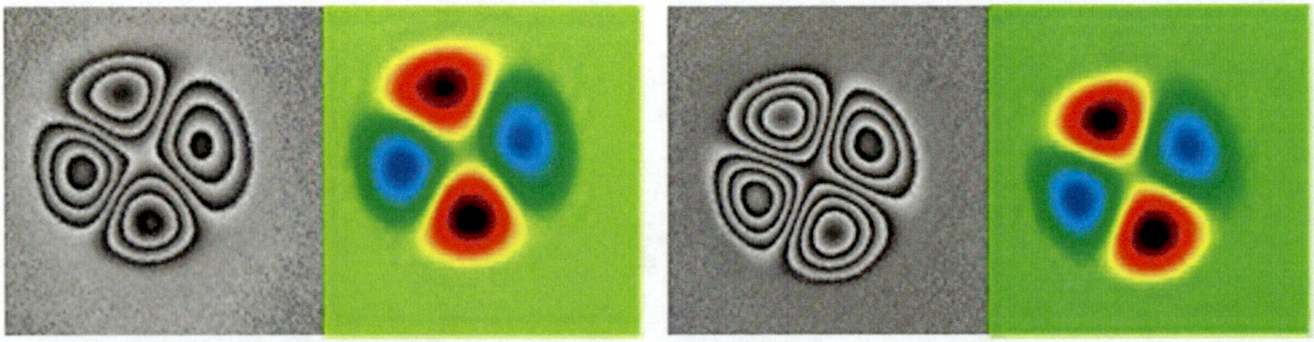

(a) (b)

Fig. 4. Unwapped and wrapped optical phase image of the displacement of a copper foil excited at 13.5 kHz. OH is supported by an optical post (a) and the PCPP (b).

3.2. Mechatronic otoscope positioner

For the implementation of the OH system in clinical environment the MOP was designed based on a commercially available microscope positioner whose mechanism was altered to fit the OH and it was also retrofitted with adjustable braking system. The mechanical modifications were based on measurements and observations of the trajectories of typical otology clinical exams and more specifically the required degrees of freedom (DOF) and workspace to position the otoscope in the patients' ear. The intuitive control system of the MOP allows it to be controlled by the examiner with a single button.

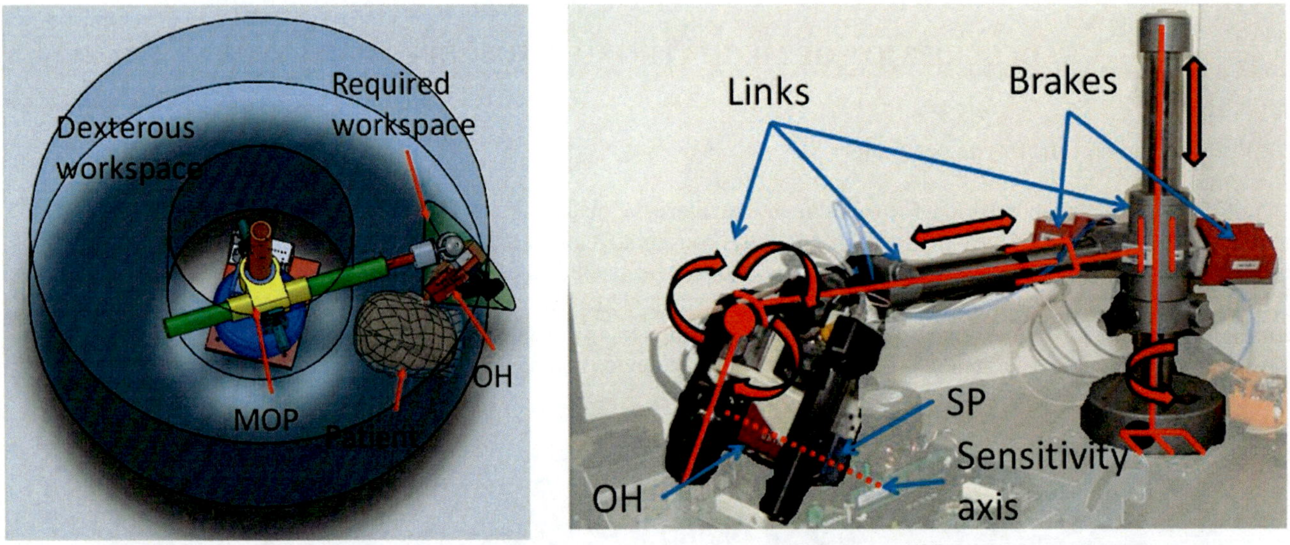

(a) (b)

Fig. 5. CAD model (a) of the required workspace for one ear and the full workspace of the positioned – generic model of patients head is included for scale; MOP (b) parts (links are red lines) and kinematic model.

The control elements in the MOP have been designed so that they are all part of a parallel control system that is easily expandable without any change in the existing hardware of the MOP. This allows for similar control system to fit large number of existing medical positioners that expands the possibilities for implementation of the DOEHS in various clinical settings.

4. RESULTS

The system has several modes [7] of measurements – time-averaged, double exposure and stroboscopic mode. Below are representative results of the measurement capabilities of the system in stroboscopic mode during measurements of a human TM, while the OH is being supported by the MOP:

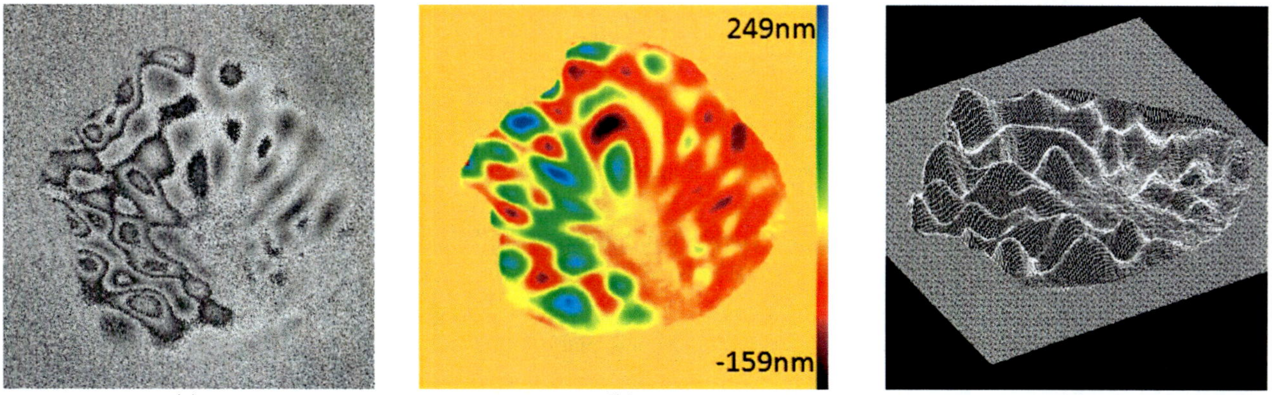

(a) (b) (c)

Fig. 6. Phase map (a) of the amplitude distribution of the TM of a postmortem chinchilla at an excitation signal frequency of 3489 Hz; A full field of view displacement map (b) across the TM indicates a 408nm p-p maximum amplitude; An isometric view (c) of the 3D shape of the displacement map.

Various acquisition techniques have been incorporated in the software to account for noise in the measurements in the system due to relative motion of the OH and the object of interest. The MOP and the PCPP have been equipped with various types of sensors to monitor the mechanical vibrations of the positioners. These signals are read into the control software and are used to synchronize the image acquisition with the mechanical disturbances of the system to minimize measurement noise.

5. CONCLUSIONS AND FUTURE WORK

The presented PCPP successfully proves the concept of non-actuated active positioner and it suggests that such control approach allows for easier implementation in existing medical environments than equivalent active positioners.

The deployment and use of the MOP in clinical environment proves completely new grounds for the applications of holography and interferometry. The system allows for large versatility of measurements and a large amount of a completely new type and much richer information for the properties of the TM and the state of patients' hearing. It provides opportunity for a completely new set of ear examinations that will greatly improve the quality of the treatments and operations of the human ear.

Future work is focused on optimization of the positioning system to maximize its stability as well as designing and implementing various acquisition techniques for minimizing measurement noise. Further stability optimizations are to be made using topology optimization methods [8] in order to minimize the weight to stiffness ratio of the MOP system. Another area of future work is the development of a suitable head restraining system for minimization of the relative motion of the head and the OH during measurements. Future research is also aimed at interferometric compensation [9] for heartbeat induced vibrations of the eardrum as well as identification of suitable methods for monitoring and compensation of the motion of a patient's head relative to the MOP.

6. ACKNOWLEDGEMENTS

This work has been funded by the National Institute on Deafness and Other Communication Disorders (NIDCD), the Massachusetts Eye and Ear Infirmary (MEEI), and the Mittal Fund. The authors also gratefully acknowledge the support of the NanoEngineering, Science, and Technology (NEST) program at the Worcester Polytechnic Institute, Mechanical Engineering Department. We would like to acknowledge our colleagues – J. M. Flores Moreno, E. J. Harrington and C. Scarpino, for their support during the developments presented in this paper.

REFERENCES

[1] Rosowski, J. J., Mehta, R. P., and Saumil N. Merchant, “Diagnostic Utility of Laser-Doppler Vibrometry in Conductive Hearing Loss with Normal Tympanic Membrane,” *Otol. Neurotol.*, 25(3): 323–332, 2004.

[2] Rosowski, J.J., Nakajima H. H., and Merchant S. N., “Clinical Utility of Laser-Doppler Vibrometer Measurements in Live Normal and Pathologic Human Ears,” *Ear Hear.*, 29(1): 3–19, 2008.

[3] Hernández-Montes, M., Furlong, C., Rosowski, J. J., Hulli, N., Harrington, E., Cheng, J. T., Ravicz, M. E., and Santoyo, F. M., “Optoelectronic holographic otoscope for measurement of nanodisplacements in tympanic membranes,” *J Biomed Opt.*; 14(3), 2009.

[4] Kolenovic, E., Furlong, C., Jüptner, W., “Inspection of micro-optical components by novel digital holographic techniques,” *Proc. SEM: 470–475,* 2004.

[5] Dobrev, I., Mauricio Flores, J. M., Furlong, C., Harrington, E., Rosowski, J. J., Scarpino, C., "Design of a positioning system for a holographic otoscope", *Proc. SPIE* 7791, 77910D; *doi:10.1117/12.862354,*2010.

[6] Harrington, E., Dobrev, I., Bapat, N., Mauricio Flores, J. M., Furlong, C., Rosowski, J. J., Cheng, J. T., Scarpino, C., Ravicz, M., "Development of an optoelectronic holographic platform for otolaryngology applications", *Proc. SPIE* 7791, 77910J; *doi:10.1117/12.862130,* 2010.

[7] Hulli, N., *“Development of an optoelectronic holographic otoscope system for characterization of sound-induced displacements in tympanic membranes”*, Worcester, Mass: Worcester Polytechnic Institute., 2009. http://www.wpi.edu/Pubs/ETD/Available/etd-011309-095519/

[8] Bendsøe, M.P., “Optimal shape design as a material distribution problem,” *Structural Optimization*, 1: 193-202, 1989.

[9] Kreis, T., *Handbook of holographic interferometry*, Wiley-VCH, Weinheim, Germany, 2005.

Characterization of Shape and Deformation of Tympanic Membranes by Dual-wavelength Lensless Digital Holography

Weina Lu*[1], Cosme Furlong[1,2], John J. Rosowski[2,3], Jeffrey T. Cheng[2], and Saumil N. Merchant[2,3]

[1] Center for Holographic Studies and Laser micro-mechaTronics, and Department of Mechanical Engineering, Worcester Polytechnic Institute, Worcester, MA, USA

[2] Eaton-Peabody Laboratory, Massachusetts Eye and Ear Infirmary, Boston, MA 02114, USA

[3] Speech and Hearing Bioscience and Technology Program, MIT-Harvard Division of Health Sciences and Technology, Cambridge, MA, USA

Corresponding author*:

weinalu@wpi.edu

ABSTRACT

We are developing an advanced computer-controlled digital optoelectronic holographic system (DOEHS) with the ability to measure both shape and acoustically induced deformations of the tympanic membrane of several species, including humans.

The DOEHS consists of laser delivery (LD), optical head (OH), and computing platform (CP) subsystems. The LD subsystem houses two tunable near-infrared laser sources, an acoustic modulator to enable stroboscopic illumination, and laser-to-fiber mechanisms to couple light into one single mode polarization maintaining fiber. The output of the LD subsystem is delivered into the OH subsystem through the polarized maintaining fiber (PM), which is arranged to perform imaging and measurements by phase-shifting lensless digital holography. A high-resolution digital camera, contained within the OH subsystem, is used for image recording at high-rates while the CP subsystem acquires and processes images in either time-averaged or double-exposure holographic modes. Shape measurements are performed by double-exposure mode with a

T. Proulx (ed.), *Mechanics of Biological Systems and Materials, Volume 2*, Conference Proceedings of the Society for Experimental Mechanics Series 9999, DOI 10.1007/978-1-4614-0219-0_28,

different laser source used in each one of the exposures whereas deformation measurements are obtained by using a single laser source.

We present preliminary results of our investigations of acoustically induced motions on tympanic membrane and the transfer of these motions to the ossicular chain, including results on sample preparation for measurements with the DOEHS.

Keywords: dual-wavelength lensless digital holography, shape and deformation measurements, tympanic membrane

1. INTRODUCTION

The tympanic membrane (TM) is the most peripheral structure in the middle ear and is the first ear structure set in motion by airbone sound [1]. Determination of the hearing loss caused by disease or trauma, which related with the sensitivity of motion, and the effect of the TM shape related with different age or sex are big concerns from doctors. In the previous work, digital optoelectronic holographic system (DOEHS) has been used to enable full field-of-view (FOV) measurement of the acoustically induced deformations of the tympanic membranes (TM) of chinchilla and human. However, limitations are also present: (1) the deformation information obtained from this system is only the relative deformation with respect to the reference and only along the optical axis, (2) no capability of shape measurement.

We are developing an advanced computer-controlled digital optoelectronic holographic system (DOEHS), the dual-wavelength lensless DOEHS, with the capabilities to measure both shape and acoustically induced deformations of the tympanic membrane qualitatively and quantitatively in a full field-of-view and non-contact mode. In order to obtain 3D deformation information, conventional methods [2, 3, and 4] consist of multiple configurations including at least three observation or three illumination systems concurrently. The goal of this new version is that, by using only one illumination system and one observation system, we can calculate the deformation information based on surface normal vector in each point of the contouring.

2. DESCRIPTION OF THE DUAL-WAVELENGTH DOEHC

2.1 Experiment Arrangement

Figures 1 and 2 depict the dual-wavelength DOEHS currently in development for high resolution shape and deformation full FOV measurements. The experiment setup mainly contains the laser delivery system and optical head system. In the laser delivery system, the coherent light sources are two tunable external cavity laser diodes (LD) with the operational wavelength from 780.04nm to 780.36nm for LD1 and from 779.54nm to 779.86nm for LD2. The output light from LD1 which is only utilized to the shape measurement is directly launched into a single mode polarized maintaining fiber (PM) and then connected with the optical switch (OS) used for switching optical signals. The output light from LD2 is first directed to the mirror (M) which changes the light angle by 90 degrees for alignment purposes. Then the light passes through the accousto-optical modulator (AOM) and joins into the fiber coupler assembly (FCA). The AOM is utilized in the stroboscopic mode of the system for deformation measurements by synchronizing it with the camera signal. The FCA is comprised of a λ/2 wave plate for maintaining the polarization direction, a focusing lens, and a fiber-optic aligner. After FCA, the light is directed through a single mode polarized maintaining fiber from the FCA to the optical switch (OS).

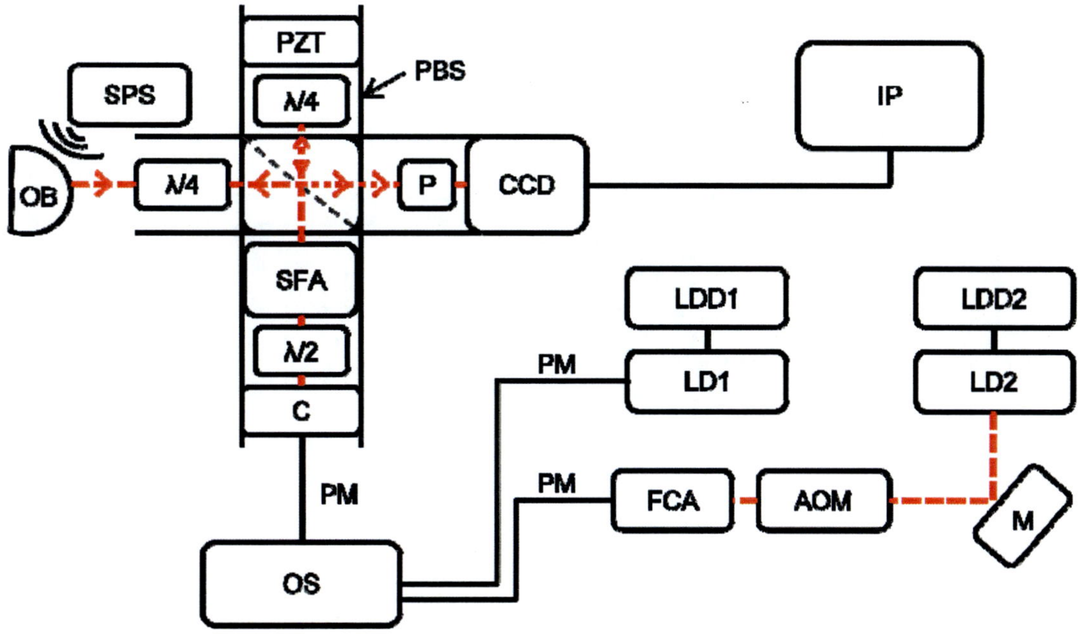

Fig. 1 Diagram of the set up for dual-wavelength DOEHS: LDD is the laser diode driver, LD is laser diode, M is the mirror, AOM is the accousto-optical modulator, FCA is the fiber coupler assembly, PM is the polarized maintaining fiber, OS is the optical switch, C is the collimator, λ/2 is the λ/2 wave plate, SFA is the spatial filter assembly, PBS is the polarizing beamsplitter cube, λ/4 is the λ/4 wave plate, PZT is the piezoelectric transducer, P is the polarizer, CCD is the CCD camera, SPS is the sound pressure system, OB is the object, and IP is the image and video processing computer.

The optical head (OH) system arrangement is shown in Figure 2. This system is a Michelson configuration setup. The light coming from the OS is directed into a collimator (C), a λ/2 wave plate and the spatial filter assembly (SFA). The function of the SFA is to filter high spatial frequencies of the beam and get a clean Gaussian beam profile. After the SFA, the light is directed through a polarized beamsplitter cube (PBS) which splits light into reference beam and object beam. The reference beam passes through a λ/4 wave plate and is directed to a mirror (M) connected with a piezoelectric transducer (PZT) that is controlled by the IP and used for application of phase-shifting algorithms. The object beam is directed through the λ/4 wave plate onto the object. The reflected object and reference beams are directed through the PBS again and combined at the surface of the CCD camera (CCD). Then the spatial intensity distributions recorded by the CCD are processed by the computer.

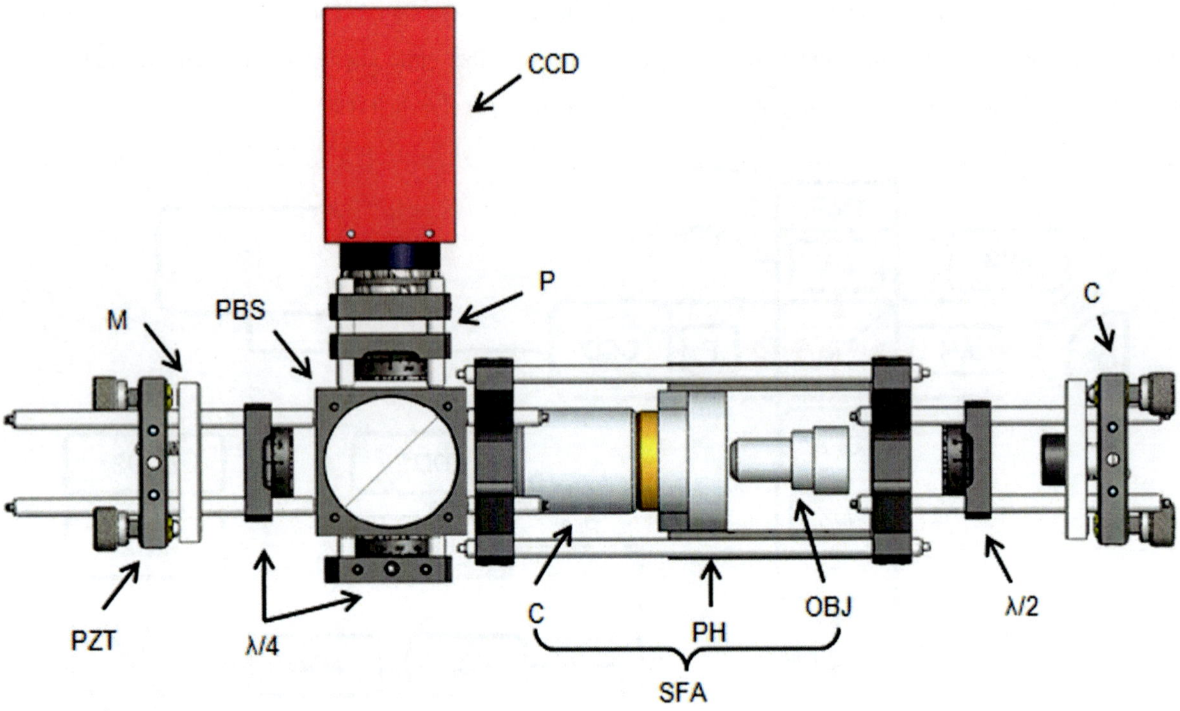

Fig. 2 CAD model of the OH cage system. C is the collimator, λ/2 is the λ/2 wave plate, SFA is the spatial filter assembly which contains OBJ, the objective, PH, the pinhole and C, PBS is the polarizing beamsplitter cube, λ/4 is the λ/4 wave plate, M is the mirror, PZT is the piezoelectric transducer, P is the polarizer, and CCD is the CCD camera.

2.2 Shape measurement with the two-wavelength contouring technique

In order to determine the shape of the object, a dual-wavelength holographic contouring technique [5] is applied by generating depth contours related with geometry of the object. This requires acquisitions of a set of amplitude and phase information at wavelength λ_1, considered as the reference or undeformed state, as well as acquisitions of a second set of amplitude and phase information at wavelength λ_2, considered as the modified or deformed state. The phase difference [6] of the two sets of data, $\Delta\phi$, is given by

$$\Delta\phi = \phi_2 - \phi_1 = \frac{2\pi}{\Lambda} OPL \tag{1}$$

where ϕ_1 is the phase of the optical path length recorded at the first wavelength λ_1, ϕ_2 is the phase of the optical path length recorded at the second wavelength λ_2, and OPL is the optical path difference defined as the distance from an illumination point to an object point and to an observation point. Λ is the synthetic wavelength given by

$$\Lambda = \frac{\lambda_1 \lambda_2}{|\lambda_1 - \lambda_2|} \tag{2}$$

The phase difference [7] related to depth contours $\Delta\phi$ can also be represented by the fringe-locus function Ω,

$$\Omega(x,y,z) = 2\pi n(x,y,z) = \left[\overline{\mathrm{K}}_2(x,y,z) - \overline{\mathrm{K}}_1(x,y,z)\right] \cdot \overline{\mathrm{L}}(x,y,z) = \overline{\mathrm{K}}(x,y,z) \cdot \overline{\mathrm{L}}(x,y,z) \tag{3}$$

where $n(x,y,z)$ is the fringe order at the (x,y,z) location of the image plane, $\overline{\mathrm{K}}(x,y,z)$ is the sensitivity vector, $\overline{\mathrm{K}}_1(x,y,z)$ is the propagation vector of light from a point source of illumination to a point on the object, $\overline{\mathrm{K}}_2(x,y,z)$ is the propagation vector of light from a point on the object to a point on the CCD sensor, and $\overline{\mathrm{L}}(x,y)$ is the displacement vector.

2.3 Deformation measurement in the stroboscopic mode

Stroboscopic mode is used for deformation measurements of the TM at a controlled sound input. This mode gives information about the phase difference [8] $\phi'(x,y)$ which is related to the complex distribution phase of the reference and deformed states respectively, given by

$$\phi'(x,y) = \arctan \frac{a_2'(x,y)a_3'(x,y) - a_1'(x,y)a_4'(x,y)}{a_2'(x,y)a_4'(x,y) + a_1'(x,y)a_3'(x,y)} \tag{4}$$

where $a_1'(x,y)$, $a_2'(x,y)$ represent the real and imaginary part of the reference state, and $a_3'(x,y)$, $a_4'(x,y)$ represent the real and imaginary part of the deformed state.

The out of plane displacement after image processing is a relative value with respect to the reference measurement. However, no displacement information along x and y axes from the processed image is obtained. In order to calculate these displacements, we need to first decompose the normal vector of the object surface based on the shape measurement shown in Figure 3. Then using the sensitivity vector $\overline{\mathrm{K}}(x,y,z)$, we can calculate the displacements along x and y axes [9].

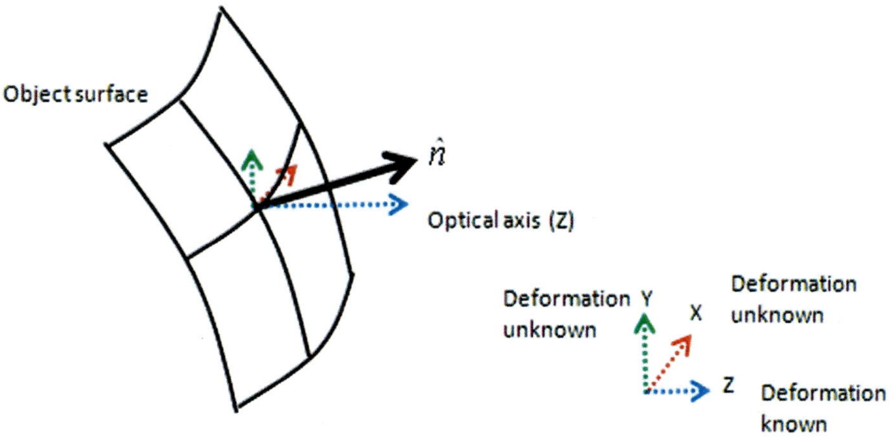

Fig. 3 Schematic drawing of the decomposition of normal vector along three axes.

3. RESULTS

3.1 Shape measurement

Fig. 4 Picture of the conical latex membrane model painted with holographic dust. The flat latex membrane was first clamped along the circular boundary of the hollow metal tube, and then was deformed by pushing with a rod at the center of the membrane.

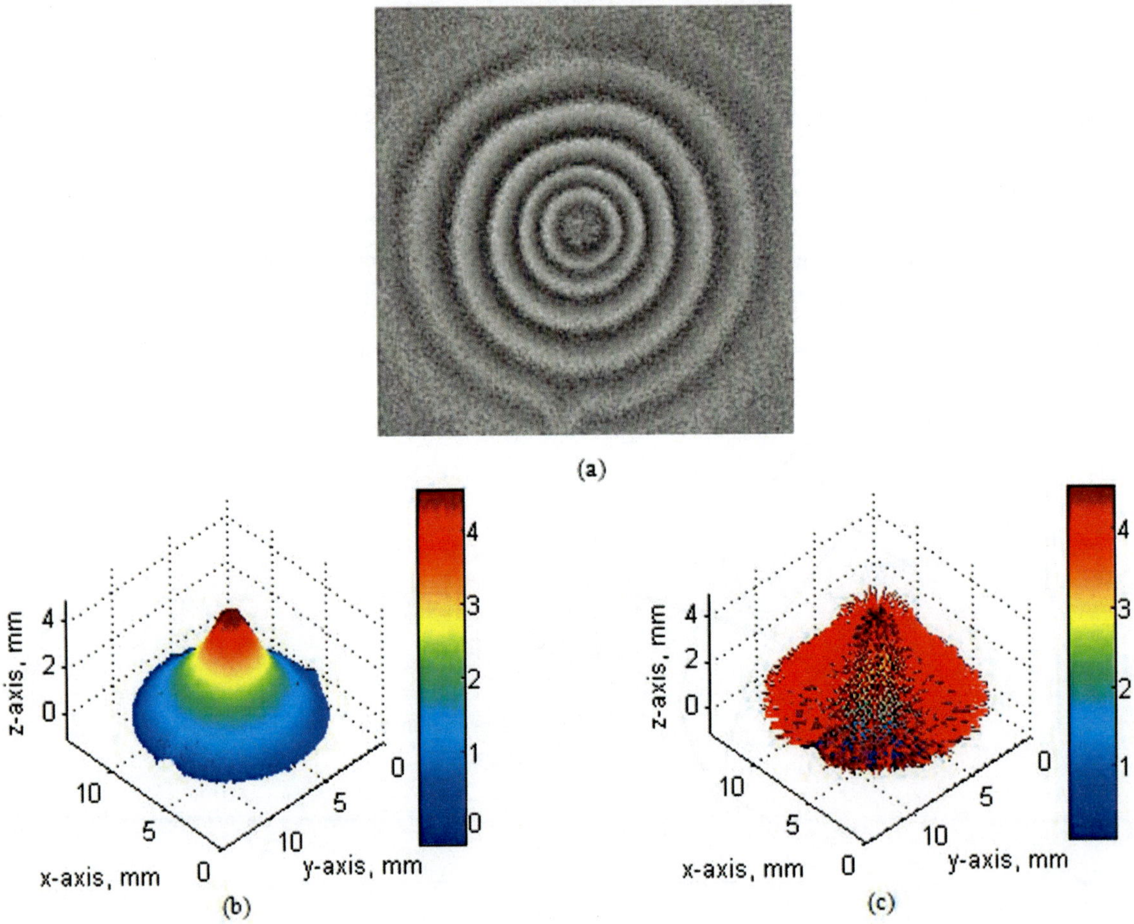

Fig. 5 Shape measurement on conical latex membrane: (a) optical phase, (b) 3D plot of the real shape, and (c) 3D plot of the shape with normal vector.

Figures 4 and 5 represent the shape measurements on a conical shape latex membrane. The object boundary conditions were controlled by the following procedure: a piece of latex membrane was clamped on the circular boundary of the hollow metal tube. And then it was deformed to a conical shape similar to the shape of tympanic membrane by pushing with a rod at the center of the membrane. The position of the rod along the optical axis is precisely controlled by the micrometer. Finally the surface of the membrane was painted with the holographic dust in order to have better contrast. The diameter of the conical shaped membrane is 12.73mm, and the height is 5mm.

3.2 Deformation measurement

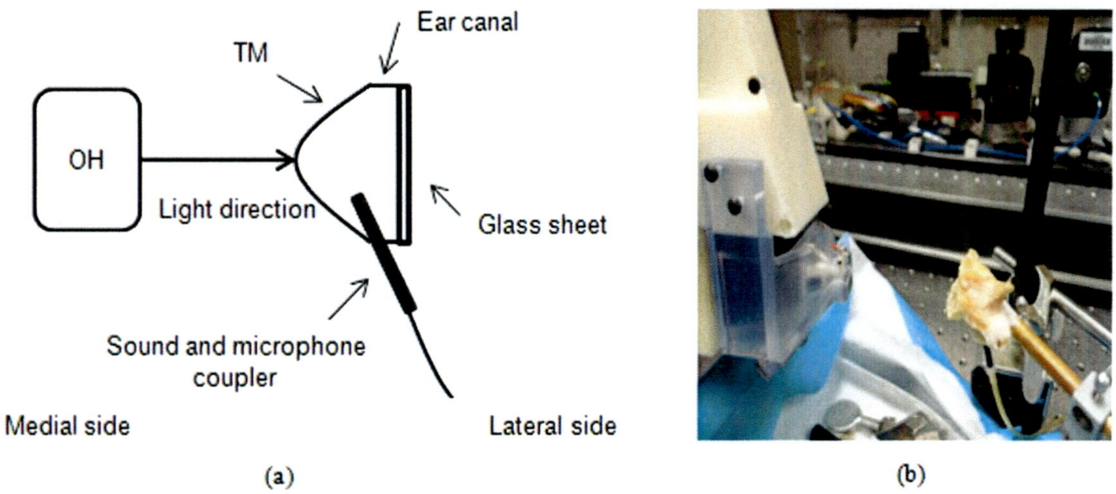

Fig. 6 Configuration of the deformation measurement in MEEI: (a) diagram of the optical head (OH) and object configuration, and (b) real configuration.

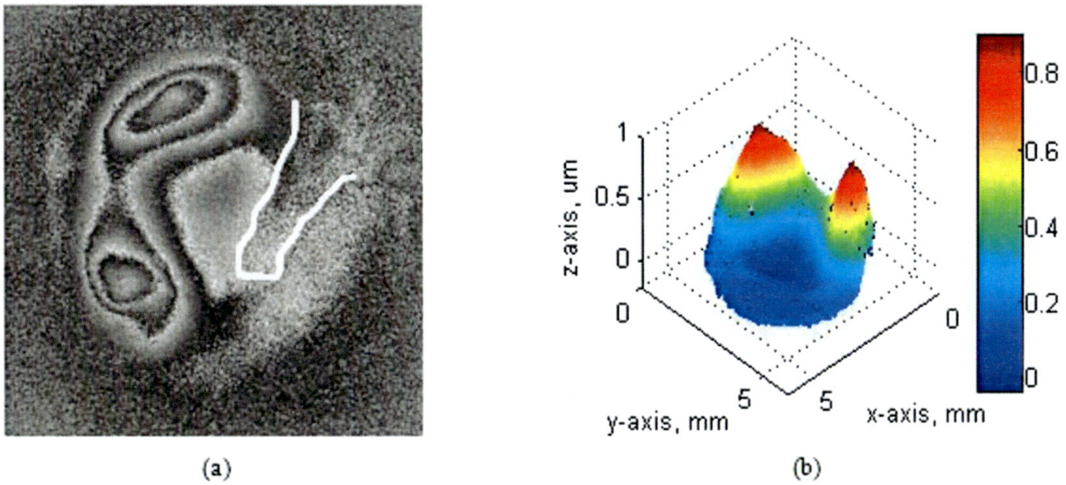

Fig. 7 The deformation of the medial side of TM at phase 180 at the sound excitation of 1 KHz: (a) optical phase with location of umbo, and (b) perspective plot of the deformation.

The deformation measurement mode was tested in Massachusetts Ear and Eye Infirmary (MEEI) in Boston. The temporal bone was from the left ear of a 65 year old male without history of otologic disease. The incus, the extended facial recess, the roof of the epitympanum, the Eustachian tube and the jugular bulb were all drilled away to expose the medial side of the TM. In Figure 6, the OH illuminated the medial side of TM with the back surface of TM perpendicular to the beam. The sound and microphone coupler was placed in the lateral side of TM through the wall of ear canal. A glass sheet was used to seal the ear canal in order to isolate the SPS. The TM was painted with ZnO to increase the contrast and mounted with the dental cement onto a hollow metal tube.

Figure 7 shows the deformation of the medial side of TM at phase 180 at the sound excitation of 1 KHz. The location of the umbo is shown in the first image. It is clear to see there are two peaks generated and the scale of the deformation is around 1 micrometer.

4. CONCLUSIONS AND FUTURE WORK

In this study, we are developing the dual-wavelength lensless digital holography system to measure both the shape and acoustically induced deformation of TMs. The typical feathers: (1) it is a collimated con-axial system, and (2) 3D displacements are calculated base on the single axis deformation and shape measurement. However, further research need to be completed in the future, such as the validation, repeatability and error estimation of the system, and simplifying and packaging the system.

5. ACKNOWLEDGMENT

This work is supported by National Institutes of Health (NIH), Massachusetts Ear and Eye Infirmary (MEEI) and Worcester Polytechnic Institute (WPI). Thanks for the help from J. M. Flores-Moreno, Ellery Harrington, Ivo Dobrev, Michael Zervas, Peter Hefti and other group members in Center for Holographic Studies and Laser micro-mechaTronics (CHSLT).

6. REFERENCE

[1] Hernandez-Montes, Maria del Socorro et al. "Optoelectronic holographic otoscope for measurement of nano-displacements in tympanic membranes." Journal of Biomedical Optics 14.3 (2009): 034023-9.

[2] C. Furlong and R. J. Pryputniewicz, "Electro-optic holography method for determination of surface shape and deformation," in Laser interferometry IX: Techniques and Analysis, M. Kujawinska, G. M Brown, and M. Takeda, Eds., *Proc. SPIE* 3478, 86–97 (1998).

[3] I. Yamaguchi, T. Ida, M. Yokota, and K. Yamashita, "Surface shape measurement by phase shifting digital holography with a wavelength shift," *Appl. Opt.* 45, 7610-7616 (2006).

[4] W. Osten, "Application of optical shape measurement for the nondestructive evaluation of complex objects," *Opt. Eng.* 39(1), 232–243 (2000) .

[5] C. Wagner, W. Osten and S. Seebacher, "Direct shape measurement by digital wavefront reconstruction and multiwavelength contouring", *Opt. Eng.* 39, 79 (2000).

[6] C. Furlong and R. J. Pryputniewicz, "Absolute shape measurements using high-resolution optoelectronic holography methods," *Opt. Eng.* 39(1), 216–223 (2000).

[7] R. J. Pryputniewicz, *Holographic numerical analysis*, Worcester Polytechnic Institute, Worcester, MA, 1996.

[8] I. Dobrev, J. M. Flores Moreno, C. Furlong, E. J. Harrington, J. J. Rosowski and C. Scarpino, "Design of a positioning system for a holographic otoscope", *Proc. SPIE* 7791, 77910D (2010).

[9] S. Seebacher, W. Osten, and W. Jüptner, "Measuring shape and deformation of small objects using digital holography," *Proc. SPIE* 3479, 104–115 (1998).

Investigating Mechanical Properties of Porcine Articular Cartilage by Flat Plate Compression Tests

[1]Liou*, N.-S., [2]Jeng, Y.-R., [1]Yen, S.-H., [1]Chen, S.-F. and [2]Wu, K.-T.
[1]Department of Mechanical Engineering, Southern Taiwan University
No. 1, Nan-Tai Street, Yongkang Dist., Tainan City 710, Taiwan R.O.C.
[2]Department of Mechanical Engineering, National Chung Cheng University
168 University Road, Minhsiung Township, Chiayi County 621, Taiwan, ROC
email: nliou@mail.stut.edu.tw

Abstract
An inverse iterative finite element analysis procedure was developed to investigate mechanical properties of porcine articular cartilage under compression load. Specimen consists of articular cartilage and bone was cut from porcine femur for this study. The specimen was mounted on metal base plate and the compression tests were performed by using material test machine with flat compression plate. The ramp-hold compression tests with different ramping displacement rates and long-term compression test with very low displacement rate were performed on the specimens. The force and compression displacement as function of time were recorded. The profile of specimen was obtained from for generating the geometry model of specimen. The finite element model which consists of bone and cartilage parts was generated from the geometry model of specimen. In this study, bone and cartilage were considered as linear elastic material and hyperviscoelastic material respectively. The generalized pattern search method was used for the optimization process to find the long-term hyperelastic and viscoelastic parameters of cartilage by minimize the experimental loading forces and the reaction forces computed form finite element analysis. The results shows 2nd order reduce polynomial hyperelastic model with two viscoelastic time constants can well describe the mechanical response of articular cartilage and bone specimen under flat plate compression.

Introduction
The articular cartilage plays important roles in the biomechanics of knee. The functionality of the articular cartilage and its tasks in biomechanics of knee such as lubrication and load transmission has been broadly discussed to better recognize traumatic and pathologic processes [1-3]. Furthermore, the studies about the effect of stresses and strains on articular cartilage reveal that the biomechanical properties of articular cartilage are anisotropic, inhomogeneous and time dependent. There are different categories of constitutive models such as poroelastic [4-5], viscoelastic [6-8], biphasic [9-10] or poroviscoelastic [11-12] models developed for articular cartilage. However, the use of many sophisticated constitutive equations of articular cartilage are limited due to they are computational expensive or difficult to be implemented into numerical tools such as finite element program. Besides, due to the limited size and non-uniform shape, it is not easy to obtain articular cartilage specimens with well defined simple geometry for standard mechanical tests. In this study, the hybrid experimental and numerical analysis was developed to investigate mechanical properties of porcine articular cartilage under compression. Inverse iterative finite element analysis procedure was used to obtain the material parameters of articular cartilage which are suitable for finite element study of knee biomechanics.

Material and method
The hyperviscoelastic model was used to describe the time dependent mechanical properties of porcine articular cartilage. The polynomial strain energy function of the hyperviscoelastic model can be written as following:

$$W = \int_0^t \left\{ \sum_{i+j=1}^{N} \left[C_{ij0} \left(1 - \sum_{k=1}^{n} g_k \left(1 - e^{-(t-\tau)/\tau_k} \right) \right) \right] \times \frac{d}{d\tau} \left[(J_1 - 3)^i (J_2 - 3)^j \right] \right\} d\tau. \qquad (1)$$

T. Proulx (ed.), *Mechanics of Biological Systems and Materials, Volume 2*, Conference Proceedings of the Society for Experimental Mechanics Series 9999, DOI 10.1007/978-1-4614-0219-0_29, © The Society for Experimental Mechanics, Inc. 2011

where g_k and τ_k are relaxation coefficients and characteristic times respectively, N is the order of polynominal in strain invariants, J_1, J_2 are strain invariants, $C_{ij0}'s$ are the instantaneous hyperelastic parameters. The least square fit method was used to obtain parameters of hyperviscoelastic constitutive model for describing mechanical properties of porcine articular cartilage under compression load. Second order reduce polynomial form were used for the energy function and cartilage were assumed to be non-compressible material. Two time-dependent terms were used and the two time constants were chosen ($\tau_1 = 5\ \mathrm{sec}$, $\tau_1 = 50\ \mathrm{sec}$). These assumptions left C_{10}, C_{20}, g_1 and g_2 to be determined by least square fit method. In the study, bone was considered as linear elastic material. The elastic modulus Poisson's ratio of bone was assumed to be constant. The hyperviscoelastic material parameters of articular cartilage were determined by using optimization-based inverse iterative finite element analysis procedure [13]. The generalized pattern search method [14-16] was used for the optimization process. The finite element analyses were performed by using ABAQUS^{TM} and the iteration loop were performed in MATLAB^{TM}.

The specimen, cut from medial condyle portion of femur, used for current study is shown in Fig. 1. The thickness of specimen is 4mm and the specimen was cut from the location where the curvature of specimen in the thickness direction can be minimized so that 2D FE simulation can be used to approximate the specimen under flat plate compression.

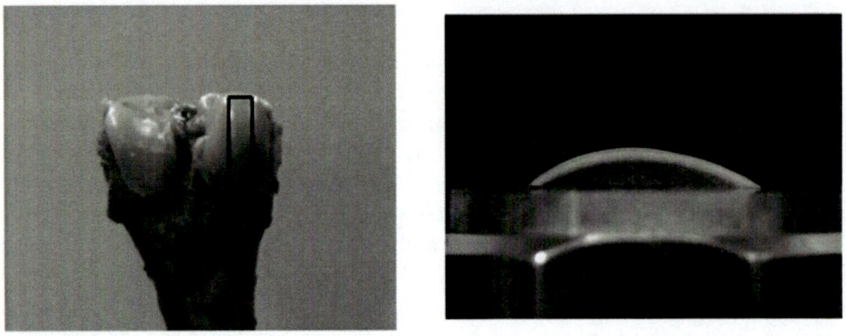

(a) (b)

Figure 1: Specimen used for flat plate compression test

In order to obtain the viscoelastic and long-term hyperelastic prameters of constitutive model, ramp-hold compression tests and low speed continuous compression test were performed on the specimen. The displacement rates used for ramp-hold compression tests are 0.1 and 0.01 mm/sec and the total compression displacement is 0.1mm. The displacement rate used for low speed continuous compression test is 0.0001mm/sec. The force response under low speed continuous compression was assumed to be solely come from the long-term response of hyperviscoelastic constitutive model. The specimen under flat plate compression test and the corresponding finite element model are shown in Fig. 2.

(a) (b)

Figure 2: Specimen under flat plate compression test and the corresponding finite element model

The hyperviscoelastic material parameters of cartilage were determined in two steps. Firstly, the long-term hyperelastic parameters were determined by minimizing the squared error between the simulated and experimental compression forces of long-term equilibrium force-displacement curve under long-term flat plate compression test. Secondly, the time dependent parameters were determined by minimizing the squared error between the simulated and experimental forces of ramp-hold compression with different compression rates.

Results and discussion

The fitted parameters $C_{10\infty}$ and $C_{20\infty}$ for long-term second order reduce polynomial hyperelastic model are 0.0673 MPa and -0.0181 MPa respectively, and the corresponding force-displacement curves of specimens under low speed continuous flat plate compression are shown in Fig. 3a.

The fitted parameters g_1 and g_2 for time-dependent terms fitted from the ramp-hold compression experimental force-displacement curves are 0.55 and 0.4 respectively, and the corresponding force-time curves of specimens under ramp-hold compression with 0.01 and 0.1 mm/sec displacement rates are shown in Fig. 3b,c respectively.

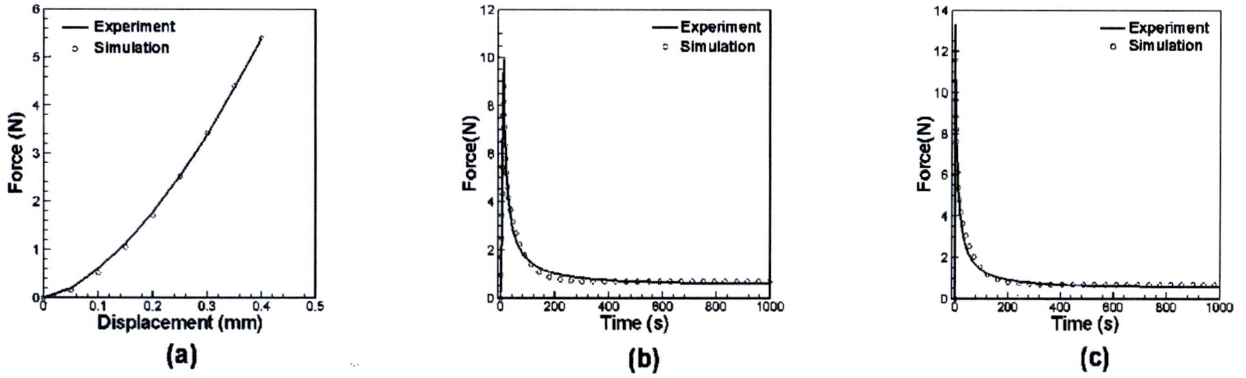

Figure 3: (a) Force-displacement curves of specimens under low speed continuous flat plate compression. (b)(c) force-time curves of specimens under ramp-hold compression with 0.01 and 0.1 mm/sec displacement rates

In this study, we used hybrid experimental and numerical analysis to identify material parameters of articular cartilage under slow and moderate compression rates. The constitutive model can be directly used with finite element software to investigate biomechanics of knee joint.

Acknowledgement

This research was supported by NSC 98-2221-E-218 -054 from the R.O.C government.

Reference

1. Sweigart, M.A., C.F. Zhu, D.M. Burt, P.D. deHoll, C.M. Agrawal, T.O. Clanton, and K.A. Athanasiou, *Intraspecies and interspecies comparison of the compressive properties of the medial meniscus.* Annals of Biomedical Engineering, 2004. 32(11): p. 1569-1579.
2. Pena, E., B. Calvo, M.A. Martinez, D. Palanca, and M. Doblare, *Finite element analysis of the effect of meniscal tears and meniscectomies on human knee biomechanics.* Clinical Biomechanics, 2005. 20(5): p. 498-507.
3. Barber, F.A., M.A. Herbert, F.A. Schroeder, J. Aziz-Jacobo, and M.J. Sutker, *Biomechanical Testing of New Meniscal Repair Techniques Containing Ultra High-Molecular Weight Polyethylene Suture.* Arthroscopy-the Journal of Arthroscopic and Related Surgery, 2009. 25(9): p. 959-967.
4. Beek, M., J.H. Koolstra, and T.M.G.J. van Eijden, *Human temporomandibular joint disc cartilage as a poroelastic material.* Clinical Biomechanics, 2003. 18(1): p. 69-76.
5. Gupta, S., J. Lin, P. Ashby, and L. Pruitt, *A fiber reinforced poroelastic model of nanoindentation of porcine costal cartilage: A combined experimental and finite element approach.* Journal of the Mechanical Behavior of Biomedical Materials, 2009. 2(4): p. 326-338.
6. Park, S. and G.A. Ateshian, *Dynamic response of immature bovine articular cartilage in tension and compression, and nonlinear viscoelastic modeling of the tensile response.* Journal of Biomechanical Engineering-Transactions of the Asme,

2006. 128(4): p. 623-630.
7. Fulcher, G.R., D.W.L. Hukins, and D.E.T. Shepherd, *Viscoelastic properties of bovine articular cartilage attached to subchondral bone at high frequencies.* Bmc Musculoskeletal Disorders, 2009. 10: p. -.
8. Egan, J.M., *A viscoelastic analysis of the tensile weakening of deep femoral head articular cartilage.* Proceedings of the Institution of Mechanical Engineers Part H-Journal of Engineering in Medicine, 2000. 214(H3): p. 239-247.
9. Ehlers, W. and B. Markert, *A linear viscoelastic two-phase model for soft tissues: Application to articular cartilage.* Zeitschrift Fur Angewandte Mathematik Und Mechanik, 2000. 80: p. S149-S152.
10. Garcia, J.J., N.J. Altiero, and R.C. Haut, *Estimation of in situ elastic properties of biphasic cartilage based on a transversely isotropic hypo-elastic model.* Journal of Biomechanical Engineering-Transactions of the Asme, 2000. 122(1): p. 1-8.
11. DiSilvestro, M.R., Q.L. Zhu, and J.K.F. Suh, *Biphasic poroviscoelastic simulation of the unconfined compression of articular cartilage: II - Effect of variable strain rates.* Journal of Biomechanical Engineering-Transactions of the Asme, 2001. 123(2): p. 198-200.
12. DiSilvestro, M.R., Q.L. Zhu, M. Wong, J.S. Jurvelin, and J.K.F. Suh, *Biphasic poroviscoelastic simulation of the unconfined compression of articular cartilage: I - Simultaneous prediction of reaction force and lateral displacement.* Journal of Biomechanical Engineering-Transactions of the Asme, 2001. 123(2): p. 191-197.
13. Liu, K.F., M.R. VanLandingham, and T.C. Ovaert, *Mechanical characterization of soft viscoelastic gels via indentation and optimization-based inverse finite element analysis.* Journal of the Mechanical Behavior of Biomedical Materials, 2009. 2(4): p. 355-363.
14. Torczon, V., *ON the convergence of pattern search algorithms.* Siam Journal on Optimization, 1997. 7(1): p. 1-25.
15. Audet, C. and J.E. Dennis, *Analysis of generalized pattern searches.* Siam Journal on Optimization, 2003. 13(3): p. 889-903.
16. Lewis, R.M. and V. Torczon, *Pattern search methods or linearly constrained minimization.* Siam Journal on Optimization, 2000. 10(3): p. 917-941.

Characterization of Tendon Mechanics Following Subfailure Damage

Sarah E Duenwald-Kuehl; Jaclyn Kondratko; Ray Vanderby, Jr.; Roderic Lakes*
University of Wisconsin-Madison, *1500 Engineering Drive, Madison, WI, 53706

Abstract
Tendon functionality depends on its mechanical properties. Therefore, a tendon's mechanical properties are essential to define and understand its normal and damaged states. In this study, we characterized porcine digital flexor tendons via stress relaxation and cyclic testing prior to and following overstretch damage. Overstretch damage was induced by pulling tendons to 6.5, 9.0, or 13.0% strain. Both elastic and viscoelastic parameters were measured pre- and post-damage for comparison. Elastic parameters such as the *maximum stress reached during relaxation* and *peak stress during cyclic testing* decreased following damage and viscoelastic parameters such as *stress reduction during relaxation* and *decrease in peak stress from the first to last cycle during cyclic testing* also decreased following damage. Therefore, sub-failure damage induced by overstretching tendons changes both the elastic and viscoelastic properties, and these changes are more pronounced as the damage level increases.

Introduction
Tendons function to move joints, absorb impacts to protect muscle, and store energy during loading to facilitate movement. Because these functions are dictated by their mechanical properties, a robust characterization of tendon's mechanical properties provides great insight into physiological behaviors. Compositionally, these properties are largely a function of their main structural component collagen, as well as other constituents of the extracellular matrix including proteoglycans, glycoproteins, and water. Properties are also a function of the microstructural organization of these components. Tendon collagen consists mainly of type I collagen fibers which are predominantly oriented parallel to the direction of load, or the long axis of the tendon [1; 2]. The viscoelastic, or time-dependent, properties of the collagen fibers, as well as their interaction with the other constituents of the extracellular matrix, lead to tendons exhibiting viscoelastic behavior. Thus, it is important to analyze both elastic and viscoelastic properties to more completely understand tendon mechanical function.

Tendons with sub-failure damage have different biomechanical behavior than normal tissues. These differences include both elastic and viscoelastic changes; for example, damage can lead to a drop in load for a given strain [3; 4]. Also, Perry *et al.* (2009) found changes in the supraspinatus tendon following rotator cuff injury were quantifiable in viscoelastic but not elastic properties, highlighting the importance of including viscoelastic characterization in damage studies [5]. "Damage" in this study was characterized as changes in the microstructure of tendon that leads to reduction in the mechanical strength [3; 4], and caused by prolonged cyclic loading or the application of stress and strain states that exceed the tendon's elastic limit [6]. The healing response to this damage could limit movement of the joint, potentially resulting in tearing of the tendon, insertion sites, or even muscle under normal movement conditions. Conversely, increased laxity from overstretch could allow hypermobility of the joint or require altered muscle contraction to facilitate movement, potentially resulting in damage of surrounding tissue. When considering the time-dependent nature of these tissues, changes in viscoelastic properties could result in differences in how the tendon copes with instantaneous or prolonged loading scenarios. Even subtle differences in tendon mechanical behavior can lead to accumulation of damage in the tendon and the surrounding tissues, which can eventually lead to catastrophic failure or chronic degeneration of the joint.

In this study we set out to examine the effects of damage on elastic and viscoelastic behavior of tendon by subjecting tendons to cyclic and stress relaxation testing prior to and following overstretch damage. The overall goal was to examine which viscoelastic and elastic mechanical parameters undergo changes following sub-failure damage induced by overstretching the tendon, and the extent to which these changes were affected by amount of overstretch strain.

Materials and Methods
Thirty digital flexor tendons were dissected from porcine lower limbs; all muscle and extraneous tissue was removed with the bony insertion left intact at the distal end of the tendons. Bony ends were potted in lightweight polyester resin filler, with care taken to ensure tendon tissue was not encased in the filler, inside a mold with the same dimensions as the testing grip to

T. Proulx (ed.), *Mechanics of Biological Systems and Materials, Volume 2*, Conference Proceedings of the Society for Experimental Mechanics Series 9999, DOI 10.1007/978-1-4614-0219-0_30,

ensure a tight fit with no movement. Unloaded area was measured by assuming an elliptical shape. Long and short axis measurements were taken at three points along the specimen with a digital caliper and averaged. Specimens were kept hydrated with physiologic buffered saline throughout the dissection and preparation process.

All mechanical testing was executed using a servohydraulic mechanical test system (Bionix 858; MTS, Minneapolis, Minnesota). Tendons were loaded into one soft-tissue grip (muscle end) capable of holding dry ice (freezing assists in gripping) and one bone grip (potted bone end) in the test frame (see Fig 1). Displacement was controlled by the servohydraulic system and load was measured via a 1000lb load cell (Honeywell, Morristown, New Jersey). Data were captured on a PC equipped with Labtech Notebook (Laboratory Technology Corporation, Fort Collins, Colorado).

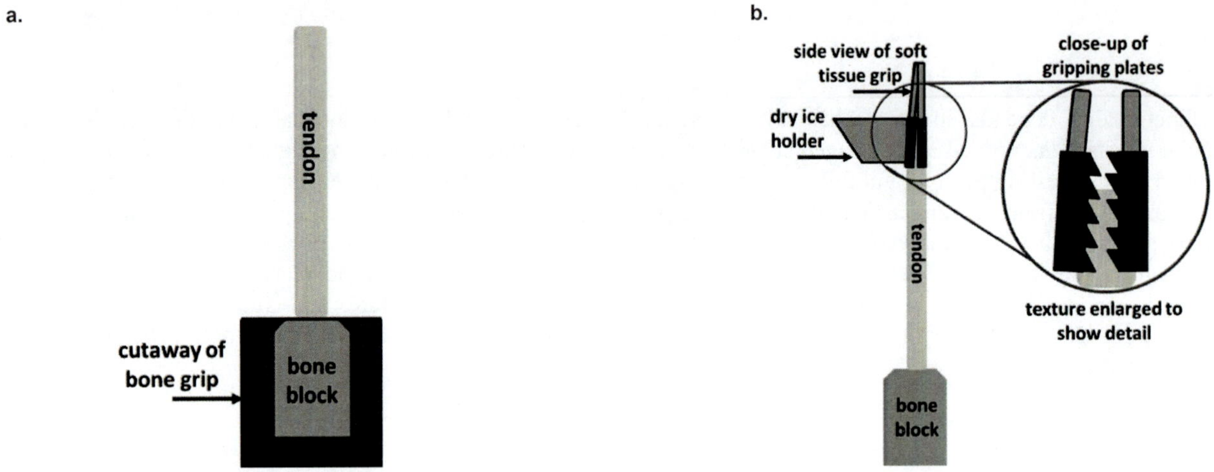

Figure 1. Schematic view of a) bone and b) soft tissue grips. The potted bone fits snugly in the bone grip while the muscle end of the tendon is gripped by the rough surface of the soft tissue grip; gripping is assisted by freezing with dry ice.

Once specimens were mounted in the mechanical test frame, a preload of 1N was applied to remove slack in the tendon. The preload was applied slowly such that the tendons effectively remained at equilibrium, rather than rapid application which would result in substantial relaxation from the 1N preload. Initial tendon length, L_0, was measured for strain calculations. Following the preload, tendons were preconditioned using a sinusoidal wave (0 to 2% strain) at 0.5Hz for 20s. A rest period of 1000s was granted prior to further mechanical testing.

Stress relaxation (step strain input held for 100s) and cyclic testing (sinusoidal strain input at 0.5Hz for 20s) to 4% strain were performed on tendons. The rise time for the nominal step stretch input was 40ms. Two relaxation and two cyclic tests were performed with 1000s of rest between tests. Following the completion of these four tests, tendons were damaged via an overstretch pull (ramp to final strain in 1s) to one of three subfailure damage strains. Maximum strains of 5-6% [7; 8] were reported in the literature during normal movement, and strains of approximately 15-20% [9; 10] were reported at tendon failure. Therefore the strains of 6.5% (n=10), 9.0% (n=10), or 13% (n=10) strain were chosen to fall outside the range of normal movement yet below the failure strain. Tendons were allowed to rest for 1000s prior to repeating stress relaxation and cyclic testing.

Force data acquired by the servohydraulic system during stress relaxation testing were used to calculate stress data (by dividing by cross-sectional area), $\sigma(t)$. Stress data were then used to calculate the *maximum stress reached during stress relaxation* (Fig 2a), σ_{max}, as well as the *reduction in stress during relaxation* (Fig 2a), σ_{decay}, both prior to and following damage. The maximum stress was measured at a time 0.1s on the time scale (the zero point taken halfway through the rise time), or $2.5t_r$ (where t_r is the rise time) as suggested by Lakes [11]. Force data acquired during cyclic testing were also used to calculate stress data, and grip-to-grip displacement data were used to calculate strain as a function of time (by dividing grip-to-grip displacement by initial length), such that stress-strain curves could be generated prior to and following damage. Stress data were then used to calculate the *peak stress reached during cyclic testing* (Fig 2b), σ_{peak}, as well as the *decrease in peak stress from the first to last cycle* (Fig 2b), $\sigma_{decrease}$, prior to and following damage. Post- to pre-damage ratios were then calculated for each parameter.

Post- to pre-damage ratios of parameters following overstretch to 6.5, 9, and 13% strain were compared using a repeated measures ANOVA. Statistical significance set at $p = 0.01$.

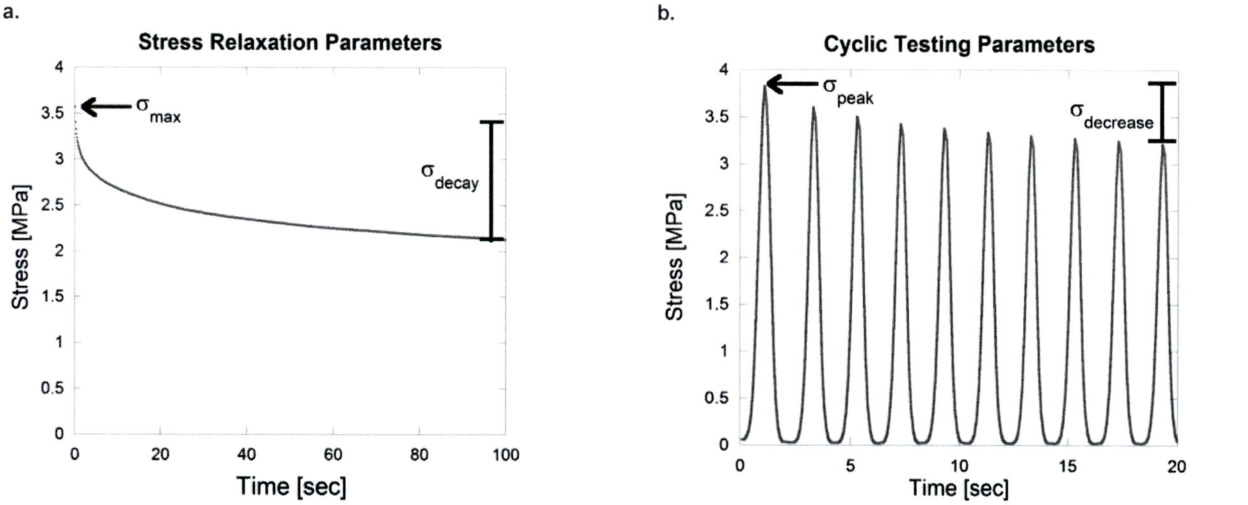

Figure2. Mechanical testing parameters for a) stress relaxation testing (σ_{max}, σ_{decay}) and b) cyclic testing (σ_{peak}, $\sigma_{decrease}$).

Results

Diffuse damage caused by overstretch lead to an increased laxity, indicated by a reduced stress at a given strain level and marked by a rightward shift in the stress-strain curve (Fig 3a). This softening was also manifested by reduced stresses during cyclic testing (Fig 3b) and relaxation testing (Fig 3c).

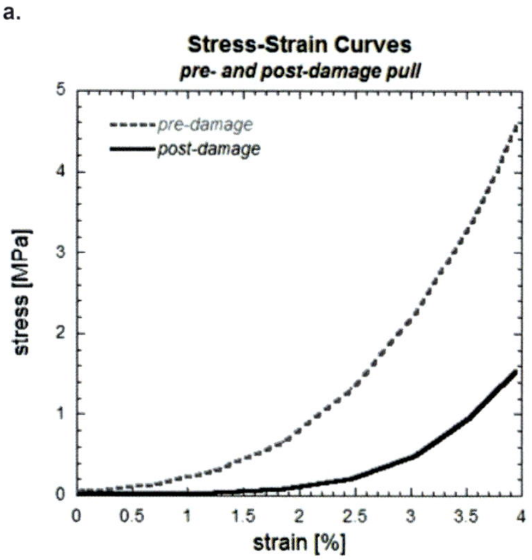

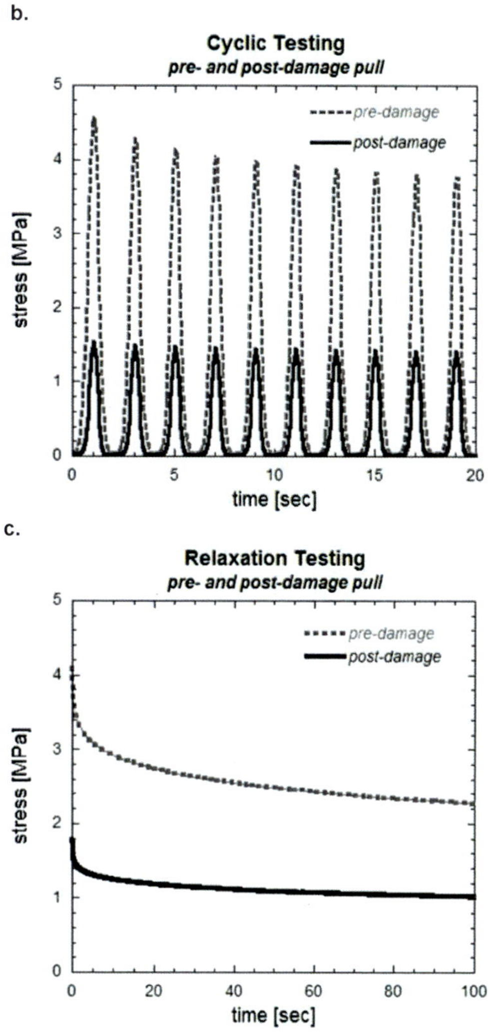

Figure 3. Representative pre- and post-damage data for a) stress-strain curves, b) cyclic testing, and c) stress relaxation testing from a specimen subjected to an overstretch of 9% strain.

The elastic parameters σ_{max} (maximum stress reached during relaxation) and σ_{peak} (peak stress reached during cyclic testing) decreased following damage (Fig 4a). The post-damage to pre-damage ratios (mean +/- standard deviation) of σ_{max} following overstretch to 6.5, 9, and 13% strain were 0.66+/- 0.08, 0.43 +/- 0.10, and 0.14 +/- 0.07, respectively, and each overstretch level was significantly different than the others (p=0.0031). Likewise, the post- to pre-damage ratios of σ_{peak} following 6.5, 9, and 13% strain were 0.56+/- 0.09, 0.34 +/- 0.10, and 0.12 +/- 0.05, respectively, and each overstretch level was significantly different than the others (p=0.0039).

The viscoelastic parameters σ_{decay} (reduction in stress during 100s of stress relaxation) and $\sigma_{decrease}$ (decrease in peak stress from cycle 1 to cycle 10) also decreased following damage (Fig 4b). The post-damage to pre-damage ratios (mean +/- standard deviation) of σ_{decay} following overstretch to 6.5, 9, and 13% strain were 0.64 +/- 0.10, 0.41+/- 0.10, and 0.14 +/- 0.09, respectively. Each overstretch strain level was significantly different than the others (p=0.0037). The post- to pre-damage ratios of $\sigma_{decrease}$ following 6.5, 9, and 13% strain were 0.23 +/- 0.07, 0.15 +/- 0.05, and 0.04 +/- 0.02, respectively. Again, each overstretch level was statistically distinct (p=0.0094).

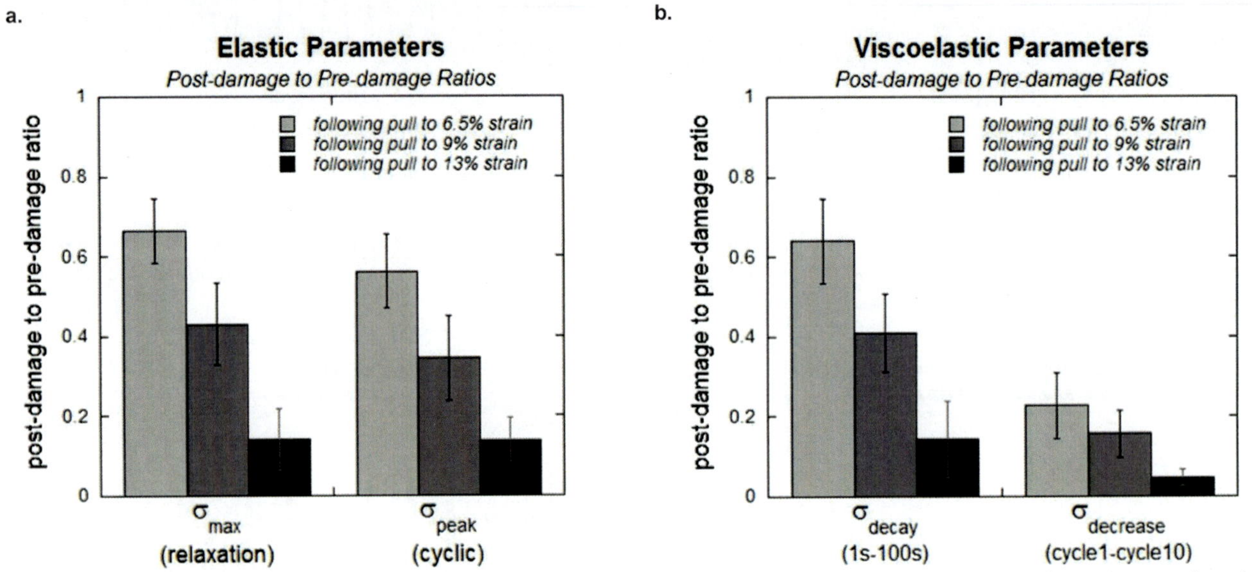

Figure 4. Post-damage to pre-damage ratios of a) elastic and b) viscoelastic parameters calculated from stress relaxation and cyclic testing data. Error bars indicate one standard deviation. The post- to pre-damage ratio gets smaller, and hence the effect of damage gets larger, as the strain during the overstretch pull gets larger.

Discussion

Though simple in appearance, tendons are intricate natural composite materials with complex mechanical properties. Analysis of their mechanical behavior describes their functional state, and knowledge of how their mechanical behavior is altered following damage is important for understanding mechanical function, quantifying compromise with subfailure damage, and benchmarking normal functional behaviors to evaluate the efficacy of injury treatments. Understanding the changes in laxity experienced by the tendon post-damage better help us understand how the muscle must compensate, as well as anticipate the potential for abnormal joint movement.

In this study, we found that the elastic parameters σ_{peak} and σ_{max} were decreased following overstretch damage. This indicates that the tendon has less resistance to motion after it has sustained subfailure damage. Normal tendon has a strain-stiffening stress-strain curve, which leads to greater resistance to deformation as strain in the tendon increases; this behavior allows tendons and ligaments to resist excessive movement of the joint and more efficient movement transfer from muscle to bone. Increased laxity due to subfailure damage therefore reduces the tendon's ability to resist excessive movement, which has the potential to cause increased damage, even ultimate failure, with time.

We also found that the viscoelastic parameters σ_{decay} and $\sigma_{decrease}$ were decreased following overstretch damage. This means that the time-dependent properties have been altered, thus affecting the tendons' ability to adapt to loading situations over time. This also indicates that the tendon's ability to absorb energy, store energy, and return energy to the system has been significantly altered. Functional implications of this are substantial. From a microstructural and molecular perspective, alterations in viscoelastic properties indicate changes in the interactions between the components within the tendon.

The amount that each of these parameters was decreased was dependent on the strain during the overstretch damage, indicating that not only is it possible to vary the amount of diffuse damage in a tendon by varying the strain used to create it, but also that the mechanical properties of the tendon can be indicative of the level of damage in the tendon. This implies that it may be possible to correlate mechanical parameters, both viscoelastic and elastic, to the damage state of the tissue. In this manner, it would be possible to diagnose diffuse damage in the tissue while it is still intact and predict risk of additional injury with continued function. With the increasing popularity and capabilities of noninvasive force measurements, there is potential for measurement of diffuse damage *in vivo*, which is currently not possible using imaging modalities such as conventional MRI or ultrasound (unlike a focal defect damage, which is readily visualized by such technologies).

We therefore conclude that subfailure diffuse damage induced by deforming the tendons to strains outside of their normal activity (but less than their failure strain) affects both the elastic and viscoelastic parameters. Both elastic and viscoelastic parameters, obtained during stress relaxation and cyclic testing, were significantly decreased following damage. Furthermore, these effects become more pronounced as the strain during the overstretch pull increases. By performing these controlled damage experiments *in vitro*, we can develop a better understanding of tendon mechanics and better anticipate the sequelae of such events *in vivo*.

References

1. Elliot, DM; Robinson, PS; Gimbel, JA; Sarver, JJ; Abboud, JA; Iozzo, RV; Soslowsky, RJ. Effect of altered matrix proteins on quasilinear viscoelastic properties in transgenic mouse tail tendons. Ann Biomed Eng, 31, 599-605, 2003.

2. Provenzano, P; Heisey, D; Hayashi, K; Lakes, R; Vanderby, R. Subfailure damage in ligament: a structural and cellular evaluation. J Appl Phys, 92, 362-371, 2002.

3. Hokanson, J; Yazdani, S. A constitutive model of the artery with damage. Mech Res Commun, 24(2), 151-159, 1996.

4. Simo, JC. On a fully three-dimensional finite-strain visco-elastic damage model: formulation and computational aspects. Comput Method Appl M, 60, 153-157, 1987.

5. Perry, SM; Getz, CL; Soslowsky, LJ. After rotator cuff tears, the remaining (intact) tendons are mechanically altered. J Shoulder Elbow Surgery, 18, 52-57, 2009.

6. Natali, AN; Pavan, PG; Carniel, EL; Lucisano, ME; Taglialavor, G. Anisotropic elasto-damage constiutive model for the biomechanical analysis of tendons. Med Eng & Phys, 27, 209-214, 2005.

7. Gardener, JC, Weiss, JA and Rosenburg, TD. Strain in the human medial collateral ligament during valgus loading of the knee. Clin Orthop Relat Res, 391, 266-274, 2001.

8. Lochner, FK; Milne, DW; Milles, EJ; Groom, JJ. In vivo and in vitro measurement of tendon strain in the horse. Am J Vet Res, 41, 1929-1937, 1980.

9. Johnson, GA; Tramaglini, DM; Levine, RE; Ohno, K; Choi, N-Y; Woo, SL-Y. Tensile and viscoelastic properties of human patellar tendon. J Ortho Res, 12, 796-803, 1994.

10. Shadwick, RE. Elastic energy storage in tendons: mechanical differences related to function and age. J Appl Physiol, 68(3), 1033-1040, 1990.

11. Lakes, RS. Viscoelastic Materials. Cambridge: Cambridge University Press, 2009.

Modeling sickle hemoglobin fibers as one chain of coarse-grained particles

He Li, Ha Vi and George Lykotrafitis*
Department of Mechanical Engineering, University of Connecticut, 191 Auditorium Road, Unit 3139, Storrs, CT 06269-3139, United States. Tel.: +1(860)486-2439, fax: +1(860)486-5088.
*Corresponding author at: E-mail address: gelyko@engr.uconn.edu (G. Lykotrafitis)

ABSTRACT

Sickle cell disease is a genetic disorder most commonly found in people of African descent and it is caused by the presence of abnormal hemoglobin S (HbS) in the patient's red blood cells (RBCs). In the deoxygenated state, the defective hemoglobin tetramers polymerize forming stiff fibers which distort the cell and change its biomechanical properties. Because the HbS fibers play a vital role in the formation of the sickle-shaped RBC, the material properties and biomechanical behaviors of polymerized HbS fibers is a subject of intense research interest. Here, we introduce a solvent-free coarse-grain molecular dynamics (CGMD) model to simulate a single hemoglobin fiber as a chain of coarse-grained particles. A finitely extensible nonlinear elastic (FENE) potential is applied between consecutive particles. Meanwhile, a FENE-type bending potential is employed to model the bending resistance of HbS fibers. The parameters of the potentials are identified via comparison between the simulation results and the experimentally measured values of bending rigidity of single HbS fibers. The Langevin thermostat is employed to control the system temperature. This model will greatly facilitate future studies on the HbS polymerization, fiber bundle and gel formation as well as the interaction of between the HbS fiber bundles and the RBC membrane. In addition, the model can be easily adapted to study other filamentous protein assembles.

Key words: Elastic properties, zippering mechanisms, Van der Waals and depletion forces, HbS fiber model, Langevin thermostat.

1. Introduction

Hemoglobin in RBCs mediates the transfer of oxygen from the lungs to tissues. The erythrocytes in patients suffering from sickle cell disease contain, instead of normal adult hemoglobin (HbA), the defective hemoglobin S (HbS) in which a charged surface group glu at β6 is replaced by a hydrophobic group val inducing polymerization of deoxygenated HbS at high enough concentrations [1-2]. The polymerized hemoglobin molecules inside RBC form fibers and gel, which result in distortion of RBCs to irregular shapes as well as decrease deformability and increase viscosity. Because of increased stiffness and cell adherence to the endothelium, the circulation of sickle cells through the body's narrow blood vessels, such as arterioles, venules, and capillaries, is often obstructed resulting in infarctions and organ damage [3].

T. Proulx (ed.), *Mechanics of Biological Systems and Materials, Volume 2*, Conference Proceedings of the Society for Experimental Mechanics Series 9999, DOI 10.1007/978-1-4614-0219-0_31, © The Society for Experimental Mechanics, Inc. 2011

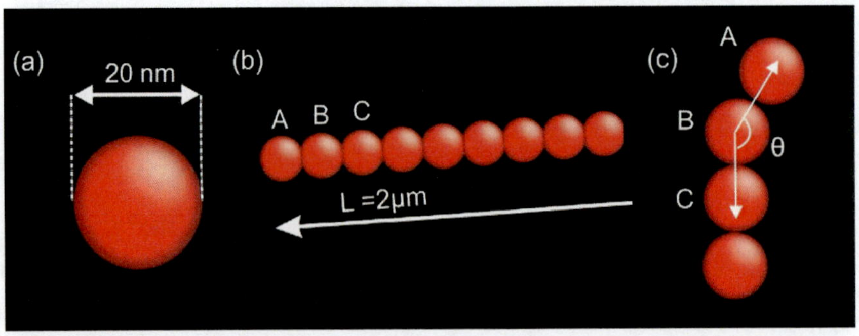

Figure 1 (Color online) (a) Top view of the HbS fiber model. (b) Side view of the HbS fiber model. (c) Bending angle measured from the HbS fiber model

Since the distortion and rigidification of abnormal RBC are caused by the formation of HbS fibers and fiber domains, the properties of the HbS fiber have been extensively studied experimentally in the past two decades. The bending and torsional rigidities of approximately 1 μm long single HbS fibers were obtained by cryo-electron microscopy [4-5]. The measurements showed that the bending rigidity of a single HbS fiber is approximately $\kappa = 5.2\times10^{-25}$ N m^2, which is consistent with the value measured in the previous study [6], whereas the torsional rigidity is approximately 6×10^{-27} J m [5]. Meanwhile, different types of fiber cross-links can be found in gel and the process of fibers zippering, such as X-shaped cross-links and Y-shaped cross-links. Afterwards, theoretical models were developed to estimate the attractive binding energy between two zippered HbS fibers, which turned out to be $u = 7.2\ k_BT$ μm^{-1} with a corresponding characteristic length $l_c = k_BT/u \approx 140$ nm [7]. It is significantly smaller than the persistence length of a single fiber, indicating that single HbS fibers are very unlikely to spontaneously separate due to thermal vibrations.

Previous research on the HbS fibers was mainly based on theoretical method and experimental observations, with only limited literatures studying HbS fibers through molecular dynamics, which investigated the contact points of the double strands when formatting HbS fibers and in general the crystal structure of HbS [8-9]. A coarse-grain (CG) model of a single HbS fiber is introduced in this work. The parameterization scheme for the CG potentials employed in this study fits this CG fiber model to the experimentally measured bending rigidities of HbS fibers[5-6].

2. Model and method

In the paper, a 2 μm long HbS fiber with diameter of 20 nm is coarse-grained into one chain of 100 soft particles, as shown in Fig. 1a and Fig. 1b. The diameter of each particle, d_0, is 20 nm and the distance between the centers of neighboring particles is also 20 nm. The translational motion of the particles are governed by the Langevin equation [10]

$$m_i \frac{d^2\boldsymbol{r}_i}{dt^2} = \boldsymbol{F}_i - f\frac{d\boldsymbol{r}_i}{dt} + \boldsymbol{F}_i^B \tag{1}$$

where m_i represents the mass of the ith particle, f stands for the friction coefficient and it is identified to be 100 m_i/t_s, where t_s is the time unit. $\boldsymbol{r}_i$ is the position vector of the ith particle, and t is time. $\boldsymbol{F}_i = -\partial U/\partial \boldsymbol{r}_i$ is the deterministic force produced by the implemented total potential U. $\boldsymbol{F}_i^B$ signifies Gaussian white noise from the environment and it obeys the fluctuation-dissipation theorem [11-12]

$$\langle \boldsymbol{F}_i^B \rangle = 0, \tag{2}$$

$$\langle \boldsymbol{F}_i^B \boldsymbol{F}_j^B \rangle = \frac{2k_BTf\delta_{ij}}{\Delta t}, \tag{3}$$

where k_B is the Boltzmann's constant, T is the absolute environmental temperature, δ_{ij} is the Kronecker delta, and Δt is the time-step. The friction coefficient, f, is selected to maintain the environmental temperature at 300 K. The time step for the numerical solution of the Langevin equation is chosen to be $\Delta t = 0.001t_s$.

In the simulations, the energy unit is $\varepsilon/k_BT = 1$. It is known that the molecular weight of each hemoglobin tetramer is approximately 68,000 Da, thus each coarse-grained particle possesses a mass of about $m_i \approx 1.58032\times10^{-21}$kg. The time scale of the simulation can be calculated to be $t_s = (m_i\sigma^2/\varepsilon)^{1/2} \approx 1.100\times10^{-8}$ s and it is set by the motion due to the deterministic force [13].

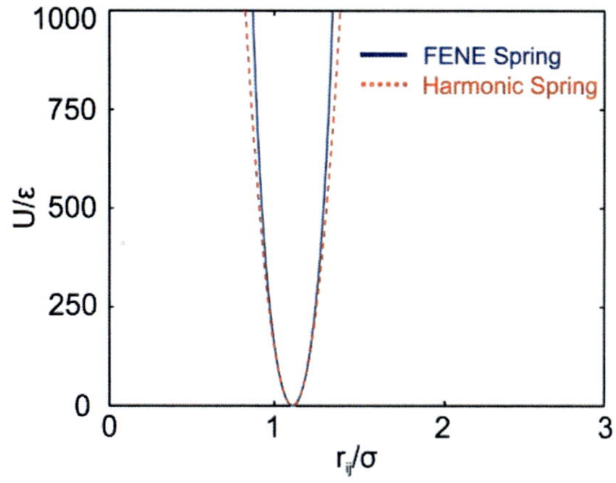

Figure 2 (Color online) The finitely extendable nonlinear elastic (FENE) potential and the corresponding harmonic potential in the HbS fiber model

A finitely extendable nonlinear elastic (FENE) potential (see Fig. 2) is introduced between adjacent particles that belong to the same chain (e.g., A and B in Fig. 1b). FENE potential is harmonic at its minimum but the bonds cannot be stretched beyond a maximum length. The FENE potential is expressed as

$$U_s = -\frac{1}{2}K_s\Delta d_{\max}\ln[1-(\frac{d_{ij}-d_0}{\Delta d_{\max}})^2] \tag{4}$$

where K_s is a parameter related to the constant $K_{sp} = K_s/\Delta d_{max}$ of the harmonic potential which has the same stiffness with the FENE potential at equilibrium, d_{ij} and d_0 are the distance and equilibrium distance between particles i and j, respectively. The bonds between two particles cannot be stretched more than Δd_{max}. In these simulations, the maximum extension allowed between two particles is set to be $\Delta d_{max} = 0.3\ d_0$. It can be showed that show that the value of K_s that relates the Young's modulus E which is consistent with the bending rigidity of the HbS fiber, is approximately $K_s = K_{sp}\Delta d_{max} = 28800\ \varepsilon/d_0$.

Bending rigidity of the HbS fiber is described by a bending FENE potential

$$U_b = -\frac{1}{2}K_b\Delta\theta_{\max}\ln[1-(\frac{\theta-\theta_0}{\Delta\theta_{\max}})^2] \tag{5}$$

where K_b is the parameter that directly regulates the bending stiffness of the hemoglobin fiber, and θ is the angle formed by three consecutive particles in the same chain, as illustrated in Fig. 1c. θ_0 is the equilibrium angle and it is chosen to be π, meaning that the three consecutive particles are initially located in-line. $\Delta\theta_{max}$ is the maximum allowed bending angle between two particles and is set to be $0.3\ \theta_0$. K_b is selected via a trial and error process in order to produce a bending rigidity identical to the experimental value and it is determined to be $K_b = 8\times10^3\varepsilon$.

The numerical method employed for the integration of equation (1) is a modified version of the leapfrog algorithm and it is similar to the approach that was employed for the modification of the velocity-Verlet algorithm used in [14].

3. Results and discussion

3.1 Measurements of bending rigidity of HbS fiber

First, we measure the bending rigidity κ, of the fiber model and prove that by tuning the parameters K_b, the model reproduced the experimental values of the bending rigidity of a HbS fiber [5]. In the measurements, one end of the HbS fiber is fixed and it is used as the reference point for the central axis of the fiber. The bending moduli of semi-flexible fibers can be derived from thermally driven fluctuations [15]. A method of measuring bending rigidity of HbS fibers under thermal fluctuation is introduced by [5-6, 16], which relates the normal deviation of the fiber middle point to its original position with fiber bending rigidity as the stiffer fibers have less fluctuations than softer ones. A similar method is adopted by [7, 17], but they measured the fluctuations at the end of the HbS fibers

$$\left\langle \delta u(L)^2 \right\rangle = \frac{k_B T L^3}{3\kappa} \tag{6}$$

and the bending rigidity of the HbS fiber is calculated from the equation

$$\kappa = \frac{k_B T L^3}{3\left\langle \delta u(L)^2 \right\rangle}\ , \tag{7}$$

where L is the total length of the HbS fiber and $\delta u(L)$ is the normal deviation of the fiber end and it is equal to $\delta u(L) = u(L) - \langle u(L) \rangle$. $\langle u(L) \rangle$ is the mean position of the fiber end. Once the bending rigidity is obtained, the persistence length can be easily calculated by the expression

$$l_p = \frac{\kappa}{k_B T} \tag{8}$$

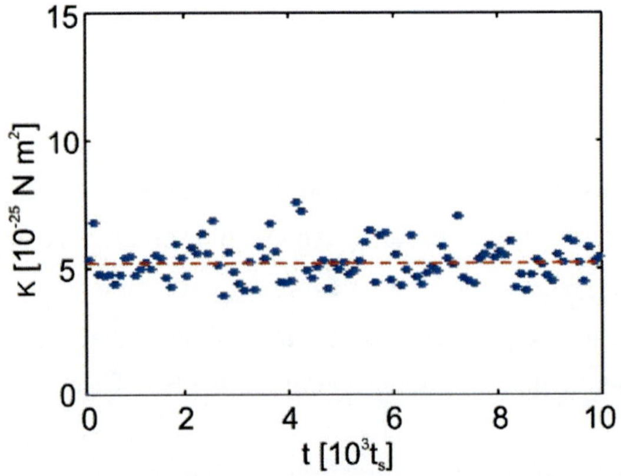

Figure 3 (Color online) Measurements of the bending rigidity of the HbS fiber model. The red dot line highlights the mean value of measured κ and it corresponds to 5.3×10^{-25} N m^2

The measured bending rigidity of HbS fiber with respect to time is plotted in Fig. 3. The measured averaged bending rigidity is in agreement with the experimentally measured value $\kappa = 5.2\times10^{-25}$ N m^2 and it corresponds to a persistence length of approximately $l_p \cong 121$ μm [5]. Due to the large persistence length, the numerical results show that a single HbS fiber subjected to thermal forces behaves similarly to a stiff rod as it fluctuates about its central axis (see Fig. 4).

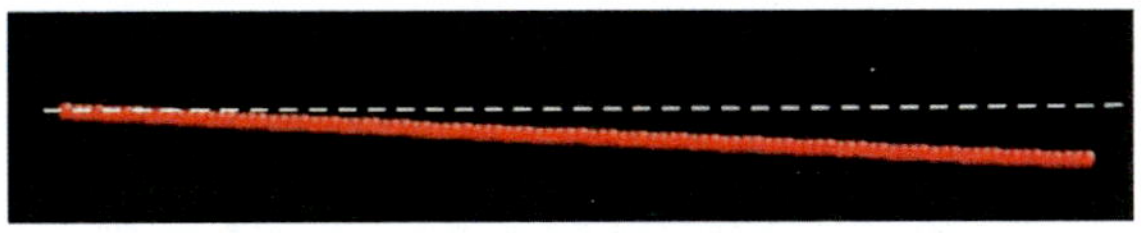

Figure 4 (Color online) Characteristic configuration of the HbS fiber model.

4. Conclusions

In this paper, we model single HbS fibers using one chain of particles and the hexagonally shaped cross-section of the HbS fiber is coarse grained into one particle. The motions of all the particles are governed by the Langevin equation. In order to simulate the thermal behaviors of HbS fibers, two interaction potentials are applied between the particles, namely a FENE potential and a bending potential. By employing these potentials, the proposed model is able to derive the experimentally measured bending rigidity of a single HbS fiber. The developed model will greatly facilitate the future study on the HbS polymerization, fiber bundle and gel formation as well as the interaction of between the HbS fiber bundles and the RBC membrane.

References

1. Ferrone, F.A., *Polymerization and sickle cell disease: A molecular view.* Microcirculation, 2004. **11**: p. 115-128.
2. Christoph, G.W., J. Hofrichter, and W.A. Eaton, *Understanding the shape of sickled red cells.* Biophys J, 2005. **88**: p. 1371-6.
3. Aprelev, A., et al., *The Effects of Erythrocyte Membranes on the Nucleation of Sickle Hemoglobin.* Biophysical Journal, 2005. **88**: p. 2815-2822.
4. Lewis, M.R., L.J. Gross, and R. Josephs, *Cryoelectron Microscopy of Deoxy-Sickle Hemoglobin Fibers.* Microscopy Research and Technique, 1994. **27**: p. 459-467.
5. Turner, M.S., et al., *Anisotropy in Sickle Hemoglobin Fibers from Variations in Bending and Twist.* Journal of Molecular Biology, 2006. **357**: p. 1422-1427.
6. Wang, J.C., et al., *Micromechanics of isolated sickle cell hemoglobin fibers: bending moduli and persistence lengths.* Journal of Molecular Biology, 2002. **315**: p. 601-612.
7. Jones, C.W., et al., *Interactions between sickle hemoglobin fibers.* Faraday Discussions, 2003. **123**: p. 221-236.
8. Prabhakaran, M. and E.J. Michael, *Molecular dynamics of sickle and normal hemoglobins.* Biopolymers, 1993. **33**: p. 735-742.
9. Roufberg, A. and F.A. Ferrone, *A model for the sickle hemoglobin fiber using both mutation sites.* Protein Science, 2000. **9**: p. 1031-1034.
10. Reif, F., ed. *Fundamentals of Statistical and Thermal Physics.* 1965, McGraw-Hill, New York, NY.
11. Noguchi, H. and G. Gompper, *Meshless membrane model based on the moving least-squares method.* Physical Review E, 2006. **73**: p. 021903.
12. Underhill, P.T. and P.S. Doyle, *On the coarse-graining of polymers into bead-spring chains.* Journal of Non-Newtonian Fluid Mechanics, 2004. **122**: p. 3-31.
13. Allen, M. and D. Tildesley, *Computer Simulations of Liquids.* 1987, New York: Clarendon Press.
14. Groot, R.D. and P.B. Warren, *Dissipative particle dynamics: Bridging the gap between atomistic and mesoscopic simulation.* Journal of Chemical Physics, 1997. **107**: p. 4423-4435.
15. Boal, D., *Mechanics of the cell.* 2002, Cambridge, United Kingdom: Cambridge University Press.
16. Turner, M.S., et al., *Fluctuations in self-assembled sickle hemoglobin fibers.* Langmuir, 2002. **18**: p. 7182-7187.

17. Jones, C.W., et al., *Measuring Forces between Protein Fibers by Microscopy*. Biophysical Journal, 2005. **88**: p. 2433-2441.